Communications in Computer and Information Science 822

Commenced Publication in 2007
Founding and Former Series Editors:
Phoebe Chen, Alfredo Cuzzocrea, Xiaoyong Du, Orhun Kara, Ting Liu,
Dominik Ślęzak, and Xiaokang Yang

More information about this series at http://www.springer.com/series/7899

Leonid Kalinichenko · Yannis Manolopoulos
Oleg Malkov · Nikolay Skvortsov
Sergey Stupnikov · Vladimir Sukhomlin (Eds.)

Data Analytics and Management in Data Intensive Domains

XIX International Conference, DAMDID/RCDL 2017
Moscow, Russia, October 10–13, 2017
Revised Selected Papers

Editors
Leonid Kalinichenko
Federal Research Center
"Computer Science and Control"
Russian Academy of Sciences
Moscow
Russia

Yannis Manolopoulos
Open University of Cyprus
Latsia
Cyprus

Oleg Malkov
Institute of Astronomy
Russian Academy of Sciences
Moscow
Russia

Nikolay Skvortsov
Federal Research Center
"Computer Science and Control"
Russian Academy of Sciences
Moscow
Russia

Sergey Stupnikov
Federal Research Center
"Computer Science and Control"
Russian Academy of Sciences
Moscow
Russia

Vladimir Sukhomlin
Moscow State University
Moscow
Russia

ISSN 1865-0929 ISSN 1865-0937 (electronic)
Communications in Computer and Information Science
ISBN 978-3-319-96552-9 ISBN 978-3-319-96553-6 (eBook)
https://doi.org/10.1007/978-3-319-96553-6

Library of Congress Control Number: 2018948633

This Springer imprint is published by the registered company Springer Nature Switzerland AG
The registered company address is: Gewerbestrasse 11, 6330 Cham, Switzerland

Preface

This CCIS volume published by Springer contains the proceedings of the XIX International Conference Data Analytics and Management in Data-Intensive Domains (DAMDID/RCDL 2017) that took place during October 9–13 in the Lomonosov Moscow State University at the Department of Computational Mathematics and Cybernetics. The DAMDID series of conferences was planned as a multidisciplinary forum of researchers and practitioners from various domains of science and research, promoting cooperation and exchange of ideas in the area of data analysis and management in domains driven by data-intensive research. Approaches to data analysis and management being developed in specific data-intensive domains (DID) of X informatics (such as X = astro, bio, chemo, geo, med, neuro, physics, chemistry, material science etc.), social sciences, as well as in various branches of informatics, industry, new technologies, finance, and business contribute to the conference content.

Traditionally DAMDID/RCDL proceedings are published locally before the conference as a collection of full texts of all regular and short papers accepted by the Program Committee as well as, abstracts of posters and demos. Soon after the conference, the texts of regular papers presented at the conference are submitted for online publishing in a volume of the European repository of the CEUR Workshop Proceedings, as well as for indexing the volume content in DBLP and Scopus. Since 2016, a DAMDID/RCDL volume of post-conference proceedings with up to one third of the submitted papers that were previously published in CEUR Workshop Proceedings have been published by Springer in their *Communications in Computer and Information Science* (CCIS) series. Each paper selected for the CCIS post-conference volume should be modified as follows: the title of each paper should be a new one; the paper should be significantly extended (with at least 30% new content); the paper should refer to its original version in the CEUR Workshop Proceedings. CCIS is abstracted/indexed in DBLP, Google Scholar, EI-Compendex, Mathematical Reviews, SCImago, and Scopus.

The program of DAMDID/RCDL 2017, as with the previous editions of these conferences, alongside the traditional data management topics reflects a rapid move into the direction of data science and data-intensive analytics. The program this year included carefully selected invited keynote talks related to rapidly developed DID. The respective plenary sessions were also aimed at attracting the attention of researchers in the selected DID. A preconference plenary session on October 9 included two talks: the keynote talk by Stefano Ceri, Professor of Database Systems at Dipartimento di Elettronica, Informazione e Bioingegneria (DEIB) of Politecnico di Milano, and the invited talk by Zoltan Szallasi, MD, senior research scientist, the Children's Hospital Informatics Program, Harvard Medical School. The session was devoted to the development of methods and techniques for genomes and diagnostics in various application domains (from health care to criminalistics). Stefano Ceri considered the implementation issues of the new-generation DNA sequencing techniques in the

European project GeCo applying big data technologies; in the talk by Zoltan Szallasi, an overview of approaches to the genomic-based diagnostics in various application domains was given. In more detail, in the tutorial given by Zoltan Szallasi on October 10 the application of genomic diagnostics in cancer immunotherapy was presented. The problems of data deluge in astronomy and approaches to their solution were considered in the keynote talk by Giuseppe Longo (Professor of Astrophysics at the University of Naples Federico II). On the basis of their talks, Zoltan Szallasi, Stefano Ceri with co-authors, and Giuseppe Longo with co-authors provided invited full papers for this CCIS volume.

The conference Program Committee reviewed 75 submissions for the conference and eight submissions for the PhD workshop. For the workshop, five papers were accepted and three were rejected. For the conference, 47 submissions were accepted as full papers, 12 as short papers, two as posters, and two as demos, whereas 12 submissions were rejected. According to the conference program, these 59 oral presentations (of the full and short papers) are structured into 19 sessions including: Data Analysis Projects in Astronomy; Semantic Web Techniques in DID; Special Purpose DID Infrastructures (two sessions); Distributed Computing; System Efficiency Evaluation; Data Analysis Projects in Neuroscience; Specific Data Analysis Techniques; Ontological Models and Applications (two sessions); Heterogeneous Database Integration; Text Analysis in Humanities (two sessions); Data Analysis Projects in Various DID; Organization of Experiments in Data-Intensive Research; Digital Library Projects; Knowledge Representation and Discovery; Approaches for Problem Solving in DID; and Applications of Machine Learning. Although most of the presentations are dedicated to the results of research conducted in organizations in the territory of the Russian Federation including Kazan, Moscow, Novosibirsk, Obninsk, Omsk, Orel, Pereslavl-Zalessky, Saint Petersburg, Tomsk, Yaroslavl, Zvenigorod, the DAMDID/RCDL 2017 conference also had international features. This move is witnessed by 12 talks (four of them are invited) prepared by the notable foreign researchers from such countries as Armenia (Yerevan), Bahrain (Manama), Belarus (Minsk), Bulgaria (Sofia), Germany (Dusseldorf, Kiel), UK (Harvel), Greece (Thessaloniki), Italy (Milan, Naples), and the USA (Harvard).

For the proceedings 19 papers were selected by the Program Committee (16 peer reviewed and three invited papers) and after careful editing they are included in this volume structured into seven sections comprising Data Analytics: two papers; Next-Generation Genomic Sequencing (Challenges and Solutions): two papers; Novel Approaches to Analyzing and Classifying of Various Astronomical Entities and Events: six papers; Ontology Population in Data-Intensive Domains: three papers; Heterogeneous Data Integration Issues: four papers; Data Curation and Data Provenance Support: one paper; Temporal Summaries Generation: one paper. Of these, eight papers (more than one third of the total number of the papers selected) were prepared by foreign researchers (from Bulgaria, Germany, Greece, Italy, UK, USA).

DAMDID/RCDL 2017 would not have been possible without the support of the Russian Foundation for Basic Research, the Federal Agency of Scientific Organizations of the Russian Federation and the Federal Research Center Computer Science and Control of the Russian Academy of Sciences. Finally, we thank Springer for publishing this proceedings volume, containing the invited and selected research papers, in their

CCIS series. The Program Committee of the conference appreciates the possibility to use the Conference Management Toolkit (CMT) sponsored by Microsoft Research, which provided great support during various phases of the paper submission and reviewing process.

May 2018

Leonid Kalinichenko
Yannis Manolopoulos
Oleg Malkov
Nikolay Skvortsov
Sergey Stupnikov
Vladimir Sukhomlin

Organization

General Chair

Igor Sokolov	Federal Research Center Computer Science and Control of RAS, Russia

Program Committee Co-chairs

Leonid Kalinichenko	Federal Research Center Computer Science and Control of RAS, Russia
Yannis Manolopoulos	Aristotle University of Thessaloniki, Greece

PhD Workshop Co-chairs

Sergey Stupnikov	Federal Research Center Computer Science and Control of RAS, Russia
Sergey Gerasimov	Lomonosov Moscow State University, Russia

Organizing Committee Co-chairs

Vladimir Sukhomin	Lomonosov Moscow State University, Russia
Victor Zakharov	Federal Research Center Computer Science and Control of RAS, Russia

Organizing Committee

Elena Zubareva	Lomonosov Moscow State University, Russia
Dmitry Briukhov	Federal Research Center Computer Science and Control of RAS, Russia
Nikolay Skvortsov	Federal Research Center Computer Science and Control of RAS, Russia
Dmitry Kovalev	Federal Research Center Computer Science and Control of RAS, Russia
Evgeny Morkovin	Lomonosov Moscow State University, Russia
Irina Karzalova	Federal Research Center Computer Science and Control of RAS, Russia
Yulia Trusova	Federal Research Center Computer Science and Control of RAS, Russia
Evgeniy Ilyushin	Lomonosov Moscow State University, Russia
Dmitry Gouriev	Lomonosov Moscow State University, Russia
Vladimir Romanov	Lomonosov Moscow State University, Russia

Supporters

Russian Foundation for Basic Research
Federal Agency of Scientific Organizations of the Russian Federation
Federal Research Center "Computer Science and Control" of the Russian Academy of Sciences (FRC CSC RAS)
Moscow ACM SIGMOD Chapter

Coordinating Committee

Igor Sokolov (Co-chair)	Federal Research Center Computer Science and Control of RAS, Russia
Nikolay Kolchanov (Co-chair)	Institute of Cytology and Genetics, SB RAS, Novosibirsk, Russia
Leonid Kalinichenko (Deputy Chair)	Federal Research Center Computer Science and Control of RAS, Russia
Arkady Avramenko	Pushchino Radio Astronomy Observatory, RAS, Russia
Pavel Braslavsky	Ural Federal University, SKB Kontur, Russia
Vasily Bunakov	Science and Technology Facilities Council, Harwell, Oxfordshire, UK
Alexander Elizarov	Kazan (Volga Region) Federal University, Russia
Alexander Fazliev	Institute of Atmospheric Optics, RAS, Siberian Branch, Russia
Alexei Klimentov	Brookhaven National Laboratory, USA
Mikhail Kogalovsky	Market Economy Institute, RAS, Russia
Vladimir Korenkov	JINR, Dubna, Russia
Mikhail Kuzminski	Institute of Organic Chemistry, RAS, Russia
Sergey Kuznetsov	Institute for System Programming, RAS, Russia
Vladimir Litvine	Evogh Inc., California, USA
Archil Maysuradze	Moscow State University, Russia
Oleg Malkov	Institute of Astronomy, RAS, Russia
Alexander Marchuk	Institute of Informatics Systems, RAS, Siberian Branch, Russia
Igor Nekrestjanov	Verizon Corporation, USA
Boris Novikov	St. Petersburg State University, Russia
Nikolay Podkolodny	ICaG, SB RAS, Novosibirsk, Russia
Aleksey Pozanenko	Space Research Institute, RAS, Russia
Vladimir Serebryakov	Computing Center of RAS, Russia
Yury Smetanin	Russian Foundation for Basic Research, Moscow
Vladimir Smirnov	Yaroslavl State University, Russia
Sergey Stupnikov	Federal Research Center Computer Science and Control of RAS, Russia
Konstantin Vorontsov	Moscow State University, Russia
Viacheslav Wolfengagen	National Research Nuclear University MEPhI, Russia

Victor Zakharov	Federal Research Center Computer Science and Control of RAS, Russia

Program Committee

Karl Aberer	EPFL, Lausanne, Switzerland
Plamen Angelov	Lancaster University, UK
Alexander Afanasyev	Institute for Information Transmission Problems, RAS, Russia
Arkady Avramenko	Pushchino Observatory, Russia
Ladjel Bellatreche	LIAS/ISAE-ENSMA, Poitiers, France
Pavel Braslavski	Ural Federal University, Yekaterinburg, Russia
Vasily Bunakov	Science and Technology Facilities Council, Harwell, UK
Evgeny Burnaev	Skoltech, Russia
George Chernishev	St. Petersburg State University, Russia
Yuri Demchenko	University of Amsterdam, The Netherlands
Boris Dobrov	Research Computing Center of MSU, Russia
Alexander Elizarov	Kazan Federal University, Russia
Alexander Fazliev	Institute of Atmospheric Optics, SB RAS, Russia
Sergey Gerasimov	Lomonosov Moscow State University, Russia
Vladimir Golenkov	Belarusian State University of Informatics and Radioelectronics, Belarus
Vladimir Golovko	Brest State Technical University, Belarus
Olga Gorchinskaya	FORS, Moscow, Russia
Evgeny Gordov	Institute of Monitoring of Climatic and Ecological Systems SB RAS, Russia
Valeriya Gribova	Institute of Automation and Control Processes FEBRAS, Far Eastern Federal University, Russia
Maxim Gubin	Google Inc., USA
Natalia Guliakina	Belarusian State University of Informatics and Radioelectronics, Belarus
Ralf Hofestadt	University of Bielefeld, Germany
Leonid Kalinichenko	FRC CSC RAS, Moscow, Russia
George Karypis	University of Minnesota, Minneapolis, USA
Nadezhda Kiselyova	IMET RAS, Russia
Alexei Klimentov	Brookhaven National Laboratory, USA
Mikhail Kogalovsky	Market Economy Institute, RAS, Russia
Vladimir Korenkov	Joint Institute for Nuclear Research, Dubna, Russia
Sergey Kuznetsov	Institute for System Programming, RAS, Russia
Sergei O. Kuznetsov	National Research University Higher School of Economics, Russia
Dmitry Lande	Institute for Information Recording, NASU, Russia
Giuseppe Longo	University of Naples Federico II, Italy
Natalia Loukachevitch	Moscow State University, Russia
Ivan Lukovic	University of Novi Sad, Serbia
Oleg Malkov	Institute of Astronomy, RAS, Russia

Yannis Manolopoulos	School of Informatics of the Aristotle University of Thessaloniki, Greece
Manuel Mazzara	Innopolis University, Russia
Alexey Mitsyuk	National Research University Higher School of Economics, Russia
Xenia Naidenova	S. M. Kirov Military Medical Academy, Russia
Dmitry Namiot	Lomonosov Moscow State University, Russia
Igor Nekrestyanov	Verizon Corporation, USA
Gennady Ososkov	Joint Institute for Nuclear Research, Russia
Dmitry Paley	Yaroslav State University, Russia
Nikolay Podkolodny	Institute of Cytology and Genetics SB RAS, Russia
Natalia Ponomareva	Scientific Center of Neurology of RAMS, Russia
Alexey Pozanenko	Space Research Institute, RAS, Russia
Andreas Rauber	Vienna TU, Austria
Roman Samarev	Bauman Moscow State Technical University, Russia
Timos Sellis	RMIT, Australia
Vladimir Serebryakov	Computing Centre of RAS, Russia
Nikolay Skvortsov	FRC CSC RAS, Russia
Vladimir Smirnov	Yaroslavl State University, Russia
Manfred Sneps-Sneppe	AbavaNet, Russia
Valery Sokolov	Yaroslavl State University, Russia
Sergey Stupnikov	FRC CSC RAS, Russia
Alexander Sychev	Voronezh State University, Russia
Dmitry Tsarkov	Google, USA
Bernhard Thalheim	University of Kiel, Germany
Dmitry Tsarkov	Manchester University, UK
Alexey Ushakov	University of California, Santa Barbara, USA
Natalia Vassilieva	Hewlett-Packard, Russia
Pavel Velikhov	Finstar Financial Group, Russia
Alexey Vovchenko	FRC CSC RAS, Moscow, Russia
Peter Wittenburg	MPI for Psycholinguistics, Germany
Vladimir Zadorozhny	University of Pittsburgh, USA
Yury Zagorulko	Institute of Informatics Systems, SB RAS, Russia
Victor Zakharov	FRC CSC RAS, Russia
Sergey Znamensky	Institute of Program Systems, RAS, Russia

Contents

Data Curation and Data Provenance Support

Temporal Summaries Generation

Data Analytics

Deep Model Guided Data Analysis

Yannic Ole Kropp(✉) and Bernhard Thalheim

Department of Computer Science,
Christian Albrechts University Kiel, 24098 Kiel, Germany
{yk, thalheim}@is.informatik.uni-kiel.de

Abstract. Data mining is currently a well-established technique and supported by many algorithms. It is dependent on the data on hand, on properties of the algorithms, on the technology developed so far, and on the expectations and limits to be applied. It must be thus matured, predictable, optimisable, evolving, adaptable and well-founded similar to mathematics and SPICE/CMM-based software engineering. Data mining must therefore be systematic if the results have to be fit to its purpose. One basis of this systematic approach is model management and model reasoning. We claim that systematic data mining is nothing else than systematic modelling. The main notion is the notion of the model in a variety of forms, abstraction and associations among models.

Keywords: Data mining · Modelling · Models · Framework · Deep model
Normal model · Modelling matrix

1 Introduction

Data mining and analysis is nowadays well-understood from the algorithms side. There are thousands of algorithms that have been proposed. The number of success stories is overwhelming and has caused the big data hype. At the same time, brute-force application of algorithms is still the standard. Nowadays data analysis and data mining algorithms are taken for granted. They transform data sets and hypotheses into conclusions. For instance, cluster algorithms check on given data sets and for a clustering requirements portfolio whether this portfolio can be supported and provide as a set of clusters in the positive case as an output. The Hopkins index is one of the criteria that allow to judge whether clusters exist within a data set or not. A systematic approach to data mining has already been proposed in [3, 16]. It is based on mathematics and mathematical statistics and thus able to handle errors, biases and configuration of data mining as well. Our experience in large data mining projects in archaeology, ecology, climate research, medical research etc. has however shown that the just described situation of ad-hoc and brute-force mining is their main approach. The results are taken for granted and believed despite the modelling, understanding, flow of work and data handling pitfalls. So, the results often become dubious due to these misconceptions and pitfalls.

Data are the main source for information in data mining and analysis. Their quality properties have been neglected for a long time. At the same time, modern data management allows to handle these problems. In [15] we compare the critical findings or

L. Kalinichenko et al. (Eds.): DAMDID/RCDL 2017, CCIS 822, pp. 3–18, 2018.
https://doi.org/10.1007/978-3-319-96553-6_1

pitfalls of [20] with resolution techniques that can be applied to overcome the crucial pitfalls of data mining in environmental sciences reported there. The algorithms themselves are another source of pitfalls that are typically used for the solution of data mining and analysis tasks. It is neglected that an algorithm also has an application area, application restrictions, data requirements, results at certain granularity and precision. These problems must be systematically tackled if we want to rely on the results of mining and analysis. Otherwise analysis may become misleading, biased, or not possible. Therefore, we explicitly treat properties of mining and analysis. A similar observation can be made for data handling.

Data mining is typically not only based on one model but rather on a model ensemble or model suite. The association among models in a model suite must be explicitly specified. These associations provide an explicit form via model suites. Reasoning techniques combine methods from logics (deductive, inductive, abductive, counter-inductive, etc.), from artificial intelligence (hypothetic, qualitative, concept-based, adductive, etc.), computational methods (algorithmics [6], topology, geometry, reduction, etc.), and cognition (problem representation and solving, causal reasoning, etc.).

In this paper, we use a model-based approach towards data mining and data analysis. The models function as mediators in investigation, reasoning, communication and understanding processes. In this context, each model consists of a *'normal model'* and a *'deep model'*. Normal models represent the parts of the models, which their users are aware of, while deep models compound of the users implicit influences, assumptions, prejudices and expert knowledge given by their culture, (academic) domain or personal beliefs. Normally only normal models are in focus of presentations and discussions. They are therefore mostly well-understood and elaborated. Deep models in contrast, are addressed rarely. They form an implicit (typically discipline specific) consensus for the ways of handling and interpreting the normal models. In effect, it can be hard for researchers to interpret normal models of foreign domains in the way the models creators expected. In addition, certain reasons for (design) decisions might be unreproducible for people outside the domain. We therefore propose an approach for data mining and analysis, which explicitly includes the handling of deep models in its design.

First, we introduce our notion of the model. Next we show how data mining can be designed. We apply this investigation to systematic modelling and later to systematic data mining. It is our goal to develop a holistic and systematic framework for data mining and analysis. Many issues are left out of the scope of this paper such as a literature review, a formal introduction of the approach, and a detailed discussion of data mining application cases.

Remark. A previous version of this paper has been presented on the 19th '*Data Analytics and Management in Data Intensive Domains*' conference (DAMDID) in 2017. Under the title '*Data Mining Design and Systematic Modelling*' that version has also already been published in the corresponding CEUR workshop proceedings [34].

In contrast to the previous version, this version has been revised. Section 5 was added to strengthen the practical relevance of the approach and other parts were shortened to keep the paper in a comfortable length for reading. In result the focus of this version lies a bit more on deep models.

2 Models and Modelling

Models are principle instruments in mathematics, data analysis, modern computer engineering (CE), teaching any kind of computer technology, and also modern computer science (CS). They are built, applied, revised and manufactured in many CE&CS sub-disciplines in a large variety of application cases with different purposes and context for different communities of practice. It is now well understood that models are something different from theories. They are often intuitive, visualisable, and ideally capture the essence of an understanding within some community of practice and some context. At the same time, they are limited in scope, context and the applicability. There is however a ***general notion*** of a model and of a conception of the model:

> *A **model** is a well-formed, adequate, and dependable instrument that represents origins* [9, 28, 29].

Its criteria of well-formedness, adequacy, and dependability must be commonly accepted by its community of practice within some context and correspond to the functions that a model fulfills in utilization scenarios.

2.1 Generic and Specific Models

The general notion of a model covers all aspects of adequateness, dependability, well-formedness, scenario, functions and purposes, backgrounds (grounding and basis), and outer directives (context and community of practice). The models themselves *function as instruments* or tools. Typically, instruments come in a variety of forms and fulfill many different functions. Instruments are partially independent or autonomous of the thing they operate on. Models are however special instruments. They are used with a specific intention within a utilization scenario. The quality of a model becomes apparent in the context of this scenario.[1]

It might thus be better to start with generic models. A ***generic model*** [4, 25, 30, 31] is a model which broadly satisfies the purpose and broadly functions in the given utilization scenario. It is later tailored to suit the particular purpose and function. It generally represents origins of interest, provides means to establish adequacy and dependability of the model, and establishes focus and scope of the model. Generic models should satisfy at least five properties: (i) they must be accurate; (ii) the quality of generic models allows that they are used consciously; (iii) they should be descriptive, not evaluative; (iv) they should be flexible so that they can be modified from time to time; (v) they can be used as a first 'best guess'.

The ***refinement of a (generic) model*** to a particular or special model provides mechanisms for model transformation along the adequacy, the justification and the sufficiency of a model. Refinement is based on *specialization* for better suitability of a model, on *removal* of unessential elements, on *combination* of models to provide a more holistic view, on *integration* that is based on binding of model components to

[1] The quality can for example be characterised through *quality characteristics* [27] such as correctness, generality, usefulness, comprehensibility, parsimony, robustness, novelty etc.

other components and on *enhancement* that typically improves a model to become more adequate or dependable.

2.2 Model Suites

Most disciplines integrate a variety of models or a *society of models*, e.g. [7, 13] Models used in CE&CS are mainly at the same level of abstraction. It is already well-known for threescore years that they form a *model ensemble* (e.g. [10, 22]) or horizontal *model suite* (e.g. [8, 26]). Developed models vary in their scopes, aspects, and facets they represent and their abstraction.

A ***model suite*** consists of a set of models $\{M_1, \ldots, M_n\}$, of an association or collaboration schema among the models, of controllers that maintain consistency or coherence of the model suite, of application schemata for explicit maintenance and evolution of the model suite, and of tracers for the establishment of the coherence.

Multi-modelling [11, 18, 23] became a culture in CE&CS. Maintenance of coherence, co-evolution, and consistency among models has become a bottleneck in development. Moreover, different languages with different capabilities have become an obstacle similar to multi-language retrieval [19] and impedance mismatches. Models are often loosely coupled. Their dependence and relationship is often not explicitly expressed. This problem becomes more complex if models are used for different purposes such as construction of systems, verification, optimization, explanation, and documentation.

2.3 Deep Models and the Modelling Matrix

Model development is typically based on an explicit and rather quick description of the *'surface'* or ***normal model*** and on the mostly unconditional acceptance of a ***deep model***. The latter one directs the modelling process and the surface or normal model. Modelling itself is often understood as development and design of the normal model. The deep model is taken for granted and accepted for a number of normal models.

The deep model can be understood as the common basis for a number of models. It consists of the grounding for modelling (paradigms, postulates, restrictions, theories, culture, foundations, conventions, authorities), the outer directives (context and community of practice), and the basis (assumptions, general concept space, practices, language as carrier, thought community and thought style, methodology, pattern, routines, commonsense) of modelling. It uses a collection of undisputable elements of the background as grounding and additionally a disputable and adjustable basis which is commonly accepted in the given context by the community of practice. Education on modelling starts, for instance, directly with the deep model. In this case, the deep model has to be accepted and is thus hidden and latent.

A ***(modelling) matrix*** is something within or from which something else originates, develops, or takes from. The matrix is assumed to be correct for normal models. It consists of the deep model and the modelling scenarios. The modelling ***agenda*** is derived from the modelling scenario and the utilization scenarios. It is something like a guideline for modeling activities and for model associations within a model suite. It improves the quality of model outcomes by spending some effort to decide what and how much reasoning to do as opposed to what activities to do. It balances resources between the data-level actions and the reasoning actions.

The modelling scenario and the deep model serve as a part of the *definitional frame* within a model development process. They define also the capacity and potential of a model whenever it is utilized.

Deep models and the modelling matrix also define some frame for adequacy and dependability. This frame is enhanced for specific normal models. It is then used for a statement in which cases a normal model represents the origins under consideration.

3 Data Mining Design

3.1 Conceptualization of Data Mining and Analysis

The data mining and analysis task must be enhanced by an explicit treatment of the languages used for concepts and hypotheses, and by an explicit description of knowledge that can be used. The algorithmic solution of the task is based on knowledge on algorithms that are used and on data that are available and that are required for the application of the algorithms. Typically, analysis algorithms are iterative and can run forever. We are interested only in convergent ones and thus need termination criteria. Therefore, conceptualization of the data mining and analysis task consists of a detailed description of *six main parameters* (e.g. for inductive learning [32]):

a. The *data analysis algorithm*: Algorithm development is the main activity in data mining research. Each of these algorithms transfers data and some specific parameters of the algorithm to a result.
b. The *concept space*: the concept space defines the concepts under consideration for analysis based on certain language and common understanding.
c. The *data space*: The data space typically consists of a multi-layered data set of different granularity. Data sets may be enhanced by metadata that characterize the data sets and associate the data sets to other data sets.
d. The *hypotheses space*: An algorithm is supposed to map evidence on the concepts to be supported or rejected into hypotheses about it.
e. The *prior knowledge space*: Specifying the hypothesis space already provides some prior knowledge. In particular, the analysis task starts with the assumption that the target concept is representable in a certain way.
f. The *acceptability and success criteria*: Criteria for successful analysis allow to derive termination criteria for the data analysis.

Each instantiation and refinement of the six parameters leads to specific data mining tasks.

The result of data mining and data analysis is described within the knowledge space. The data mining and analysis task may thus be considered to be a transformation of data sets, concept sets and hypothesis sets into chunks of knowledge through the application of algorithms.

Problem solving and modelling considers, however, typically six aspects [15]:

1. *Application, problems, and users*: The domain consists of a model of the application, a specification of problems under consideration, of tasks that are issued, and of profiles of users.
2. *Context*: The context of a problem is anything what could support the problem solution, e.g. the sciences' background, theories, knowledge, foundations, and concepts to be used for problem specification, problem background, and solutions.
3. *Technology*: Technology is the enabler and defines the methodology. It provides [22] means for the flow of problem solving steps, the flow of activities, the distribution, the collaboration, and the exchange.
4. *Techniques and methods*: Techniques and methods can be given as algorithms. Specific algorithms are data improvers and cleaners, data aggregators, data integrators, controllers, checkers, acceptance determiners, and termination algorithms.
5. *Data*: Data have their own structuring, their quality and their life span. They are typically enhanced by metadata. Data management is a central element of most problem solving processes.
6. *Solutions*: The solutions to problem solving can be formally given, illustrated by visual means, and presented by models. Models are typically only normal models. The deep model and the matrix is already provided by the context and accepted by the community of practice in dependence of the needs of this community for the given application scenario. Therefore, models may be the final result of a data mining and analysis process beside other means.

Comparing these six spaces with the six parameters we discover that only four spaces are considered so far in data mining. We miss the user and application space as well as the representation space.

3.2 Meta-models of Data Mining

An abstraction layer approach separates the application domain, the model domain and the data domain [16]. This separation is illustrated in Fig. 1.

The data mining design framework uses the inverse modeling approach. It starts with the consideration of the application domain and develops models as mediators between the data and the application domain worlds. In the sequel we are going to combine the three approaches of this section. The meta-model corresponds to other

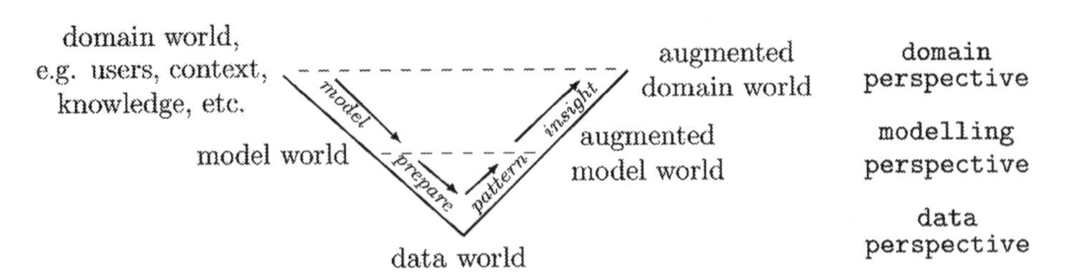

Fig. 1. The V meta-model of model-based Data Mining Design

meta-models such as inductive modelling or hypothetical reasoning (hypotheses development, experimenting and testing, analysis of results, interim conclusions, reappraisal against real world).

4 Data Mining: A Systematic Model-Based Approach

Our approach presented so far allows to revise and to reformulate the model-oriented data mining process on the basis of well-defined engineering [14, 24] or alternatively on systematic mathematical problem solving [21]. Figure 2 displays this revision. We realize that the first two phases are typically implicitly assumed and not considered. We concentrate on the non-iterative form. Iterative processes can be handled in a similar form.

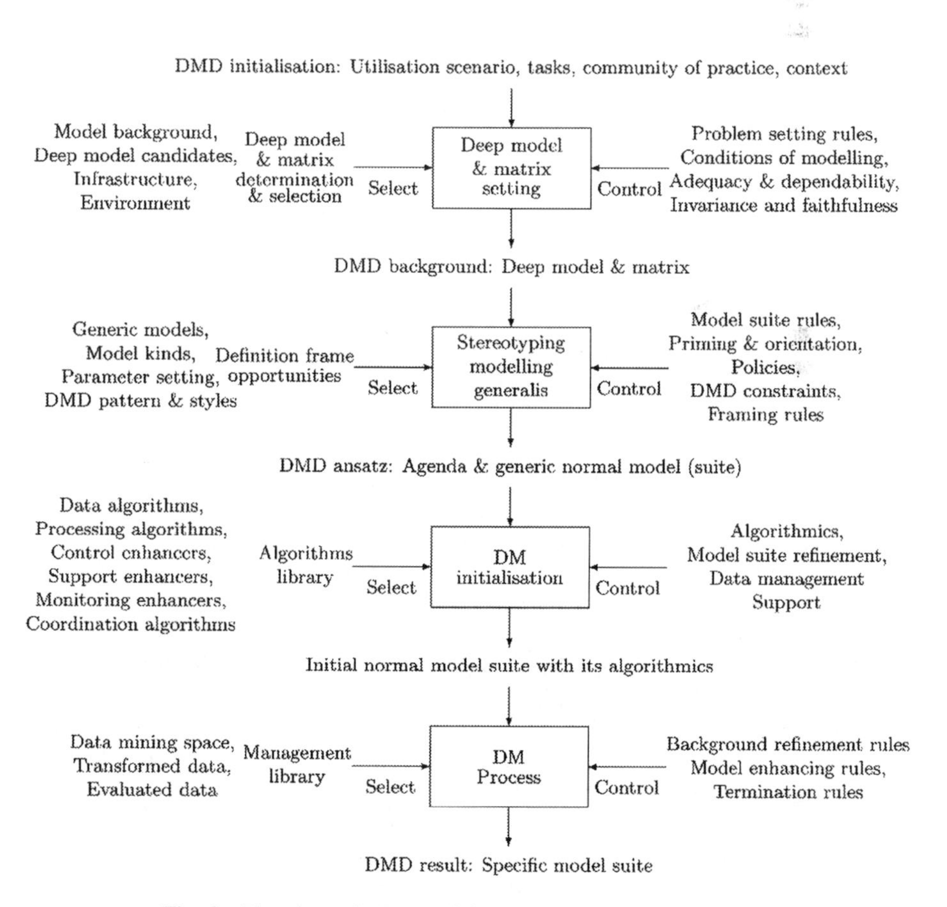

Fig. 2. The phases in Data Mining Design (Non-iterative form)

4.1 Setting the Deep Model and the Matrix

The problem to be tackled must be clearly stated in dependence on the utilization scenario, the tasks to be solved, the community of practice involved, and the given context. The result of this step is the deep model and its matrix. The first one is based on the background, the specific context parameter such as infrastructure and environment, and candidates for deep models.

The data mining tasks can be now formulated based on the matrix and the deep model. We set up the context, the environment, the general goal of the problem and also criteria for *adequateness* and *dependability* of the solution, e.g. *invariance properties* for problem description and for the task setting and its mathematical formulation and *solution faithfulness properties* for later application of the solution in the given environment. What is exactly the problem, the expected benefit? What should a solution look like? What is known about the application?

Deep models already use a background consisting of an undisputable grounding and a selectable basis. The explicit statement of the background provides an understanding of the postulates, paradigms, assumptions, conceptions, practices, etc. Without the background, the results of the analysis cannot be properly understood. Models have their profile, i.e. goals, purposes and functions. These must be explicitly given. The parameters of a generic model can be either order or slave parameters [12], either primary or secondary or tertiary (also called genotypes or phenotypes or observables) [1, 5], and either ruling (or order) or driven parameters [12]. Data mining can be enhanced by knowledge management techniques.

Additionally, the concept space into which the data mining task is embedded must be specified. This concept space is enhanced during data analysis.

4.2 Stereotyping the Process

The general flow of data mining activities is typically implicitly assumed on the basis of stereotypes which form a set of tasks, e.g. tasks of prove in whatever system, transformation tasks, description tasks, and investigation tasks. Proofs can follow the classical deductive or inductive setting. Also, abductive, adductive, hypothetical and other reasoning techniques are applicable. Stereotypes typically use model suites as a collection of associated models, are already biased by priming and orientation, follow policies, data mining design constraints, and framing.

Data mining and analysis is rather stereotyped. For instance, mathematical culture has already developed a good number of stereotypes for problem formulation. It is based on a mathematical language for the formulation of analysis tasks, on selection and instantiation of the best fitting variable space and the space of opportunities provided by mathematics.

Data mining uses *generic models* which are the basis of normal models. Models are based on a separation of concern according the problem setting: dependence-indicating, dependence-describing, separation or partition spaces, pattern kinds, reasoning kinds, etc. This separation of concern governs the classical data mining algorithmic classes: association analysis, cluster analysis, data grouping with or without classification, classifiers and rules, dependences among parameters and data subsets, predictor analysis, synergetics, blind or informed or heuristic investigation of the search space, and pattern learning.

4.3 Initialization of the Normal Data Models

Data mining ***algorithms*** have their capacity and potential [2]. Potential and capacity can be based on SWOT (strengths, weaknesses, opportunities, and threats), SCOPE (situation, core competencies, obstacles, prospects, expectation), and SMART (how simple, meaningful, adequate, realistic, and trackable) analysis of methods and algorithms. Each of the algorithm classes has its strengths and weaknesses, its satisfaction of the tasks and the purpose, and its limits of applicability. Algorithm selection also includes an explicit specification of the order of application of these algorithms and of mapping parameters that are derived by means of one algorithm to those that are an input for the others, i.e. an explicit association within the model suite. Additionally, evaluation algorithms for the success criteria are selected. Algorithms have their own obstinacy, their hypotheses and assumptions that must be taken into consideration. Whether an algorithm can be considered depends on acceptance criteria derived in the previous two steps.

So, we ask: *What kind of model suite architecture suits the problem best? What are applicable development approaches for modelling? What is the best modelling technique to get the right model suite? What kind of reasoning is supported? What not? What are the limitations? Which pitfalls should be avoided?*

The result of the entire data mining process heavily depends on the appropriateness of the ***data sets***, their properties and quality, and more generally the data schemata with essentially three components: application data schema with detailed description of data types, metadata schema [17], and generated and auxiliary data schemata. The first component is well-investigated in data mining and data management monographs. The second and third components inherit research results from database management, from data mart or warehouses, and layering of data. An essential element is the explicit specification of the quality of data. It allows to derive algorithms for data improvement and to derive limitations for applicability of algorithms. Auxiliary data support performance of the algorithms.

Therefore typical data-oriented questions are: *What data do we have available? Is the data relevant to the problem? Is it valid? Does it reflect our expectations? Is the data quality, quantity, recency sufficient? Which data we should concentrate on? How is the data transformed for modelling? How may we increase the quality of data?*

4.4 The Data Mining Process Itself

The data mining process can be understood as a coherent and stepwise refinement of the given model suite. The model refinement may use an explicit transformation or an extract-transform-load process among models within the model suite. Evaluation and termination algorithms are an essential element of any data mining algorithm. They can be based on quality criteria for the finalized models in the model suite, e.g. generality, error-proneness, stability, selection-proneness, validation, understandability, repeatability, usability, usefulness, and novelty.

Typical questions to answer within this process are: *How good is the model suite in terms of the task setting? What have we really learned about the application domain? What is the real adequacy and dependability of the models in the model suite? How these models can be deployed best? How do we know that the models in the model suite are still valid? Which data are supporting which model in the model suite? Which kind of errors of data is inherited by which part of which model?*

The final result of the data mining process is then a combination of the deep model and the normal model whereas the first one is a latent or hidden component in most cases. If we want, however, to reason on the results then the deep model must be understood as well. Otherwise, the results may become surprising and may not be convincing.

5 A Typical Application Case

The presented approach demands an explicit consideration of deep models, matrices and typical (generic) model suites when preparing data mining. It also includes the common steps of data and algorithm selection, initialization of the process and the data mining process itself. It therefore can be used as a guideline or generic process for good proceedings in executing data mining.

At first the additional workload caused by this explicit consideration and almost bureaucratic and stepwise procedure might seem over the top for simple and non-recurring data mining procedures. However, this approach brings huge benefits in research where recurring data mining is performed and especially in (collaborative) projects where multiple disciplines are involved. Multiple deep models, matrices and typical (generic) model suites do likely exist in interdisciplinary projects with multiple (scientific) domains. Also different requirements on data, algorithms and results are common. Additionally these deep models, matrices and other latent factors are normally implicitly taken for granted by the domain experts, but not known by the experts of other domains. These experts of the other domains rather take the implicit agreements of their domains for granted, what easily results in communication problems.

Besides communication, the exchange of data among project members from (academic) disciplines might cause problems too. The data acquisition is also affected by implicit assumptions. This concerns for example the granularity and quality of data. Measuring methods of data might not be adequate for all purposes of all domains. A hypothetical example are the differences in standard procedures, like the computation of the average of a set of numeric values. If the corresponding data is just labelled with '*average*', their integration into data mining processes could affect the integrity of the analysis negatively, as the data set might not contain the expected data. There are huge differences in mode, median and arithmetic mean. Within a discipline the usual method is known, but for other disciplines their usual methods can differ. Applying our approach and explicitly consider the domain specific characteristics can help to reduce such obstacles in communication.

5.1 Collaborative Research Cluster 1266

Let us consider an application case. The CRC 1266[2,3] investigates processes of transformation from 15,000 BCE to 1 BCE. As we are currently involved in this multi-domain interdisciplinary collaborative research project, we notice various of those implicit assumptions. These may ease the communication within a single domain, as they represent the best practices and '*normal*' proceedings of this domain, but they cause problems in cross-disciplinary collaborations. Still it is utopian to try to force the researchers to get rid of their implicit assumptions. First it is unlikely to achieve this goal as many assumptions are latent and only subliminal, second they are directly connected to the academic fields and working environments of the researchers and third there is mostly no perception of an urgent need to do so. It is therefore crucial to explicitly identify those *deep models*, *matrices* and typical (generic) *model suites*, to enable unproblematic collaboration by making them public to all project partners. When those latent structures become know, it is possible to select adequate data for analyses cross-disciplinary. The same applies for data which can be adapted to become adequate. In other words does the awareness of the own latent assumptions and their comparison with the latent assumptions of other project members ease collaboration.

The *deep models*, *matrices* and (generic) *model suites* are (fortunately) quite stable. In contrast to the dynamics in research itself, the latent structures change rarely. Therefore, it is possible to reuse the collected information about those structures. The results of the first steps of our approach can be stored and used for reruns or similar data mining activities. Especially in research it is common to rerun certain procedures at different points in time due to new or revised data or with slightly different parameters due to new knowledge. It is also common for researchers to perform research on the same subject in varying contexts. Typical examples for such varying contexts in archaeology are different regions or different time-frames. In a project like the *CRC 1266* it is therefore reasonable to store, maintain and manage the results of the '*DM Initialisation*' phase, presented in Sect. 4.3, for later reuse. The results combine the information about the disciplinary perspectives (*deep models* etc.), the connections between those perspectives and the requirements for specific research intentions. Especially such a result contains information about adequate data for the research in a specific context and background. In other words it summarizes for a specific research (or data mining) intention, which data within the project can be used and eventually how it has to be preprocessed beforehand. Storing this information saves much time for later and similar research activities. Also this information are valuable meta information, concerning the origins of used data and provenance in a data mining process. In general this eases the usage of this information and therefore encourages its consideration. We suspect that this has a positive impact on the overall quality of interdisciplinary research.

---

[2] https://www.sfb1266.uni-kiel.de/en.

[3] *"Scales of Transformation - Human Environmental Interaction in Prehistoric and Archaic Societies"*.

In the *CRC 1266* we are currently developing and deploying a method to support the approach of this paper by implementing such a storage possibility. The next paragraph will shortly describe its main idea, based on so called '*viewpoints*'.

5.2 Excursus: Viewpoints in CRC 1266

To manage the different perspectives within a project, multiple *viewpoints* are created. A *viewpoint* represents a specific perspective on the available data in the project. It is defined by the domains and the research intentions of its users. It basically models the results of the first three phases of the systematic approach presented in Sect. 4.[4] The central questions, which are addressed by a *viewpoint* include: *Is there relevant data for my research, created by other project members?* and *How to preprocess it to make it fit my perspective?*

When accessing the data pool of the project in context of a viewpoint, only the relevant data for this viewpoint will be shown. The data will also be structured according to the viewpoints specifications. On a technical level this is implemented by storing all project data inside of an abstract universal central database system[5] and generating (database) views on top of it. Each viewpoint is represented by a set of views, which are dynamically constructed according to its specifications. The members of the project access the database and its data via those views. Also interactions like *insert, update* and *delete* is performed via these views. As our community of practice mostly does not interact with databases via raw-sql on a daily basis, a web-portal basing on the *Django* web framework[6] will offer an interface for these services. The specifications for creating the viewpoint and its views are created using the results of the '*DM Initialization*' phase of Sect. 4.3 (which bases on the results of the previous phases). In detail the specifications of a viewpoint do provide information about:

1. *the label* -The most basic information provided is a name (or label) to be able to identify and address a viewpoint and its views.
2. *the description* - Another pretty basic information, is a short description of the viewpoint. This information is not necessary for the generation of the viewpoint, but helps users to put the viewpoint into proper context. (Especially when the *label* is not expressive.) Typical information contained in the *description* would be the recommended users, the domain, the focus of research and the research subject.
3. *the projection (of structures)* - A way more important information are the structural details about the kinds of data objects, which belong to the viewpoint. This includes the (relational) types and the properties/attributes of these types, as well as the possible connections/relations between types (and their properties). In short this represents the information *what* kind of views are in the viewpoint and *how* they compound. For example do the experts for *ancient DNA*[7] collect huge amounts of

[4] the phases before the actual data mining process.

[5] The structure of that database is not topic of this paper, but roughly bases on the schema presented in [33].

[6] https://www.djangoproject.com/.

[7] short: *aDNA*.

DNA sequence data, but these are hard to interpret by non-physicians. In a viewpoint for archaeologists, this raw sequence data might be needless and therefore not integrated. The other way round, in the research of aDNA experts might the exact stratigraphy of excavation sites be needless too. Another example could be scientists who investigate distribution patterns of pottery. They do not need all available information of the found pottery (like weight, dimensions, description of ornaments etc.), but just the coordinates of its excavation and its classification.

This description of the view structures eventually also includes viewpoint specific labels for types, properties etc. Such a specification might be necessary as in different academic disciplines and domains the same labels for different objects/subjects or different labels for same objects/subjects can be in use.

4. *the selection (of instances)* - Another quite important information concerns the data in the views. In contrast to the previous point, this one deals with instances. Even if a type is adequate for a viewpoint, because it fits to the research focus, this does not necessarily imply that also all data tuples of this type are adequate. Further filtering of the available data might be necessary. Such requirements are typically based on thresholds on properties/attributes or metadata. Common examples from archaeology originate from restrictions of the research focus. Referring back to the previous example about the scientists who investigate distribution patterns of pottery, these researchers might for a specific context just be interested in pottery from a specific region or (ancient) time.
5. *the adaptation/adjustment (of values)* - Finally there can be need for (minor) transformations concerning the instance values. This can be necessary if the values from different sources are in different forms and shall be mapped to the standard form of the viewpoints domain. Such mapping or transformation rules can for instance be used to change the unit of a property, by adapting its values (e.g. the unit of an attribute is transformed from *centimeter* to *meter* by multiplying its values with 0.01). Another example could be the transfer of coordinates to a different coordinate reference system.

To create such specifications for viewpoints it is useful to use the systematic approach presented in Sect. 4 for collecting the required information. This standardizes their construction and eases the process.

6 Conclusion

The literature on data mining is fairly rich. Mining tools have already gained the maturity for supporting any kind of data analysis if the data mining problem is well understood, the intentions for models are properly understood, and if the problem is professionally set up. Data mining aims at development of model suites that allows to derive and to draw dependable and thus justifiable conclusions on the given data set. Data mining is a process that can be based on a framework for systematic modelling that is driven by a deep model and a matrix. Textbooks on data mining typically explore algorithms as blind search. Data mining is a specific form of modeling. Therefore, we can combine modeling with data mining in a more sophisticated form.

Models have however an inner structure with parts which are given by the application, by the context, by the commonsense and by a community of practice. These fixed parts are then enhanced by normal models. A typical normal model is the result of a data mining process.

The current state of the art in data mining is mainly technology and algorithm driven. The problem selection is made on intuition and experience. So, the matrix and the deep model are latent and hidden. The problem specification is not explicit. Therefore, this paper aims at the entire data mining process and highlights a way to leave the ad-hoc, blind and somehow chaotic data analysis. The approach we are developing integrates the theory of models, the theory of problem solving, design science, and knowledge and content management. We realized that data mining can be systematized. The framework for data mining design exemplarily presented is an example in Fig. 2.

Acknowledgement. This research was supported by the CRC 1266 '*Scales of Transformation - Human-Environmental Interaction in Prehistoric and Archaic Societies*' which is funded by the DFG. We thank both institutions for enabling this work. We are also very thankful for the fruitful discussions with the members of the CRC.

References

1. Bell, G.: The Mechanism of Evolution. Chapman and Hall, New York (1997)
2. Berghammer, R., Thalheim, B.: Methodenbasierte mathematische Modellierung mit Relationenalgebren. In: Wissenschaft und Kunst der Modellierung: Modelle, Modellieren, Modellierung, pp. 67–106. De Gryuter, Boston (2015)
3. Berthold, M.R., Borgelt, C., Höppner, F., Klawonn, F.: Guide to Intelligent Data Analysis. Springer, London (2010). https://doi.org/10.1007/978-1-84882-260-3
4. Bienemann, A., Schewe, K.-D., Thalheim, B.: Towards a theory of genericity based on government and binding. In: Embley, David W., Olivé, A., Ram, S. (eds.) ER 2006. LNCS, vol. 4215, pp. 311–324. Springer, Heidelberg (2006). https://doi.org/10.1007/11901181_24
5. Booker, L.B., Goldberg, D.E., Holland, J.H.: Classifier systems and genetic algorithms. Artif. Intell. **40**(1–3), 235–282 (1989)
6. Brassard, G., Bratley, P.: Algorithmics - Theory and Practice. Prentice Hall, London (1988)
7. Coleman, A.: Scientific models as works. Cat. Classif. Q. **33**, 3–4 (2006). Special Issue: Works as Entities for Information Retrieval
8. Dahanayake, A., Thalheim, B.: Co-evolution of (information) system models. In: Bider, I., et al. (eds.) BPMDS/EMMSAD -2010. LNBIP, vol. 50, pp. 314–326. Springer, Heidelberg (2010). https://doi.org/10.1007/978-3-642-13051-9_26
9. Embley, D., Thalheim, B. (eds.): The Handbook of Conceptual Modeling: Its Usage and Its Challenges. Springer, Heidelberg (2011). https://doi.org/10.1007/978-3-642-15865-0
10. Gillett, N.P., Zwiers, F.W., Weaver, A.J., Hegerl, G.C., Allen, M.R., Stott, P.A.: Detecting anthropogenic influence with a multi-model ensemble. Geophys. Res. Lett. **29**, 31–34 (2002)
11. Guerra, E., de Lara, J., Kolovos, D.S., Paige, R.F.: Inter-modelling: From Theory to Practice. In: Petriu, Dorina C., Rouquette, N., Haugen, Ø. (eds.) MODELS 2010. LNCS, vol. 6394, pp. 376–391. Springer, Heidelberg (2010). https://doi.org/10.1007/978-3-642-16145-2_26
12. Haken, H., Wunderlin, A., Yigitbasi, S.: An introduction to synergetics. Open. Syst. Inf. Dyn. **3**(1), 1–34 (1994)

13. Hunter, P.J., Li, W.W., McCulloch, A.D., Noble, D.: Multiscale modeling: physiome project standards, tools, and databases. IEEE Comput. **39**(11), 48–54 (2006)
14. ISO/IEC 25020: Software and system engineering - software product quality requirements and evaluation (square) - measurement reference model and guide. ISO/IEC JTC1/SC7 N3280 (2005)
15. Jaakkola, H., Thalheim, B., Kidawara, Y., Zettsu, K., Chen, Y., Heimbürger, A.: Information modelling and global risk management systems. In: Information Modeling and Knowledge Bases XX, pp. 429–446. IOS Press (2009)
16. Jannaschk, K.: Infrastruktur für ein Data Mining Design Framework. Ph.D. thesis, Christian-Albrechts University, Kiel (2017)
17. Kramer, F., Thalheim, B.: A metadata system for quality management. In: Information Modelling and Knowledge Bases, pp. 224–242. IOS Press (2014)
18. Nakoinz, O., Knitter, D.: Modelling Human Behaviour in Landscapes. Springer, Heidelberg (2016). https://doi.org/10.1007/978-3-319-29538-1
19. Pardillo, J.: A systematic review on the definition of UML profiles. In: Petriu, Dorina C., Rouquette, N., Haugen, Ø. (eds.) MODELS 2010. LNCS, vol. 6394, pp. 407–422. Springer, Heidelberg (2010). https://doi.org/10.1007/978-3-642-16145-2_28
20. Petrelli, D., Levin, S., Beaulieu, M., Sanderson, M.: Which user interaction for cross-language information retrieval? Design issues and reflections. JASIST **57**(5), 709–722 (2006)
21. Pilkey, O.H., Pilkey-Jarvis, L.: Useless Arithmetic: Why Environmental Scientists Can't Predict the Future. Columbia University Press, New York (2006)
22. Podkolsin, A.S.: Computer-based modelling of solution processes for mathematical tasks (in Russian). ZPI at Department of Mechanics and Mathematics, Lomonosov Moscow State University, Moscow (2001)
23. Pottmann, M., Unbehauen, H., Seborg, D.E.: Application of a general multi-model approach for identification of highly nonlinear processes – a case study. Int. J. Control **57**(1), 97–120 (1993)
24. Rumpe, B.: Modellierung mit UML. Springer, Heidelberg (2012). https://doi.org/10.1007/978-3-642-22413-3
25. Samuel, A., Weir, J.: Introduction to Engineering: Modelling Synthesis and Problem Solving Strategies. Elsevier, Amsterdam (2000)
26. Simsion, G., Witt, G.C.: Data Modeling Essentials. Morgan Kaufmann, San Francisco (2005)
27. Skusa, M.: Semantische Kohärenz in der Softwareentwicklung. Ph.D. thesis, CAU, Kiel (2011)
28. Thalheim, B.: Towards a theory of conceptual modelling. J. Univ. Comput. Sci. **16**(20), 3102–3137 (2010)
29. Thalheim, B.: The conceptual model an adequate and dependable artifact enhanced by concepts. In: Information Modelling and Knowledge Bases XXV, pp. 241–254. IOS Press (2014)
30. Thalheim, B.: Conceptual modeling foundations: the notion of a model in conceptual modeling. In: Liu, L., Özsu, M. (eds.) Encyclopedia of Database Systems. Springer, New York (2017). https://doi.org/10.1007/978-1-4899-7993-3_80780-1
31. Thalheim, B., Tropmann-Frick, M.: Wherefore models are used and accepted? The model functions as a quality instrument in utilisation scenarios. In: Comyn-Wattiau, I., du Mouza, C., Prat, N. (eds), Ingenierie Management des Systemes D'Information (2016)
32. Thalheim, B., Wang, Q.: Towards a theory of refinement for data migration. In: Jeusfeld, M., Delcambre, L., Ling, T.-W. (eds.) ER 2011. LNCS, vol. 6998, pp. 318–331. Springer, Heidelberg (2011). https://doi.org/10.1007/978-3-642-24606-7_24

33. Jannaschk, K., Rathje, C.A., Thalheim, B., Förster, F.A.: generic database schema for CIDOC-CRM data management. In: ADBIS 2011, Research Communications, Proceedings II of the 15th East-European Conference on Advances in Databases and Information Systems, CEUR Workshop Proceedings, pp. 127–136 (2011)
34. Kropp, Y.O., Thalheim, B.: Data mining design and systematic modelling. In: Selected Papers of the XIX International Conference on Data Analytics and Management in Data Intensive Domains (DAMDID/RCDL), CEUR Workshop Proceedings 2022, pp: 273–280 (2017)

Data Mining and Analytics for Exploring Bulgarian Diabetic Register

Svetla Boytcheva[1(✉)], Galia Angelova[1], Zhivko Angelov[2], and Dimitar Tcharaktchiev[3]

[1] Institute of Information and Communication Technologies, Bulgarian Academy of Sciences, 25A Acad. G. Bonchev Street, 1113 Sofia, Bulgaria
svetla.boytcheva@gmail.com, galia@lml.bas.bg

[2] ADISS Lab Ltd., 4 Hristo Botev Blvd, 1463 Sofia, Bulgaria
angelov@adiss-bg.com

[3] University Specialized Hospital for Active Treatment of Endocrinology – Medical University Sofia, 2 Zdrave Street, 1431 Sofia, Bulgaria
dimitardt@gmail.com

Abstract. This paper discusses the need of building diabetic registers in order to monitor the disease development and assess the prevention and treatment plans. The automatic generation of a nation-wide Diabetes Register in Bulgaria is presented, using outpatient records submitted to the National Health Insurance Fund in 2010–2014 and updated with data from outpatient records for 2015–2016. The construction relies on advanced automatic analysis of free clinical texts and business analytics technologies for storing, maintaining, searching, querying and analyzing data. Original frequent pattern mining algorithms enable to discover maximal frequent itemsets of simultaneous diseases for diabetic patients. We show how comorbidities, identified for patients in the prediabetes period, can help to define alerts about specific risk factors for Diabetes Mellitus type 2, and thus might contribute to prevention. We also claim that the synergy of modern analytics and data mining tools transforms a static archive of clinical patient records to a sophisticated knowledge discovery and prediction environment.

Keywords: Big data analytics · Data mining · Frequent pattern mining Text mining · Health informatics

1 Introduction

Diabetes is an increasingly common disease and a global public health problem that places a considerable economic burden on society. The World Health Organization (WHO) reports that diabetes prevalence among adults has risen from 4.7% in 1980 to 8.5% in 2014. It is expected that diabetes will be the seventh leading cause of death in 2030 [1]. In the recent Global Report on Diabetes WHO recommends: "Strengthen national capacity to collect, analyze and use representative data on the burden and trends of diabetes and its key risk factors. Develop, maintain and strengthen a diabetes registry if feasible and sustainable" [2]. All countries in Europe have national plans for

L. Kalinichenko et al. (Eds.): DAMDID/RCDL 2017, CCIS 822, pp. 19–33, 2018.
https://doi.org/10.1007/978-3-319-96553-6_2

discovery, treatment and prevention of diabetes [3]; seven countries have diabetic registers in 2014 [4]. However, one hardly finds information about the execution of national diabetes plans, monitoring of various plan measures and evaluation of their success. Positive health outcomes are difficult to assess too, moreover this needs to be done dynamically, at national level in order to improve the treatment plans. From a technological point of view, the general impression is that healthcare authorities lack understanding about the potential of modern Information and Communication Technologies (ICT) as an enabling tool that facilitates data collection, monitoring of indicators, knowledge discovery, early alerting and automatic sending of feedbacks, evaluation of updated indicators and automatic preparation of aggregated recaps.

In this paper we discuss the automatic generation of a national Diabetic Register, using outpatient records submitted to the Bulgarian National Health Insurance Fund (NHIF) for the period 2010–2016 and present research efforts to explore the register data by extracting useful information about patients and disease development over time. Some results concern discovery of correlations among data items and have more scientific value while other outcomes are actually aggregated reports addressing the healthcare management. The authors believe that these developments, which are already integrated in the software infrastructure underlying the Diabetic Register and regularly used by the national healthcare authorities, will influence the forthcoming implementation of Electronic Health Record (EHR) system in Bulgaria.

This paper presents novel results extending [5]. Section 2 briefly overviews the need and construction of diabetic registers in Europe. Section 3 presents the Bulgarian Diabetic Register which was generated automatically using a national collection of more than 262 mln outpatient records. We emphasize on the originality of our approach: starting from a very large repository of full-text clinical records, we had to employ more sophisticated software solutions in order to cope with the input data and to provide dynamic exploration of the constantly growing archive of pseudonimyzed outpatient records. Some examples of aggregated reports, prepared by a business analytics tool, demonstrate the potential of the software behind the register. Section 4 shows another data mining tool for discovery of correlations. It sketches an original method for frequent pattern mining and discusses its application for searching of comorbidities in the register. Section 5 contains the conclusion and plans for future work.

2 Diabetic Registers in Europe

The Euro Diabetes Index 2014 compares the figures of diabetes prevalence to previous ones and concludes that prevention and screening in Europe have improved after 2008 because less people die [4]. Patient awareness is raising, devices for self-monitoring become much more accessible, and the variety of medications is growing. However, still a very high number of diabetic patients are undiagnosed and half of the European countries cannot provide reasonably good data of procedure indicators. It is claimed that “as long as important data is not systematically reported and transformed into methodology, diabetes care will remain inefficient and, at worst, haphazard” [4].

On the other hand, it is well known that availability of high-quality data is hard to achieve. Information about diabetic patients is often not collected nationally but rather in hospitals or at regional level, with limited comparability of collected indicators. Available data often come from isolated national projects with fixed duration or EU-funded initiatives like EUBIROD (European Best Information through Regional Outcomes in Diabetes, 2008–2012) [6]. After the project ends, no strategic plans are built by the respective political or governing institutions and in this way projects that started and proved to be successful remain feasibility studies without practical effects.

Seven European countries have diabetic registers in 2014: Sweden, Denmark, Norway, Netherlands, UK, Switzerland, and Hungary. Without making detailed overview of data collection procedures, we emphasize that data input to the registers listed in [4] is ensured either by self-registration or by burdening medical professionals with additional documentation tasks. However self-registration means that a significant percent of the patients remains unregistered. For instance in Sweden, which according to Euro Diabetes Index 2014 is the country with the best diabetes care delivery in Europe, the register was constructed by self-registration. During its development phase 2001–2005 the self-registration rate of patients gradually increased and reached 75% which in 2010 still remains stable and is one of the highest in the country [7]. No information is available about the procedures for register update and maintenance.

The Euro Diabetes Index 2014 summarizes the situation with the nice phrase "*No data, no cure*". Surprisingly, no attempts for automatic extraction of registers from available EHR repositories are mentioned in 2014. In the next section we present our achievement for building a national Diabetes Register as a component of the healthcare system, where clinical narratives can be reused dynamically for ensuring good diabetes care to patients, on the one hand, and reducing the documentation burden to many healthcare professionals, on the other hand.

3 Bulgarian Diabetes Register and Its Exploration

3.1 Automatic Register Generation

A pseudonymized Register of diabetic patients was generated in 2015 from the Outpatient Records (ORs), collected by the Bulgarian NHIF, in compliance with all legal requirements for safety and data protection [8]. The usual patient registration process was kept without burdening the medical experts with additional paper work. NHIF is the only obligatory Insurance Fund in Bulgaria so using ORs ensures 100% registration of all patients who contacted the healthcare system at all (however there are Bulgarian citizens who are not insured and some others who have ORs but are not properly diagnosed with diabetes). The data repository, underpinning the Register, currently contains more than 262 mln pseudonymized ORs submitted to the NHIF in 2010–2016 for more than 7.3 mln Bulgarian citizens (more than 5 mln yearly), including 483,836 diabetic patients. In Bulgaria ORs are produced by General Practitioners (GPs) and specialists from Ambulatory Care whenever they contact patients. Despite the primary accounting purpose these ORs summarize sufficiently the case and motivate the requested reimbursement. ORs are semi-structured files with predefined XML-format.

Many indicators in the Diabetic Register copy the structured data submitted to NHIF in ORs: *(i)* date and time of the visit; *(ii)* pseudonymized personal data, age, gender; *(iii)* pseudonymized visit-related information; *(iv)* diagnoses in ICD-10; *(v)* NHIF drug codes for medications that are reimbursed; *(vi)* a code if the patient needs special monitoring; *(vii)* a code concerning the need for hospitalization; *(viii)* several codes for planned consultations, lab tests and medical imaging.

The ORs contain also values of clinical tests and lab data, presented in the free text fields. Using extractors for automatic text analysis of Bulgarian texts, which have been developed in our previous projects, we mine these values from four OR fields: *(i) Anamnesis*: summarizes case history, previous treatments, often family history, risk factors; *(ii) Status*: summary of patient state, height, weight, BMI, blood pressure etc.; *(iii) Clinical tests*: values of clinical examinations and lab data; *(iv) Prescribed treatment*: codes of drugs reimbursed by NHIF, free text descriptions of other drugs.

We develop text mining tools for clinical texts in Bulgarian language since many years. The focus was placed mostly on clinical narratives discussing diabetic patients due to the social importance of this chronic disease. Initially various indicators concerning the patient status were extracted from hospital discharge letters [9], later the attention was shifted to extraction of numeric values of clinical tests and lab data from NHIF outpatient records [10, 11]. A brief overview of natural language processing (NLP) from clinical narratives is provided in [5].

3.2 Business Analytics as an Exploratory Tool

The Diabetes Register contains more than structured indicators in a database; actually data about subsequent visits of all patients to medical doctors is kept so the patient records in the Register have variable length. In addition, all underlying pseudonimyzed outpatient records for all diabetic patients in Bulgaria can be accessed in an efficient manner for detailed full text inspection. Due to this reason, the usual database functionality is insufficient to provide the necessary capacity for search and exploration of the Register repository. Moreover the archive size excludes direct observations by database tables. Our solution is based on business intelligence. As far as we know, this approach to construction and maintenance of medical Registers is unique.

The system BITool supports the Diabetes Register at the University Specialized Hospital for Active Treatment of Endocrinology “Acad. Ivan Penchev”, Medical University – Sofia, Bulgaria (authorized by the Bulgarian Ministry of Health to host the Register of diabetic patients in Bulgaria). BITool shows correlations among various indicators, significant for diabetes and its complications, and the prescribed drugs. Given detailed and semi-structured descriptions of all case histories, BITool identifies the importance of various risk factors combinations for diabetes development over time. The relatively complex business analytics functionalities with appropriate visualization extend the main Register purpose from monitoring to prevention. Some examples illustrate the services.

BITool displays the correlation between the compensation of Diabetes and Hypertension for the diabetic patients included in the Register at certain moment (Fig. 1). Age groups show clear distinction between children and adults. Here BITool operates on the structured information from the NHIF archive as patient pseudonym, age and types of diabetes using also aggregated lab test data. Further statistics of this kind might concern explorations of diabetic patients per region code, types of diabetes and diabetes complications, per GPs, per types of medication, according to visit frequency etc.

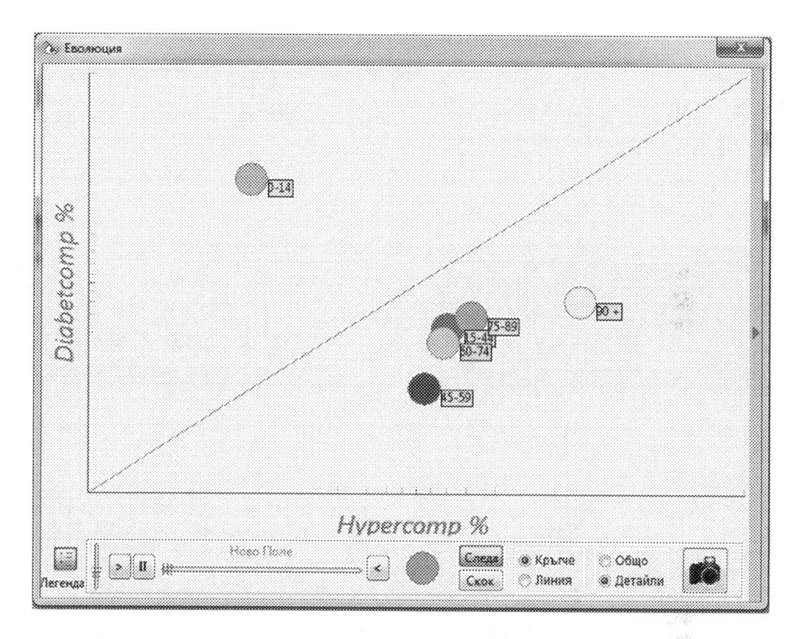

Fig. 1. Compensation of diabetes and hypertension by age groups

BITool easily finds the support (number of patients) for combinations of five risk factors for diabetes development in a cohort of patients without Diabetes (Fig. 2). The patients are outside the Register and data is extracted from the respective ORs using the same software tools that generated the Register. The latter is updated yearly with information provided within an archive of pseudonimyzed ORs for the respective year.

BITool integrates drill down functionality as well; clicking on some item, aggregating a list, moves the user to a level of greater detail. For instance, Fig. 3 shows an aggregated report about drugs prescribed to diabetic patients for 2016. Patient numbers are listed in age groups and genders. Clicking on any number (e.g. “2” in line A10AD, age 0–14, 2 boys in the second column) will open a list with these two patient identifiers and their basic Register indicators, from where access to all the information about them is provided.

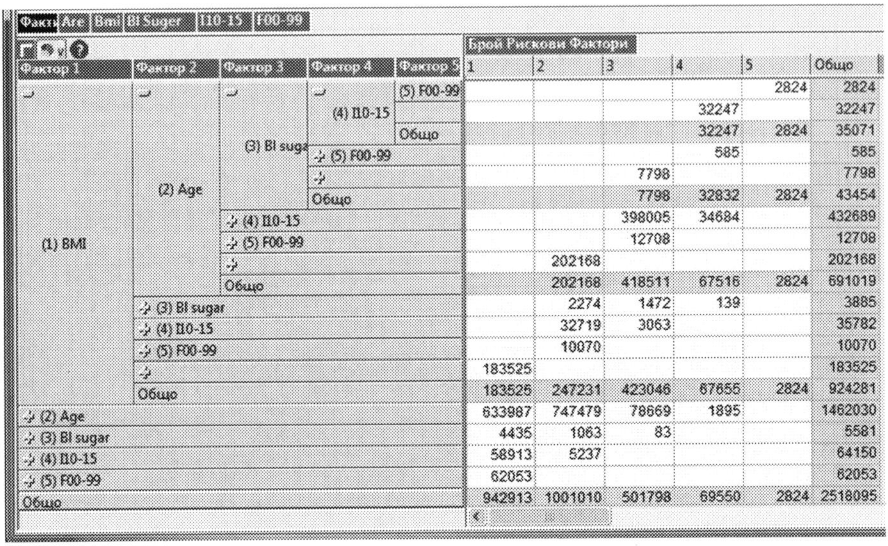

Фактор 1	Фактор 2	Фактор 3	Фактор 4	Фактор 5	1	2	3	4	5	Общо
(1) BMI	(2) Age	(3) Bl sugar	(4) I10-15	(5) F00-99					2824	2824
								32247		32247
				Общо				32247	2824	35071
			(5) F00-99					585		585
							7798			7798
			Общо				7798	32832	2824	43454
		(4) I10-15					398005	34684		432689
		(5) F00-99					12708			12708
						202168				202168
		Общо				202168	418511	67516	2824	691019
	(3) Bl sugar					2274	1472	139		3885
	(4) I10-15					32719	3063			35782
	(5) F00-99					10070				10070
					183525					183525
	Общо				183525	247231	423046	67655	2824	924281
(2) Age					633987	747479	78669	1895		1462030
(3) Bl sugar					4435	1063	83			5581
(4) I10-15					58913	5237				64150
(5) F00-99					62053					62053
Общо					942913	1001010	501798	69550	2824	2518095

Fig. 2. Monitoring number of citizens with risk factors for diabetes development

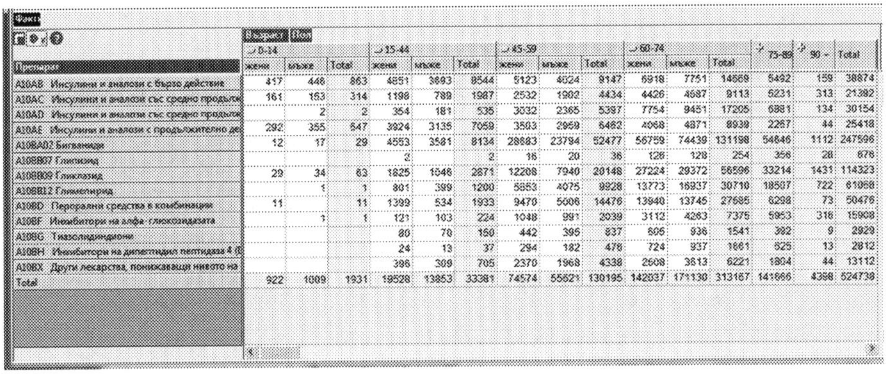

Препарат	0-14			15-44			45-59			60-74			75-89	90 -	Total
	жени	мъже	Total	жени	мъже	Total	жени	мъже	Total	жени	мъже	Total			
A10AB Инсулини и аналози с бързо действие	417	448	863	4851	3893	8544	5123	4024	9147	6918	7751	14669	5492	159	38874
A10AC Инсулини и аналози със средно продълж	161	153	314	1198	789	1987	2532	1902	4434	4426	4687	9113	6231	313	21382
A10AD Инсулини и аналози със средно продълж		2	2	354	181	535	3032	2365	5397	7754	9451	17205	6881	134	30154
A10AE Инсулини и аналози с продължително де	292	355	647	3924	3135	7059	3503	2959	6462	4068	4871	8939	2267	44	25418
A10BA02 Бигваниди	12	17	29	4553	3581	8134	28683	23794	52477	56759	74439	131198	54646	1112	247596
A10BB07 Глипизид				2		2	16	20	36	126	128	254	356	28	676
A10BB09 Гликлазид	29	34	63	1825	1046	2871	12208	7940	20148	27224	29372	56596	33214	1431	114323
A10BB12 Глимепирид		1	1	801	399	1200	5853	4075	9928	13773	16937	30710	18507	722	61068
A10BD Перорални средства в комбинации	11		11	1399	534	1933	9470	5006	14476	13940	13745	27685	6298	73	50476
A10BF Инхибитори на алфа-глюкозидазата		1	1	121	103	224	1048	991	2039	3112	4263	7375	5953	316	15908
A10BG Тиазолидиндиони				80	70	150	442	395	837	605	936	1541	392	9	2929
A10BH Инхибитори на дипептидил пептидаза 4 (D				24	13	37	294	182	476	724	937	1661	625	13	2812
A10BX Други лекарства, понижаващи нивото на				396	309	705	2370	1968	4338	2608	3613	6221	1804	44	13112
Total	922	1009	1931	19528	13853	33381	74574	55621	130195	142037	171130	313167	141866	4398	524738

Fig. 3. Groups of drugs (by ATC codes) prescribed to diabetic patients (age, gender) in 2016

4 Frequent Pattern Mining for Knowledge Discovery

4.1 Motivation and Context

The Register is pseudonimyzed, i.e. all ORs for each patient are linked in one case history. Then data mining can be used to discover unknown associations among data items in the Register. The algorithm MixCO for finding Maximal Frequent Itemsets (MFI) in Frequent Pattern Mining (FPM) has been developed [5, 11] and recently we apply it to study associations between diseases (so called comorbidities) for patients with Diabetes Mellitus Type 2 (DM2). Given the importance of early diabetes discovery and prevention, our aim is to identify risk factors using the Register data.

We consider the patients in prediabetes condition taking ORs from the period of two years preceding the onset of DM2. Below we show how retrospective analyses are done using the ORs: some comorbidities are identified for the prediabetes period, they are analyzed and given to medical experts who can define alerts about more complex risk factors for DM2. In general comorbidities are considered as frequent patterns of diagnoses.

Formally, for the collection S of ORs we extract the set of all different patient identifiers $P = \{p_1, p_2, \ldots, p_N\}$. This set corresponds to transaction identifiers (*tids*) and we call them *pids* (patient identifiers). We consider each patient visit to a doctor as a single event. For each patient $p_i \in P$ an event sequence of tuples $\langle event, timestamp \rangle$ is generated: $E(p_i) = (\langle e_1, t_1 \rangle, \langle e_2, t_2 \rangle, \ldots, \langle e_{k_i}, t_{k_i} \rangle), i = \overline{1, N}$. Let $\mathcal{E}$ be the set of all possible events and $\mathcal{T}$ be the set of all possible timestamps. Let $I = \{id_1, id_2, \ldots, id_p\}$ be the set of all diseases ICD-10[1] codes, which we call *items*. Each subset $X \subseteq I$ is called an *itemset*. We define a projection function $\pi{:}(\mathcal{E} \times \mathcal{T})^N \rightarrow 2^I$: $\pi(E(p_i)) = I(p_i) = (id_{1i}, id_{2i}, \ldots, id_{m_i})$, such that for each patient $p_i \in P$ the projected time sequence contains only the first occurrence (onset) of each disorder recorded in $E(p_i)$. Let $D \subseteq P \times 2^I$ be the set of all itemsets in our collection after projection π in the format $\langle pid, itemset \rangle$. We shall call D a *database*. We are looking for itemsets $X \subseteq I$ with frequency ($\sup(X)$) above given *minsup*. Let $\mathcal{F}$ denote the set of all frequent itemsets, i.e. $\mathcal{F} = \{X | X \subseteq I \text{ and } \sup(X) \geq minsup\}$. A frequent itemset $X \in \mathcal{F}$ is called *maximal* if it has no frequent supersets. Let $\mathcal{M}$ denote the set of all maximal frequent itemsets, i.e. $\mathcal{M} = \{X | X \in \mathcal{F} \text{ and } \nexists\, Y \in \mathcal{F}, \text{ such that } X \subset Y\}$. Let 2^X denote the power set (set of all subsets) of itemset X. Then each subset of $X \in \mathcal{F}$ is also a frequent itemset, i.e. $\forall Y \in 2^X \text{ implies that } Y \in \mathcal{F}$. For each item $id \in I$ we define the set called *pidset*: $p(id) = \{p_i | \langle p_i, I(p_i) \rangle \in D \text{ and } id \in I(p_i)\}$.

The majority of FPM and MFI algorithms consider no contextual information of the processed data [12]. Only few methods for contextual FPM and FSM (frequent sequence mining) use structured background knowledge: hierarchies [13] and ontologies [14], or some metrics to measure distances between the frequent patterns context [15]. Rabatel et al. [13] propose a hierarchical organization of attributes that allows different levels of abstraction. They present an application in the marketing domain based on clustering of frequent patterns of customers depending on their age, gender, etc. in contrast to the classic FSM methods. Ziembiński [15] proposes a new FSM approach for extracting small contextual models from smaller collections of data that are summarized later in generalized models using information from contextual models with common information. A metrics for measuring distance of context models is applied. Huang et al. [14] present one of the first approaches for contextual FPM in EHRs for adverse drug effect monitoring. Two algorithms are proposed: semantic hypergraph-based k-itemset generation and ontology-based k-itemset enrichment. These methods identify some complex patterns which are usually skipped by other FPM algorithms and prove to be very useful in health informatics.

---

[1] International Classification of Diseases and Related Health Problems 10th Revision. http://apps.who.int/classifications/icd10/browse/2015/en.

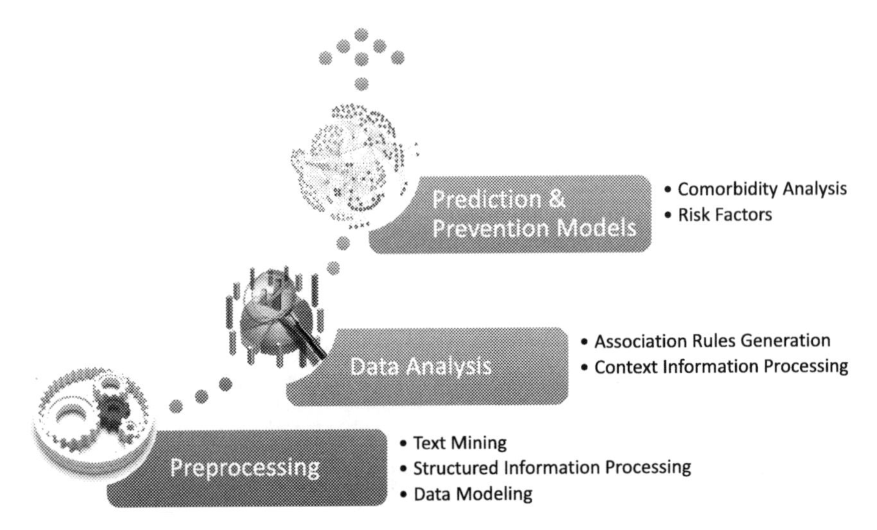

Fig. 4. System architecture

We define a set of attributes of interest $A = \{a_1, a_2, \ldots, a_k\}$. Context Q for some patient $p_i \in P$ is defined as the set of attribute-value pairs from patient profile information: $Q(p_i) = \{\langle a_1, q_1 \rangle, \langle a_2, q_2 \rangle, \ldots, \langle a_k, q_k \rangle\}$.

From $Q(p_i)$ we generate a feature vector $v(p_i) = (v_{1i}, v_{2i}, \ldots, v_{mi})$, where each attribute $a_j \in A$ with N_j possible values is represented by N_j consecutive positions in the vector. For a set of MFI $\mathcal{M}$ with cardinality $|\mathcal{M}| = \mathrm{K}$ we have K classes of comorbidities. We apply classification of multiple classes in order to generate rules for each comorbidity class. We use large scale multi class classification as we deal with a big database (millions of ORs) and a large group of comorbidity classes (ICD-10 contains approx. 12,000 four-sign codes of diagnoses). We use Support Vector Machines and optimization based on block minimization method described by Yu et al. [16].

For searching diseases comorbidities we apply the MixCO algorithm for searching MFI. We propose a cascade data mining approach for MFI enriched with context information. MixCO is a tabular method using a vertical database, depth-first traversal as well as set intersection and diffsets [11].

The architecture of the experimental workbench is shown in Fig. 4. We start with preprocessing by gathering context data and diagnosis codes for FPM. Then we provide data analysis by applying MIxCO and context based analysis. The post-processing identifies the importance of different attributes for each MFI. To study the nature of comorbidities we need to investigate the context in which they occur.

The preprocessing modules combine structured OR data (age, gender, and demographic region, clinic visits and hospitalizations, ATC codes of drugs that are reimbursed by NHIF) and perform free text analysis in order to deliver additional context attributes beyond the structured information about the patients. Text mining tools [9] extract vital parameters (BMI, blood pressure – Riva Rocci), lab tests values (HbA1c, Blood Glucose levels, etc.), and some prescribed therapy (ATC codes of drugs beyond the ones that are reimbursed by NHIF). Due to the huge number of possible distinct attribute values some aggregation is needed. WHO provides some standard aggregated categories like standard age groups, BMI classification[2] - *underweight, normal weight, overweight, obesity*. An approach for generalization of attributes related to geolocations is presented in [11], it helps for identify associations between patient attributes and the location where they live. For the status and lab test data we take the worst value for the period, according to the risk factors definition.

4.2 Experiments and Results

We discuss experimental results for patients with DM2 onset in 2015. We excerpted from the Diabetes Register the ORs of these patients for 2013–2014 when, as we assume, they were in a prediabetes condition. The idea is to check whether we can successfully discover risk factors for these patients looking only at their ORs in 2013 and 2014. Then, mapping our hypotheses to the real data in 2015, we test whether our approach is feasible (due to the short period of observation and lack of data about mortality, at the moment we cannot follow diabetes development in longer periods.)

In the Register each OR, corresponding to a single visit, contains up to four diagnoses encoded in ICD-10. Some diagnoses are presented by 4-sign encodings, i.e. in a more specific way, while others use the more general 3-sign encoding. Due to the hierarchical organization of ICD-10 we analyze individually two collections: the original one, which is more specific (with 4-sign codes) and we generalize also all diagnoses to more general classes (with 3-sign codes). The result of data analysis for patients with DM2 onset in 2015 are shown in Table 1.

Table 1. Data analysis results for patients with prediabetes in 2013–2014

Set	2013		2014		2013–2014	
Items	ICD-10 3 signs	ICD-10 4 signs	ICD-10 3 signs	ICD-10 4 signs	ICD-10 3 signs	ICD-10 4 signs
Patients	27,082	27,082	27,902	27,902	29,205	29,205
Outpatient records	267,194	267,194	296,129	296,129	556,323	556,323
ICD-10 codes	1,142	4,701	1,145	4,834	1,257	5,503
minsup	0.01	0.01	0.01	0.01	0.01	0.01
Total MFI	203	486	219	512	521	1,406
Longest MFI	5	8	5	9	6	9
Frequent itemsets	608	7,452	689	8,935	1,909	32,093
Association rules	686	58,299	810	78,052	2,722	381,012

[2] WHO, BMI Classification http://apps.who.int/bmi/index.jsp?introPage=intro_3.html.

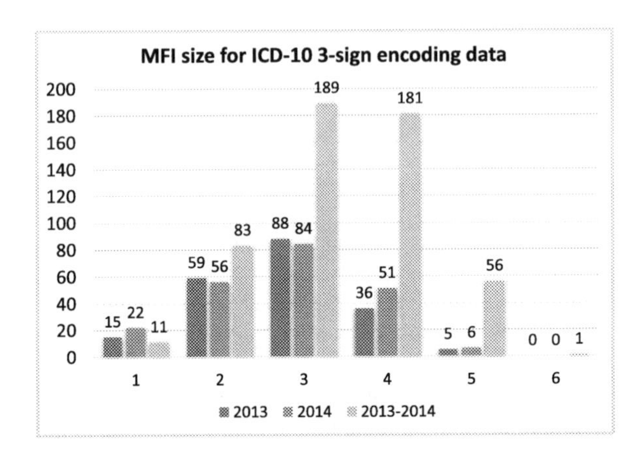

Fig. 5. Distribution of MFI by size for three OR collections with ICD-10 3-sign encodings

The distribution of MFIs by size for three collections with ICD-10 3-sign and 4-sign encodings is shown correspondingly in Figs. 5 and 6.

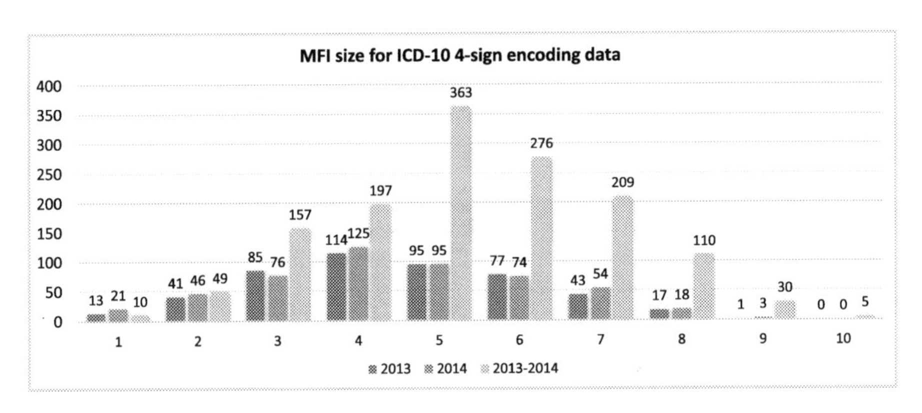

Fig. 6. Distribution of MFI by size for three OR collections with ICD-10 4- sign encodings

The top three strongest (with maximal support) MFI found by the algorithm are shown in Table 2 (ICD-10 3-sign encodings) and Table 3 (ICD-10 4-sign encodings), where the support value is denoted by #S.

Table 2. The top 3 MFI for data collection with 3-sign ICD-10 encodings

2013	2014	2013–2014
I11 I20 H25 #S:671	I11 I20 M51 #S:755	I11 I20 E78 #S:931
I10 H52 #S:667	I11 I20 H25 #SUP:748	I11 I20 J20 #S:847
I11 I20 M51 #S:628	I10 H52 #S:722	I11 I20 K29 #S:831

Table 3. The top 3 MFI for data collection with 4-sign ICD-10 encoding

2013	2014	2013–2014
I11.9 I20.8 I48 #S:583	I11.9 I20.8 I20.9 #S:740	I11.9 I20.8 I11.0 I50.0 I48 #S:400
I11.9 E04 #S:555	I11.9 I20.8 I69.8 #S:736	I11.9 I20.8 I20.9 I11.0 I50.0 #S:372
I10 I20.9 #S: 512	I11.9 I20.8 M17.0 #S:567	I11.9 I20.8 H52.4 H35.0 #S:357

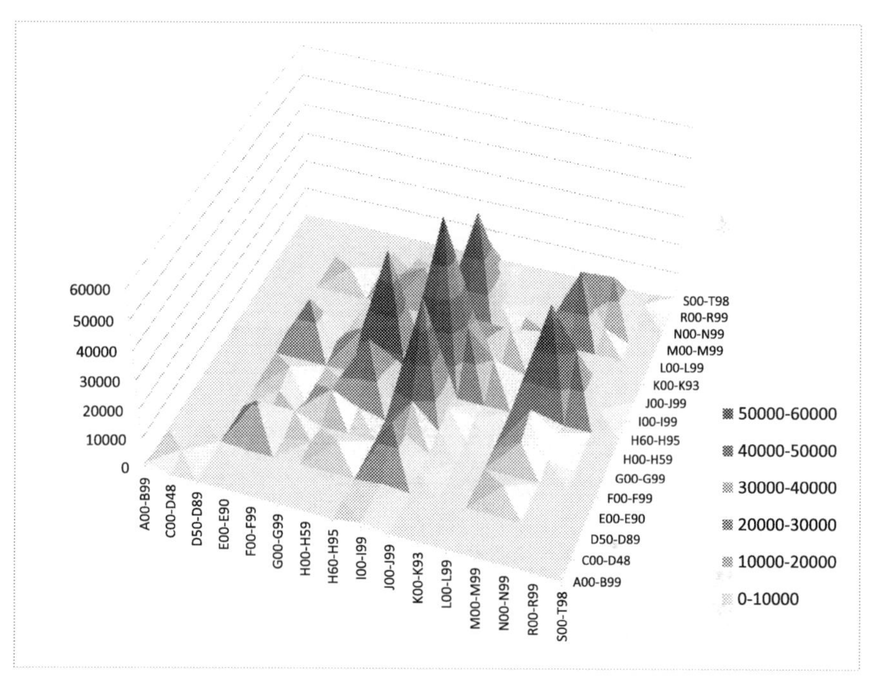

Fig. 7. Comorbidities for 2013-2014 collection of ORs grouped by classes in ICD-10

Now we need to explain why the diagnoses in the MFIs appear together. It is not surprising that the strongest top 3 MFIs in Tables 2 and 3 contain different diseases of the circulatory system, like Hypertensive diseases (I10-I15), Ischaemic heart diseases (I20-I25), Atrial fibrillation and flutter (I48), Cerebrovascular diseases (I60-I69), and other forms of heart disease (I30-I52). It is well known that diseases of the circulatory system are primary risk factors for DM2. These can be seen also as highest peaks in Fig. 7 which presents the comorbidities of different ICD-10 classes for 2013–2014 in ORs of patients with DM2 onset in 2015.

Further classes of diseases with higher frequency in the MFIs are shown in Fig. 7: Diseases of the eye and adnexa (H00-H59), Diseases of the musculoskeletal system and connective tissue (M00-M99), Diseases of the nervous system (G00-G99), Acute bronchitis (J20), and Gastritis and duodenitis (K29), all of them typical for prediabetes.

One unusual finding is the frequency of Malignant neoplasm of breast (C50) that was also identified as a maximal frequent itemset MFI#149 with a single diagnose only in all three collections. Figure 8 shows the demographic information for prevalence of

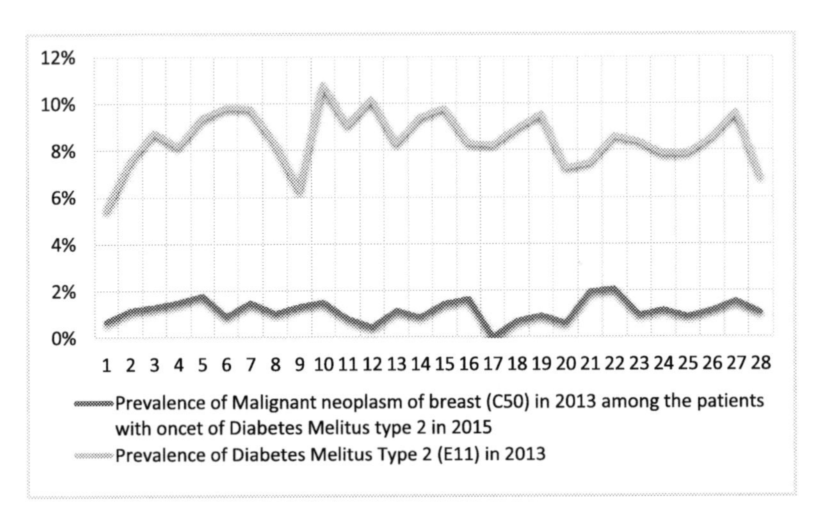

Fig. 8. Demographic data about the prevalence of C50 in prediabetes condition and E11

Maligna neoplasm of breast (C50) in prediabetes condition and DM2, with ICD-10 code E11, in 28 Bulgarian regions. There is a strong correlation of 0.93 between these two diagnoses except for two regions with ID#9 and ID#17. The latter finding is unexpected and needs further clarification; the Register shows that there are less registered diabetic patients in region #9 but this is insufficient to motivate the correlation shown in Fig. 8.

Figure 9 shows the distribution of patients in the support of "MFI#149" according to their age. The gender value in the support of "MFI#149" is female, with one exception for a male, for whom this diagnose is considered as a rare disease. The age values show that these are mainly female patients in menopause which is considered as a period with high risk for breast cancer. From the context information in the support of "MFI#149" for BMI and blood pressure we can also observe that most patients in this support set have higher risk of DM2 due to the presence of multiple risk factors as obesity (ICD-10 code I66) and hypertension (ICD-10 codes I10-I11).

Usually the association between Malignant neoplasm of breast (C50) and DM2 is studied in the opposite direction, considering the diabetes treatment as a risk factor for breast cancer [17]. However recently the association of breast cancer as a risk factor in prediabetes condition was in focus as well [18]. We note that in general the ICD-10 diagnose C50 is not considered risky for diabetes. But our algorithm reveals this unknown and latent interrelationship so it needs deeper analysis by medical experts.

Finally we briefly discuss the data quality issue and how we deal with it in our data mining approach. It is well known that missing data in medical documentation is inevitable. There are many patients for whom the available ORs contain no information about certain context attributes. Thus some attribute values are replaced by the value NA, which is considered as the most general value.

Data about HbA1c (glycated hemoglobin) are available only for 15% of patients, that is why we consider this attribute as a more general value ANY. But we note that

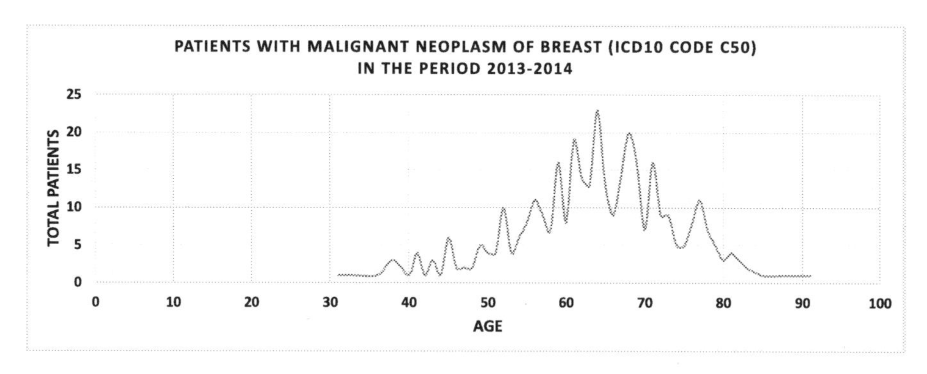

Fig. 9. Age of the patients in the support set of "MFI#149"

the lack of HbA1c measurements is not surprising because tests for HbA1c are made when the diabetes is diagnosed (and this has happened in 2015 for the selected patient cohort). Data for blood glucose are available only for 45% of these patient and for 30% of them the values were high.

5 Conclusion and Future Work

In this paper we present the national Diabetes Register, automatically generated using semi-structured patient records in Bulgarian language, and show how the stepwise integration of modern data processing technologies turn the Register to an environment for monitoring, prediction, issuing alerts, and discovery of specific risk factors. Application of automatic NLP in large scale is a real novelty in this area. Perhaps one of the most important achievement is the demonstration that reuse of available medical documentation leads to new quality when modern ICT is integrated as an enabling tool. The authors show this achievement to the national healthcare authorities whenever possible and officially propose to reuse existing clinical texts in the implementation of the Electronic Health Record (EHR) system in Bulgaria.

Future work includes further elaboration of specific algorithms that take into consideration temporal sequences of events. Developing more efficient knowledge discovery tools will provide functionality to monitor patient status over time, in the context of all available information, and to issue alerts for coincidence of risk factors that open the door to socially-important chronic diseases. In this way it will become possible to identify the Bulgarian citizens who have predisposition to various serious diseases.

Acknowledgements. This research is partially supported by grant IZIDA 02/4 (SpecialIZed Data MIning MethoDs Based on Semantic Attributes), funded by the Bulgarian National Science Fund in 2017–2019. The authors acknowledge also the support of Medical University – Sofia, the National Health Insurance Fund and the Bulgarian Ministry of Health.

References

1. WHO Diabetes Fact Sheets, November 2017. http://www.who.int/mediacentre/factsheets/fs312/en/. Accessed 20 Jan 2018
2. WHO Global Report on Diabetes (2016). http://apps.who.int/iris/bitstream/10665/204871/1/9789241565257_eng.pdf?ua=1. Accessed 20 Jan 2018. ISBN 978 924 156525 7
3. Richardson, E., (ed.): National Diabetes Plans in Europe: what lessons are there for the prevention and control of chronic diseases in Europe? Policy Brief of the Joint Action on Chronic Diseases and Promoting Healthy Ageing across the Life Cycle, WHO Regional Office for Europe (2016). ISSN 1997-8065
4. Garrofé, B., Björnberg, A., Phang, A.Y.: Euro Diabetes Index 2014. Health Consumer Powerhouse Ltd., (2014). ISBN 978-91-980687-4-0
5. Boytcheva, S., Angelova, G., Angelov, Z., Tcharaktchiev, D.: Integrating Data Analysis Tools for Better Treatment of Diabetic Patients. In: Kalinichenko, L., Manolopoulos, Y., Skvortsov, N., Sukhomlin, V. (eds.) Selected Papers of the XIX International Conference on Data Analytics and Management in Data Intensive Domains (DAMDID/RCDL 2017), CEUR Workshop Proceedings, vol. 2022, pp. 230–237 (2017). http://ceur-ws.org/Vol-2022/. Accessed 20 Jan 2018
6. European Best Information through Regional Outcomes in Diabetes (EUBIROD) homepage. http://www.eubirod.eu/. Accessed 20 Jan 2018
7. Hallgren Elfgren, I.M., Törnvall, E., Grodzinsky, E.: The process of implementation of the diabetes register in primary health care. Int. J. Qual. Health Care **24**(4), 419–424 (2012)
8. Tcharaktchiev, D., Zacharieva, S., Angelova, G., Boytcheva, S., Angelov, Z., et al.: Building a bulgarian national registry of patients with diabetes mellitus. J. Soc. Med. **2**, 19–21 (2015). ISSN 1310-1757 (in Bulgarian Language)
9. Boytcheva, S., et al.: Obtaining status descriptions via automatic analysis of hospital patient records. Informatica **34**, 269–278 (2010)
10. Boytcheva, S., Angelova, G., Angelov, Z., Tcharaktchiev, D.: Text mining and big data analytics for retrospective analysis of clinical texts from outpatient care. Cybern. Inf. Technol. **15**(4), 58–77 (2015). https://doi.org/10.1515/cait-2015-0055
11. Boytcheva, S., Angelova, G., Angelov, Z., Tcharaktchiev, D.: Mining comorbidity patterns using retrospective analysis of big collection of outpatient records. Health Inf. Sci. Syst. **5**(1), 3 (2017). https://doi.org/10.1007/s13755-017-0024-y
12. Aggarwal, C., Bhuiyan, M., Hasan, M.: Frequent pattern mining algorithms: a survey. In: Aggarwal, C., Han, J. (eds.) Frequent pattern mining, pp. 19–64. Springer, Cham (2014). https://doi.org/10.1007/978-3-319-07821-2_2
13. Rabatel, J., Bringay, S., Poncelet, P.: Mining sequential patterns: a context-aware approach. In: Guillet, F., Pinaud, B., Venturini, G., Zighed, D. (eds.) Advances in Knowledge Discovery and Management, pp. 23–41. Springer, Heidelberg (2013). https://doi.org/10.1007/978-3-642-35855-5_2
14. Huang, J., Huan, J., Tropsha, A., Dang, J., Zhang, H., Xiong, M.: Semantics-driven frequent data pattern mining on electronic health records for effective adverse drug event monitoring. In: 2013 IEEE International Conference on Bioinformatics and Biomedicine BIBM, pp. 608–611. IEEE (2013). https://doi.org/10.1109/bibm.2013.6732567
15. Ziembiński, R.Z.: Accuracy of generalized context patterns in the context based sequential patterns mining. Control Cybern. **40**(3), 585–603 (2011). http://yadda.icm.edu.pl/baztech/element/bwmeta1.element.baztech-article-BATC-0009-0001/c/httpwww_bg_utp_edu_plartcc2011ziembinski.pdf. Accessed 20 Jan 2018

16. Yu, H.F., Hsieh, C.J., Chang, K.W., Lin, C.J.: Large linear classification when data cannot fit in memory. ACM Trans. Knowl. Discov. Data **5**(4), 23 (2012). https://doi.org/10.1145/2086737.2086743
17. Pan, X.F., He, M., Yu, C., Lv, J., Guo, Y., Bian, Z., et al.: Type 2 Diabetes and risk of incident cancer in China: a prospective study among 0.5 million Chinese adults. Am. J. Epidemiol., kwx376 (2018). https://doi.org/10.1093/aje/kwx376
18. Onitilo, A.A., Stankowski, R.V., Berg, R.L., Engel, J.M., Glurich, I., Williams, G.M., Doi, S.A.: Breast cancer incidence before and after diagnosis of type 2 diabetes mellitus in women: increased risk in the prediabetes phase. Eur. J. Cancer Prev. **23**(2), 76–83 (2014). https://doi.org/10.1097/CEJ.0b013e32836162aa

Next Generation Genomic Sequencing: Challenges and Solutions

An Introduction to the Computational Challenges in Next Generation Sequencing

Zoltan Szallasi(✉)

Computational Health Informatics Program, Boston Children's Hospital, Harvard Medical School, Boston, USA
zoltan.szallasi@childrens.harvard.edu

Abstract. During the last decade next generation sequencing has become one of the research areas that poses the most significant challenges both in terms of big data handling and algorithmic problems.

In this review we will discuss those challenges with a particular emphasis on those issues where scientific innovation will be essential to make progress.

Keywords: Next generation sequencing · Big data · Bioinformatics

1 Introduction

Science collects data in order to uncover natural laws that could be used to make scientific predictions. "Hard" or "quantitative" sciences usually work with homogeneous data types, such as the location or speed of an object. This allows a relatively simple summary analysis and derivation of universal laws. Life sciences and especially medical research, however, historically suffers from relative data scarcity and the difficulties of integrating widely disparate data types. For example, oncological pathology (diagnostic evaluation of cancer) makes predictions about the chances of overall patient survival based on a histological image of the tumor combined with a mutation status of a gene combined with the number and location of tumor metastasis etc.

Since the middle of the last century it has been known, however, that biology is held together by strong unifying scientific principles as well. The genetic information of all living organisms is coded in nucleic acids, usually DNA, which is translated into RNA, that is in turn converted into the actual effector molecules, proteins. The nucleic acid code, with minor modifications, is universal across the entire living universe. This, of course, suggested that just by sequencing a large number of DNA and RNA samples one will be able to learn a lot about living organisms. The problem, until about a decade ago, was that sequencing was rather expensive and cumbersome. Truly visionary technology developers realized that the chemistry of nucleic acids offers itself to massive parallelization and thus significantly increasing throughput. This has led to the next generation sequencing revolution. The price drop of nucleic acid sequencing did not only follow Moore's rather optimistic law but starting at about 2008 it even accelerated [1]. At the writing of this article, an entire human genome can be sequenced at a 30x coverage for less than $1000, and the goal of the industry is to reach the $100 mark for the same level of sequencing coverage. This is clearly pointing to the direction

L. Kalinichenko et al. (Eds.): DAMDID/RCDL 2017, CCIS 822, pp. 37–45, 2018.
https://doi.org/10.1007/978-3-319-96553-6_3

that the already considerable speed of data accumulation will further accelerate and data management will become a true computational problem both from a scientific and an economical point of view. For the latter, one should remember that an entire human genome is on the order of 100 GB of data. The sequencing cost of this in the foreseeable future, as mentioned above, will be between $100 and $1000 but data storage for this amount of raw data is currently in the range of 40–60 USD/year. Consequently, it is becoming more and more evident that next generation sequencing based analysis will face a variety of problems, traditionally associated with "big data" issues in computer sciences.

These developments not only opened up new opportunities for computer scientists looking for new challenges but necessitated their active involvement as well. No modern research institute or biotechnology company can survive without experts skilled in the field of genomics related bioinformatics. While this ensures solid job prospects of those students specializing in the standard tool set of bioinformatics, it is less clear how much truly challenging problems exist for those inventive minds entering the field of genomics from other quantitative research fields.

In this paper, we will illustrate with examples how innovation originating in other scientific disciplines has contributed to genomics and we will list a set of open problems that are waiting for the next generation of bioinformatics researchers [2].

2 Challenge #1 – the Data Deluge

2.1 Data Types and Source of Data

One of the most appealing features of next generation sequencing is that based on the same technological principle one can gather information on a wide variety of levels from the biological organism. One can sequence DNA (nuclear or mitochondrial) and thus characterize the genome. One can also sequence RNA, characterizing the transcriptome and thus characterize the expression level of each gene in a given biological sample. DNA sequencing can be modified to query the methylated sites in the DNA. It is also possible to "pull down" chemically and sequence those regions of the genome that bind to a given protein, such as a transcription factor and thus identify transcription factor binding sites. The biochemical preparatory steps of all these various methods produce the same starting material for next generation sequencing, a DNA library [3]. The sequencing machines will process these libraries by reading out the sequence of the nucleic acids contained in those. The output of this process is a list of tags contained in standardized format, such as FASTQ files. Thus, at the end of each sequencing reaction we end up with a long list of tags, usually varying in size from a few hundred to several thousands of letters (base pairs). At this point, the biological problem is converted from biochemistry to a pure computational problem, identifying strings and counting them.

2.2 Projected Data Amount

Sequencing reactions produce a lot of data at a constantly decreasing price. As mentioned above, 100 GB of sequencing data is produced for less than $1000. This is predicting an immense speed of data accumulation [4]. In fact, genomics is estimated to produce about 1 zetta-bases/year requiring at least 2–40 EB/year storage [4]. This data accumulation rate is far outpacing several other rapidly expanding data domains such as astrophysics or the social media such as twitter or youtube.

What are the implications of this for the individual bioinformatics researcher and the research community as a whole? First, cost of data storage is rapidly becoming a sizeable factor of research budgets. Second, it is not effective, in fact nearly impossible, to download sizeable genomic data sets (a good example is the 560 breast cancer whole genome sequencing data set [5]) to each lab that wants to reanalyze it. It is much more effective to establish cloud based computational analytical resources, when the starting dataset is stored at a single site and individual researchers can run their own algorithms independently from each other on that core data set [6] (see e.g. https://gdc.cancer.gov/). While in principle this is a reasonable plan, it will take a significant amount of actual technical expertise and work invested to make such a system work in a truly effective manner that would allow all researchers to run their own analysis.

2.3 Quality of Data

Before considering the purely computational challenges of next generation sequencing, one should briefly mention several issues at the interface of the biochemical sequencing reactions and the computational data management.

Like all measurements next generation sequencing is also affected by noise that manifests in erroneous base calling. Quantitatively this error is characterized by the probability of an incorrect base call. This error rate widely varies across different technologies and usually ranges from 0.1% single pass error rate (e.g. in the case Illumina HiSeq 2500) to up to 13% single pass error rate in the case of the PacBio technology [7]. This error rate is usually presented as the "Phred quality score" (named after the Phred base calling algorithm [8]): $Q = -10 \log_{10} P$.

A Q-score of 30 (Q30) is equivalent to the probability of an incorrect base call 1 in 1000 times. These numbers will, of course, have a significant impact on next generation sequencing analysis that aims to detect mutations e.g. in tumors.

3 Challenge #2: Algorithmic Challenges in Genomics

The second source of challenges is defined by the actual analytical goal of next generation sequencing. As we outlined above, the output of all sequencing reactions is a large set of sequence tags that are essentially strings of varying lengths containing four letters A, C, G, T. Each of those tags are basically a copy of a given genomic location of the investigated organism. The sequence tags carry useful information only in the context of this genomic location. Therefore, the first computational task is to determine where that actual tag is coming from, i.e. where that actual sequence tag is mapping in the genome. This process, however, requires a complete catalogue of genomic sequences of a given organism, i.e. assembling the genome.

3.1 Assembling Genomes

Assembling genomes is not a trivial task and it has significantly benefitted from the insight of computer scientists. Next generation sequencing is a "shotgun" approach when a large number of sequence tags are produced that are more or less randomly selected across the entire genome. These randomly selected tags during shotgun sequencing then need to be assembled. In the first generation of sequence assembly algorithms this was done more or less by brute force, when all possible pairs of sequence pairs were tested for overlap and then these overlapping pieces were then assembled into a contiguous genome during the "layout" step. These early genome assembly methods had several drawbacks, the most prominent of those being computational cost and their inability to deal with repetitive sequences. This problem was elegantly overcome by the application of de Bruijn graphs by Pevzner et al. [9]. In this work, they converted the genome assembly problem from the NP-hard Hamiltonian cycle problem into the polynomial Eulerian cycle problem and they could also handle efficiently the problem of repetitive sequences. These results opened up a whole new field in applied computer science that was gradually chipping away at other real-life problems of shotgun sequencing based genomes sequencing, such as those caused by erroneous reads or multiple chromosomes [10].

As genomics is used to address more and more complex problems the computational analysis will certainly benefit from further innovation. For example, while most genomes are diploid (having two sets of chromosomes), certain species have 3, 4, 5 or 6 (or more) sets of chromosomes. This means that during sequencing the same genomics regions will be sampled over and over again with the slight variations (such as single nucleotide variations) that might be present in the various chromosomes [11]. Metagenomics, is another exciting application of next generation sequencing when a large number, hundreds or thousands, of different microorganisms are sequenced in the same sample. Genomes are assembled and identified from these "wholesale" set of sequence tags. Since the abundance of each species is different and genomic material is often exchanged between the various microorganisms, this field also faces its own well defined computational challenges [12].

Due to the above described difficulties producing a reliable, complete genome without gaps is a rather difficult task and even for such an extensively studied genome such as ours (the human genome), the work still continues and significant updates are released on a regular basis (https://www.ncbi.nlm.nih.gov/grc/human/data).

3.2 Identification of Germline Sequence Variations

Once the genome of a given species, for example that of humans, is assembled one can ask questions about the variations between individuals, as those probably underlie a significant portion of the differences between human beings such as susceptibility to various diseases. Again, this relatively simple sounding problem turns out to be rather difficult due to the particular characteristics of the human genome. This is perhaps best exemplified by the genomics based classification of HLA subtypes in a given individual. HLA molecules present epitopes on the surface of human cells and thus play a central role in cellular immunity. In order to maximize our chances to fight viral infections as a species, natural selection created an enormous sequence diversity of HLA molecules across entire humanity. Well over 10,000 different HLA sequences have been observed in the human population and the average difference between two pairs of HLA molecules can be as much as 10% (about 50 bp out of the ~500 bp relevant sequence region) [13]. Furthermore, each individual has two, often highly divergent copies of a given HLA molecule in his or her genome. Since individual sequence tags do not cover the entire polymorphic region of HLA genes, it is difficult to determine whether two polymorphic regions detected by next generation sequencing are part of the same allele or produced by two different alleles [14]. It has taken several ingenious algorithmic approaches over a decade to find reliable solutions for this problem [15, 16].

3.3 Identification of Somatic Sequence Variations

Mutation calling is one of the most important applications of next generation sequencing. Cancer is essentially a genomic disease. Specific mutations initiate and maintain neoplastic growth. Therefore, it is not surprising that one of the main sources of genomic data is sequencing the somatic mutations of tumor samples. In principle, this problem is deceptively simple. After sequencing determine the genomic location of each sequence tag, i.e. map to the human genome. Then determine, whether there are any mutations such as single nucleotide variations, short deletions or insertion present in those. In practice, however, due to the nature of human tumor biopsies and the human genome, this is a difficult task that still lacks a widely applicable reliable solution. Two examples highlight the difficulties associated with this analytical task. Genomic instability is one of the hallmarks of cancer. Cancer cells accumulate mutations at a remarkable speed. Consequently, virtually each cell has a slightly different mutational profile, leading to the possibility that a significant portion of actual mutations will be present only in a few cells resulting in a detected low mutant allele frequency. This means that an actual mutation may be represented only by a single read in a sequencing run. Considering the error rate of sequencing discussed above, this makes it difficult to distinguish between a technical sequencing error and a low allele frequency real mutation. The human genome also contains a number of analytical traps for example in the form of pseudogenes [17]. These are defined as fragments of once-functional genes that have been inactivated by nonsense, frameshift or missense mutations. When a gene and its pseudogene copy is present in the genome, the same sequence tag may correspond to the mutated form of the original functioning gene or the evolutionarily mutated form of the gene contained in the pseudogene and distinguishing between them is extremely difficult. Considering these factors, it comes as no surprise that when mutation calling algorithms are compared, only a minority of the mutations are called consistently by several algorithms [18]. Improvements in this analysis will be essential for the reliable analysis diagnostic cancer sequencing.

3.4 Mutational Signatures

The genomic instability underlying cancerous processes has posed several other important analytical problems. In normal cells a wide array of DNA repair pathways and genome integrity checkpoints ensure the faithful propagation of genetic information from parent to daughter cells. In cancer, one or more of these mechanisms maintaining genomic integrity is inactivated allowing the cancerous cells to accumulate genetic mutations rapidly. Importantly, the various DNA repair mechanism are responsible for correcting different types of mutations. Therefore, when a given DNA repair pathway is inactivated one can expect a specific mutational signature arise. However, it is not clear how many such distinct repair pathway aberrations and other mutational mechanisms are present in cancer. Therefore, an agnostic, unsupervised learning method, non-negative matrix factorization, was applied across the entire data base of all sequenced cancer samples (TCGA). As a starting point, for each cancer sequence data set the number of a given single nucleotide variation in the context of the two neighboring base pairs (96 variables in total) was counted. This created a data matrix defined by the samples and mutation types and this matrix was subjected to the above mentioned unsupervised analysis [19]. Remarkably, this approach identified several biologically explained mutational processes and helped researchers to identify which mutational process is driving a given neoplastic growth. This approach was later expanded to short deletion, insertion analysis and large scale genome rearrangements [5]. The mutational signatures identified in the original unsupervised analysis could be explained away by an associated biological mechanism only in the minority of the signatures [19]. Therefore, it is an open question how many mutational processes can be identified in human cancer by the summary analysis of genomic aberrations profiles.

4 Making Sense of Genomic Data

In the previous sections, we outlined several types of obstacles that make the accurate conversion of next generations sequence tags into genomic information. However, even if genomic analysis was virtually error free, we would still need to make sense of all the genomic information collected and this will also pose significant computational and scientific challenges for the next generation of bioinformaticians.

4.1 Predicting Function from Sequence Information

Biochemical reactions occur in three-dimensional dynamic systems and yet a significant amount of information determining those processes are coded in a simple linear sequence of base pair “letters”. For example, a protein, in most cases, is fully determined by the primary amino acid sequence and a freshly synthesized such primary amino acid sequence will fold into the appropriate three-dimensional structure by itself. Yet, predicting the same 3D structure computationally, i.e. *de novo* protein structure prediction, is far beyond the reach of current computational capabilities [20]. Since computer science considerations place this and similar problems well beyond the realm of scientific reality at the moment [21], one may want to start solving significantly simpler problems with still considerable scientific utility. For example, how can we predict from the primary sequence whether a given mutation inactivates the function of a given protein. This has immense practical significance. Let us consider, e.g. the BRCA1 gene (also referred to as the “Angelina Jolie gene” in popular culture).

This tumor suppressor gene plays an important role in repairing DNA damage and its functional loss leads to increased breast and ovarian cancer incidence. This is an enormous gene (1863 amino acids), and it is often mutated in a single amino acid location in the average population. Predicting accurately whether a given mutation leads to the inactivation of this gene has enormous impact [22] but it is obviously a rather complex problem. On one hand, a premature stop codon at the beginning of the gene will obviously inactivate the gene, on the other hand a mutation changing an alanine into a glycine in a functionally less relevant part of the gene will probably have no clinical impact. Real life mutations are on various parts on this spectrum leaving researchers with a significant number of variants of uncertain clinical significance (VUS). Recent approaches combining bioinformatics and knowledge based approaches made progress in this field [23], but we are still far from a generally applicable loss of function predictor.

Primary sequence based biological function predictors successfully contributed to many fields in biology. A salient example is immunology. We already referred to the central role HLA/epitope complexes play in activating cellular immunity. A given HLA molecule can bind a wide variety of 8–11 amino acid long sequences, epitopes, and thus presenting them to T cell receptors. Machine learning approaches were proposed and applied early on to identify the sequence characteristics of epitopes that makes them good binders to HLA [24]. Methods such as NetMHC show a remarkable accuracy to predict cytotoxic immune response induced by, e.g., novel cancer associated mutations, neoantigens [25]. While these results are impressive, there is plenty of room for improvement especially considering the fact that individualized cancer vaccines derived from next generation sequencing data are entering clinical trials [26].

4.2 Clonal Mutational Evolution in Cancer

Accumulating genetic mutations plays an important role in the development and maintenance of cancer. Different mutations have a different impact on the growth potential and survival capability of a given cancer cell. Therefore, clones with a different set of mutations will have a growth advantage or disadvantage and they will be represented with different frequencies when the tumor biopsies are sequenced. It is important to determine, which mutations and in which combination support cancer growth. This would require the identification of individual clones and their representative mutational profiles from a single whole exome or whole genome sequencing reaction. Again, this was a highly non-trivial problem, which required the implementation of a Dirichlet process based statistical model that simultaneously estimated clonal frequencies and mutation genotype from deeply sequenced somatic single nucleotide variations and copy number measurements [27].

5 The Human Resource Aspect

Technological revolutions often leave the research community unprepared. While initially the utility of cancer genome sequencing was questioned [28] it has become evident that whole exome or whole genome sequencing allows us to investigate the biology of cancer with unprecedented depths and detail [5, 19]. It seems the stream of such important research questions will continue to expand at a speed faster than the training of capable bioinformatics researchers. This shortage of experts has been noted both in industry and academia [29]. Another factor that may contribute to the current situation is that bioinformaticians often seem to be uncertain about their career path. By some estimates [29] the majority of efforts/working hours of bioinformaticians is spent in an albeit essential but still supporting role. The questions are defined by the biologists and the bioinformatics experts are expected to perform well defined steps of analysis without much intellectual contribution beyond the correct application of already established algorithms. As we outlined in this article, there have been plenty of exciting challenges in the analysis of genomic data but nevertheless, the number of those cases is much smaller than the number of bioinformaticians needed for research centers and industry to provide support. It is not clear how this dilemma will be resolved. One possibility is training more biologists in bioinformatics methods, therefore the actual researchers in charge of and performing the experiments would be able to perform the essential bioinformatics analysis. This will, however, require a significant shift in the current curricula of most universities.

References

1. Muir, P., Li, S., Lou, S., et al.: The real cost of sequencing: scaling computation to keep pace with data generation. Genome Biol. **17**(1), 53 (2016)
2. Szallasi, Z.: Development of genomic based diagnostics in various application domains. In: XIX International Conference on Data Analytics and Management in Data Intensive Domains (DAMDID/RCDL 2017), CEUR WS, vol. 2022, pp. 3–4, (2017). http://ceur-ws.org/Vol-2022/. Extended Abstract
3. Boone, M., De Koker, A., Callewaert, N.: Capturing the "ome": the expanding molecular toolbox for RNA and DNA library construction. Nucleic Acids Res. **107**, 1 (2018)
4. Stephens, Z.D., Lee, S.Y., Faghri, F., et al.: Big data: astronomical or genomical? PLoS Biol. **13**(7), e1002195 (2015)
5. Nik-Zainal, S., Davies, H., Staaf, J., et al.: Landscape of somatic mutations in 560 breast cancer whole-genome sequences. Nature **534**(7605), 47–54 (2016)
6. Reynolds, S.M., Miller, M., Lee, P., et al.: The ISB cancer genomics cloud: a flexible cloud-based platform for cancer genomics research. Cancer Res. **77**(21), e7–e10 (2017)
7. Rhoads, A., Au, K.F.: PacBio sequencing and its applications. Genomics Proteomics Bioinf. **13**(5), 278–289 (2015)
8. Liao, P., Satten, G.A., Hu, Y.-J.: PhredEM: a phred-score-informed genotype-calling approach for next-generation sequencing studies. Genet. Epidemiol. **41**(5), 375–387 (2017)
9. Pevzner, P.A., Tang, H., Waterman, M.S.: An Eulerian path approach to DNA fragment assembly. Proc. Natl. Acad. Sci. U.S.A. **98**(17), 9748–9753 (2001)
10. Compeau, P.E.C., Pevzner, P.A., Tesler, G.: How to apply de Bruijn graphs to genome assembly. Nat. Biotechnol. **29**(11), 987–991 (2011)
11. Yang, J., Moeinzadeh, M.-H., Kuhl, H., et al.: Haplotype-resolved sweet potato genome traces back its hexaploidization history. Nat. Plants **3**(9), 696–703 (2017)
12. Olson, N.D., Treangen, T.J., Hill, C.M., et al.: Metagenomic assembly through the lens of validation: recent advances in assessing and improving the quality of genomes assembled from metagenomes. Brief. Bioinf. **8**, e61692 (2017)

13. Buhler, S., Sanchez-Mazas, A.: HLA DNA sequence variation among human populations: molecular signatures of demographic and selective events. PLoS One **6**(2), e14643 (2011)
14. Szilveszter Juhos, K.R., Horváth, G.: On Genotyping Polymorphic HLA Genes — Ambiguities and quality measures using ngs. next generation sequencing - advances, applications and challenges. InTech (2016). https://doi.org/10.5772/61592
15. Szolek, A., Schubert, B., Mohr, C., et al.: OptiType: precision HLA typing from next-generation sequencing data. Bioinformatics **30**(23), 3310–3316 (2014)
16. Shukla, S.A., Rooney, M.S., Rajasagi, M., et al.: Comprehensive analysis of cancer-associated somatic mutations in class I HLA genes. Nat. Biotechnol. **33**(11), 1152–1158 (2015)
17. Goodhead, I., Darby, A.C.: Taking the pseudo out of pseudogenes. Curr. Opin. Microbiol. **23**, 102–109 (2015)
18. Krøigård, A.B., Thomassen, M., Lænkholm, A.-V., et al.: Evaluation of nine somatic variant callers for detection of somatic mutations in exome and targeted deep sequencing data. PLoS One **11**(3), e0151664 (2016)
19. Alexandrov, L.B., Nik-Zainal, S., Wedge, D.C., et al.: Signatures of mutational processes in human cancer. Nature **500**(7463), 415–421 (2013)
20. Dill, K.A., MacCallum, J.L.: The protein-folding problem, 50 years on. Science **338**(6110), 1042–1046 (2012)
21. Berger, B., Leighton, T.: Protein folding in the hydrophobic-hydrophilic (HP) model is NP-complete. J. Comput. Biol. **5**(1), 27–40 (1998)
22. Eccles, D.M., Mitchell, G., Monteiro, A.N.A., et al.: BRCA1 and BRCA2 genetic testing-pitfalls and recommendations for managing variants of uncertain clinical significance. Ann. Oncol. **26**(10), 2057–2065 (2015)
23. Li, Q., Wang, K.: InterVar: Clinical Interpretation of Genetic Variants by the 2015 ACMG-AMP Guidelines. Am. J. Hum. Genet. **100**(2), 267–280 (2017)
24. Jurtz, V., Paul, S., Andreatta, M., et al.: NetMHCpan-4.0: improved peptide-MHC Class I interaction predictions integrating eluted ligand and peptide binding affinity data. J. Immunol. **199**(9), 3360–3368 (2017)
25. Bjerregaard, A.-M., Nielsen, M., Jurtz, V., et al.: An analysis of natural T cell responses to predicted tumor neoepitopes. Front. Immunol. **8**, 1566 (2017)
26. Ott, P.A., Hu, Z., Keskin, D.B., et al.: An immunogenic personal neoantigen vaccine for patients with melanoma. Nature **547**(7662), 217–221 (2017)
27. Shah, S.P., Roth, A., Goya, R., et al.: The clonal and mutational evolution spectrum of primary triple-negative breast cancers. Nature **486**(7403), 395–399 (2012)
28. Miklos, G.L.G.: The human cancer genome project—one more misstep in the war on cancer. Nat. Biotechnol. **23**(5), 535–537 (2005)
29. Chang, J.: Core services: Reward bioinformaticians. Nature **520**(7546), 151–152 (2015)

Overview of GeCo: A Project for Exploring and Integrating Signals from the Genome

Stefano Ceri(✉), Anna Bernasconi, Arif Canakoglu, Andrea Gulino, Abdulrahman Kaitoua, Marco Masseroli, Luca Nanni, and Pietro Pinoli

Dipartimento di Elettronica, Informazione e Bioingegneria,
Politecnico di Milano, Milano, Italy
{stefano.ceri,anna.bernasconi,arif.canakoglu,andrea.gulino,
abdulrahman.Kaitoua,marco.masseroli,luca.nanni,pietro.pinoli}@polimi.it

Abstract. Next Generation Sequencing is a 10-year old technology for reading the DNA, capable of producing massive amounts of genomic data - in turn, reshaping genomic computing. In particular, tertiary data analysis is concerned with the integration of heterogeneous regions of the genome; this is an emerging and increasingly important problem of genomic computing, because regions carry important signals and the creation of new biological or clinical knowledge requires the integration of these signals into meaningful messages. We specifically focus on how the GeCo project is contributing to tertiary data analysis, by overviewing the main results of the project so far and by describing its future scenarios.

Keywords: Genomic computing · Data translation and optimization
Cloud computing · Next generation sequencing · Open data

1 Introduction

Genomics is a relatively recent science. The double helix model of DNA, due to Nobel prizes James Watson and Francis Crick, was published on Nature in April 1953 and the first draft of the human genome, produced as result of the Human Genome Project, was published on Nature in February 2001, with the full sequence completed and published in April 2003. The Human Genome Project, primarily funded by the National Institutes of Health (NIH), was the result of a collective effort involving twenty universities and research centers in the United States, the United Kingdom, Japan, France, Germany, Canada, and China.

In the last fifteen years, the technology for DNA sequencing has made gigantic steps. Figure 1 shows the cost of DNA sequencing in the last fifteen years; by inspecting the curve, one can note a huge drop around 2008, with the introduction of Next Generation Sequencing, a high-throughput, massively parallel technology based upon the use of image capturing [24]. The cost of producing a complete human sequence dropped to 1000 US$ in 2015 and is expected to drop below 100 US$ in the next two-three years.

L. Kalinichenko et al. (Eds.): DAMDID/RCDL 2017, CCIS 822, pp. 46–57, 2018.
https://doi.org/10.1007/978-3-319-96553-6_4

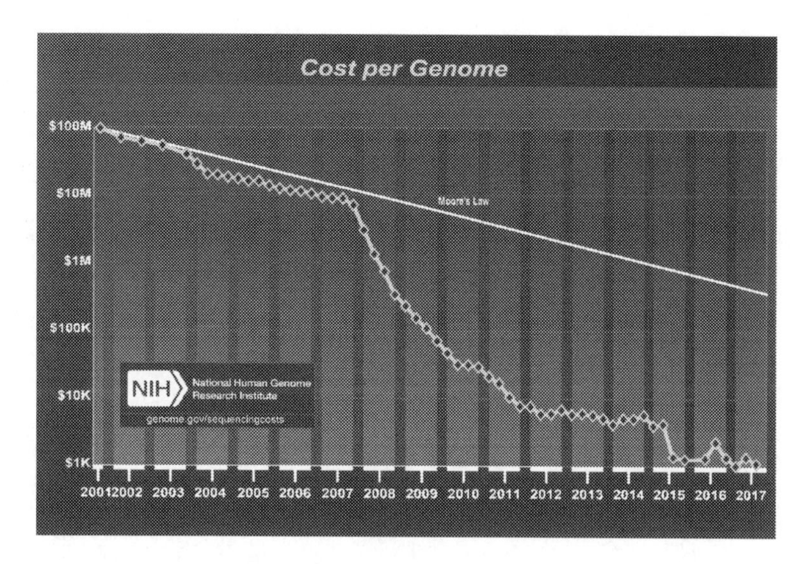

Fig. 1. Cost of DNA Sequencing, Source: NIH

Each sequencing produces a mass of information (raw data) in the form of "short reads" of genome strings. Once stored, the raw data of a single genome reach a typical size of 200 Gbyte; it is expected that between 100 million and 2 billion human genomes will be sequenced by 2025, thereby generating the biggest "big data" problem for the mankind [25].

1.1 From Strings to Signals

Technological development also marked the generation of new methods for extracting signals from the genome, and this in turn is helping us in better understanding the information that the genome is bringing to us. Our concept of genome has evolved, from considering it as a long string of 3.2 billions of base pairs, encoding adenine (A), cytosine (C), guanine (G), and thymine (T), to that of a living system producing signals, to be integrated and interpreted. The most interesting signals can be classified as:

- mutations, telling us specific positions or regions of the genome where the code of an individual differs from the expected code of the "reference" human being. Mutations are associated with genetic diseases – which are inherited and occur on specific positions of the genes – and other diseases such as cancer – which are also produced during the human life and correlate with factors such as nutrition and pollution.
- gene expression, telling us in which specific conditions genes are active (i.e. they are translated to a protein) or inactive. It turns out that the same gene may have an intense activity in given conditions and no activity in others.

– peaks of expression, indicating specific positions of the genome where there is an increase of short reads due to a specific treatment of DNA; these in turn indicate specific biological events, such as the binding of a protein to the DNA.

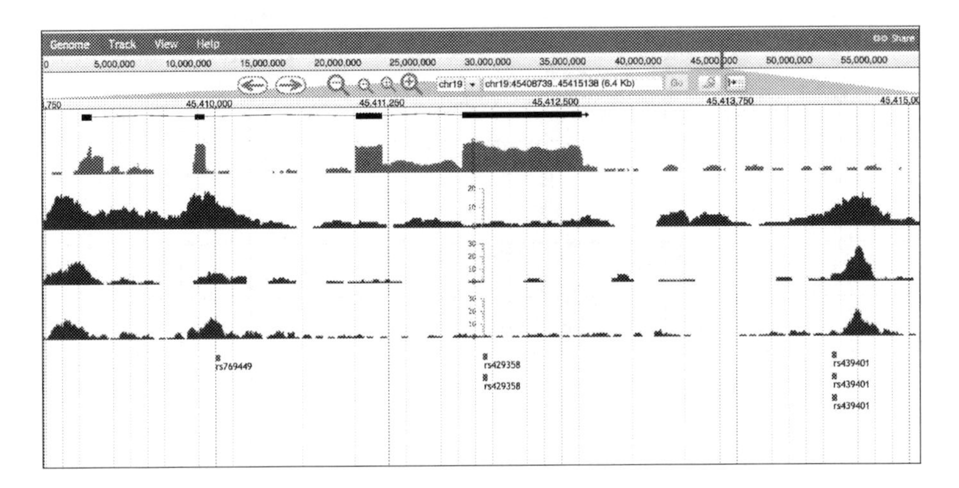

Fig. 2. Signals on a genome browser, corresponding to genes (an annotation in the header where genes are shown in black), gene expression (one track with red signal), peaks (three tracks with blue signals) and mutations (three tracks with red segments pointing to mutated positions of the genome)

These signals can be observed by using a genome browser, i.e. a viewer of the genome. All signals are aligned to a reference genome (the standard sequence characterizing human beings; such sequence is constantly improved and republished by the scientific community). The browser is open on a window of a given length (from few bases to millions of bases) and the signals are presented as tracks on the browser; each track, in turn, displays the signal – either by showing its intensity or just by showing its position.

Figure 2 presents a genome browser window; tracks of different colours describe gene expression, peaks of expressions, and mutations. The black line indicates the position four gene exons – these are known information, or "annotations", that can be included in the window. An interesting biological question could be: "find the genes which are expressed where there are three peaks (representing the fact that the specific event denoted by the peak is confirmed by all experiments) and such that the gene is close to at least one mutation". Such question would, in this specific example, extract the second exon; GeCo's objective is to allow the expression of such questions over genomic signals.

1.2 Tertiary Data Analysis

Signals can be loaded on the browser only after being produced as a result of long and complex bioinformatics pipelines. In particular, analysis of NGS

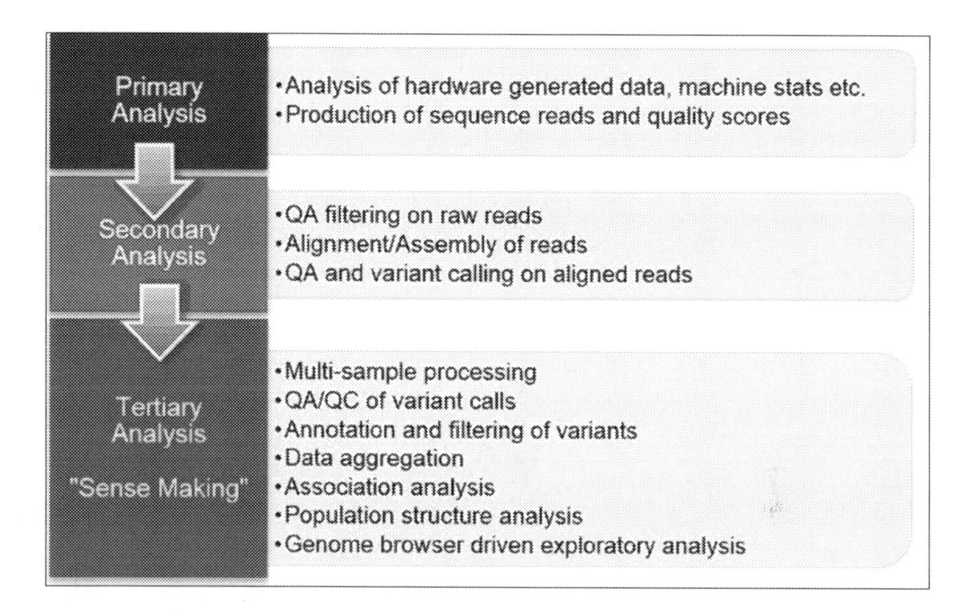

Fig. 3. Primary, secondary, and tertiary data analysis for genomics, from http://blog.goldenhelix.com/grudy/a-hitchhikers-guide-to-next-generation-sequencing-part-2/

data is classified as primary, secondary, and tertiary (see Figs. 3 and 4): primary analysis is essentially responsible of producing raw data - i.e. small sequences in output from the sequencing machine; secondary analysis is responsible of extracting ("calling") the signals from raw data and aligning them to the reference genome; tertiary analysis is responsible of data extraction, integration, and analysis. Thousands of processed datasets are being produced at large sequencing

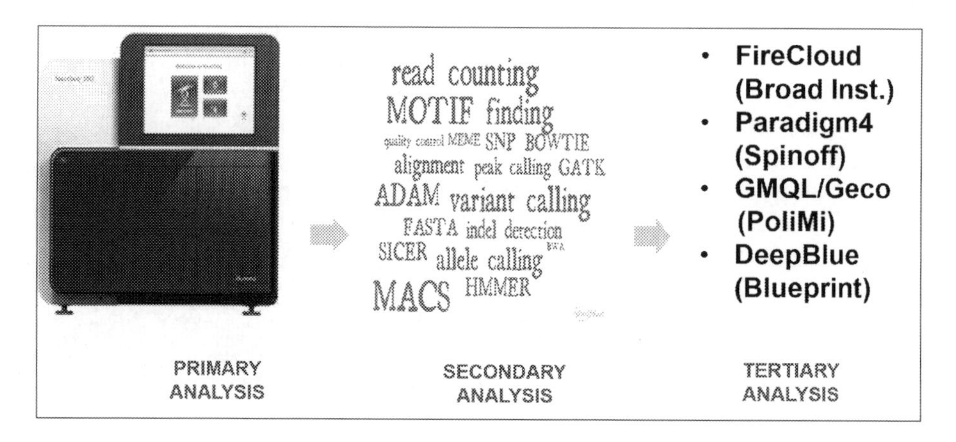

Fig. 4. The landscape of genomic computing tools; few of them are dedicated to tertiary data analysis

centers, are being assembled by large international consortia, and made available for secondary research use. International consortia include the Encyclopedia of DNA Elements (ENCODE) [13], The Cancer Genome Atlas (TCGA) [26], and the 1000 Genomes Project [1].

While many bioinformatics systems are dedicated to secondary data analysis, a few of them include tertiary analysis. Among these: SciDB, a scientific database produced by the spin-off company Paradigm4, supports a genomic addition focused on genomic data integration [3]; DeepBlue, provides easy access to datasets produced within the BluePrint consortium, with a language which is quite similar to GeCo's query language GMQL [2]; FireCloud, developed by the Broad Institute, offers an integrated platform supporting cancer research pipelines [14]. All these systems already support access to a huge number of open datasets, including ENCODE and TCGA.

2 Overview of GeCo

The introduction, which expands the abstract reported in [10], has set the stage for discussing why GeCo is an important project in the context of tertiary data analysis for genomics. We now illustrate the main results achieved by the project, providing as well a short description of the various publications where such results are presented.

2.1 Data Model

The Genomic Data Model (GDM) [19] is based on the notions of datasets and samples and on two abstractions: one for genomic regions, which represent portions of the DNA and their features, and one for their metadata. Datasets are collections of samples and each sample consists of two parts: the region data, which describe the characteristics of genomic features called during secondary analysis, and the metadata, which describe general properties of the sample.

Genomic region/feature data describe a broad variety of molecular aspects, which are individually measured, and therefore they are available in a variety of formats which hamper their integration and comprehensive assessment. GDM provides a schema to the genomic features of DNA regions; thus, it makes such heterogeneous data self-describing and interoperable. We map data from data files in their original format into the GDM format when they are used, without including them into a database, so as to preserve the possibility for biologists to work with their usual file-based tools. The provided data schema has a fixed part, which guarantees the comparability of regions produced by different kinds of processing, and a variable part reflecting the "feature calling process" that produced the regions and describing the region features determined through various processing types.

Metadata characterize the high heterogeneity of genomic data and their processing; they are collected in a broad variety of data structures and formats that constitute barriers to their use and comparison. To cope with the lack of

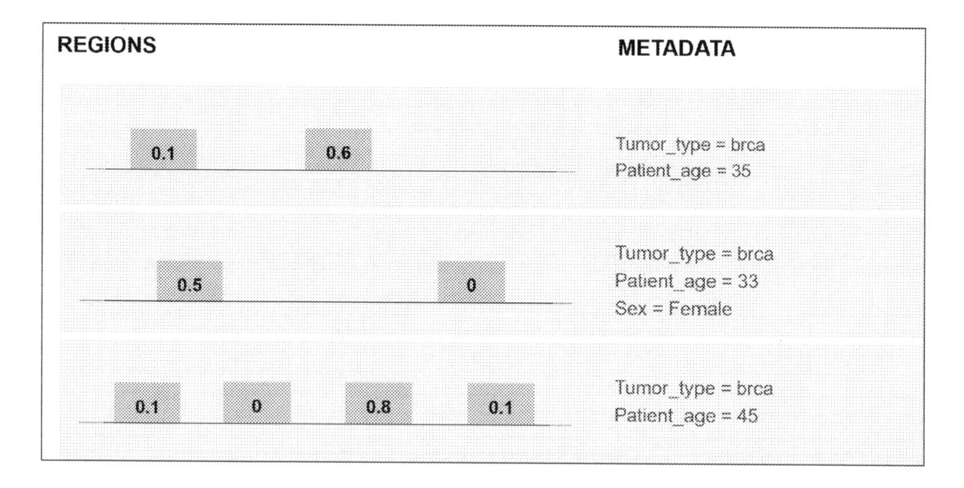

Fig. 5. Genomic Data Model

agreed standards for metadata, GDM models metadata simply as free arbitrary semi-structured attribute-value pairs, where attributes may have multiple values (e.g., the Disease attribute can have both "Cancer" and "Diabetes" values). We expect metadata to include at least the considered organism, tissue, cell line, experimental condition (e.g., antibody target – in the case of NGS ChIP-seq experiments, treatment, etc.), experiment type, data processing steps, feature calling, and analysis method used for the production of the related data; in the case of clinical studies, also individual's descriptions including phenotypes.

Figure 5 shows a GDM dataset consisting of three samples, each associated with a patient affected by Breast Cancer (BRCA). The region part has a simple schema, describing a particular value associated to each region; regions are aligned along the whole genome and belong to specific chromosomes (not shown in the figure). The metadata part includes free attribute-value pairs, describing the patient's age and sex; in GDM, attributes are freely associated to samples, without specific constraints.

2.2 Query Language

The name 'GenoMetric Query Language' (GMQL) [18] stems from the language's ability to deal with genomic distances, which are measured as nucleotide bases between genomic regions (aligned to the same reference) and computed using arithmetic operations between region coordinates. The language supports a very rich set of predicates describing distal conditions, i.e. distal properties of regions (e.g., being at minimal distance above a given threshold from given genes). Furthermore, the management of stranded regions (i.e. regions with an orientation) is facilitated by predicates that deal with such orientation (e.g., to locate the region's start and stop coordinates according to the strand, or the upstream or downstream directions with respect to the region's ends).

GMQL operations include classic algebraic operations (SELECT, PROJECT, UNION, DIFFERENCE, JOIN, SORT, AGGREGATE) and domain-specific operations (e.g., COVER deals with replicas of a same experiment; MAP refers genomic signals of experiments to user selected reference regions; GENOMETRIC JOIN selects region pairs based upon distance properties); the full description and language specification of GMQL is provided at the GMQL website[1]. A typical GMQL query starts with a SELECT operation, which creates a dataset with only the data samples filtered out from an input dataset by using a predicate upon their attributes. Then, the query proceeds by processing the selected samples in batch with operations applied on their region data and/or metadata. Finally, it ends with a MATERIALIZE operation, which stores a dataset by saving the region data of each of its samples and the related metadata. Tracing metadata provenance during the processing of each operation is a unique aspect of our approach; knowing metadata associated with resulting samples allows tracing the input samples which contributed to results.

2.3 Implementations

The development of GMQL V1, reported in [11] was relatively fast, as the implementation took about one year; Pig Latin [21] was chosen as target language because GMQL syntactically recalls Pig Latin - both languages progressively construct queries by introducing intermediate results and naming them by using variables; this style of query writing was considered similar to the scripting style which is dominant in bioinformatics, although of course GMQL scripts are high-level and not intertwined with programming language constructs.

The main problem with this approach is that the optimization strategies were hard-coded within the software produced by the translator, which in turn was focused on the specific features of our target language Pig Latin. Moreover, at the end of 2014, it became clear that Pig Latin was no longer strongly supported, while other engines were becoming much more popular; in particular Spark [6] was achieving a dominant position, Flink [4] had very interesting capabilities as a streaming engine, and SciDB [23] had a totally different approach to storage through arrays. Therefore, at the beginning of 2015, we opted for a major system redesign, starting the development of GMQL Version 2.

The design of GMQL V2 [17] has been centered around the notion of an intermediate representation for the query language, developed as an abstract operator tree (actually a directed acyclic graph or DAG, as operations can be reused), that is at the center of the software architecture. The intermediate representation carries the semantics of the query language in terms of elementary operations that are applicable to metadata and to region data separately, and opens up to various options for expressing the language's syntax (which can be expressed as relational expressions, or in embedded form within programming or workflow languages, or in logical format by using a Datalog-like style) and for

[1] http://www.bioinformatics.deib.polimi.it/geco.

deploying over different cloud-based engines. In particular, at various times we used Spark, Flink, and SciDB.

2.4 Comparison of Cloud-Based Implementations

Compared to V1, V2 has a much more complex software organization. While V1 was cooperatively developed by 3 PhD students, V2 has a larger group of developers, which included in the past several master student. At the moment, the Spark implementation is fully supported, while the Flink and SciDB implementations are only maintained for specific operations.

Comparative analysis, published in [8,9], shows that the performances of Flink and Spark are remarkably similar, while the performances of Spark and SciDB are very different, with SciDB being faster than Spark when operations involve selections and aggregates (as they are facilitated by an array organization), whereas Spark is faster than SciDB in JOIN and MAP operations (thanks to the general power of the Spark execution engine).

2.5 Data Indexing

While many solutions for efficient data management of "sequence reads" have been developed, in GeCo we concentrated on efficient data management for genomic intervals. GMQL is grounded on the efficient execution of operations for the composition and comparison of (epi)genomic regions and their associated attributes. To cope efficiently with complex region calculus, we have developed the 1D Interval Inverted Index (Di3), a multi-resolution single-dimension indexing framework [15].

Di3 is independent from the data layer and adaptable to any key-value pair persistence technology. These may range from NoSQL databases to classical B+ tree and simple in-memory key-value collections implemented by most of modern programming languages; such separation makes Di3 adaptable to a variety of application scenarios, from small scale ad-hoc solutions (using B+ tree), to large scale systemic solutions (using cloud-based key-value pair persistence technologies).

The most important contribution of Di3 is its extendable, orthogonal, and comprehensive region calculus. Region-based operations include the identification of co-occurrences or accumulations of regions, possibly from different biological tests and/or of distinct semantic types, within the same area of the DNA, sometimes within a certain distance from each other and/or from DNA regions with known structural or functional properties.

2.6 Desktop Data Analysis

We developed a rich set of abstractions for client-side data analysis, exploration and visualization, supported by a tool named GenoMetric Space Explorer (GeMSE) [16]; GeMSE supports primitives for data exploration spanning from *select*, *sort*, and *discretize*, to *clustering*, and *pattern extraction*. GeMSE leverages on GMQL as its back-end tertiary data retrieval framework, but can be used on any text files in standard BED (Browser Extensible Data), BroadPeak, NarrowPeak, GTF (General Transfer Format), or general tab-delimited format.

GeMSE applies to a *genometric space* produced by a specific GMQL operation, called `MAP` [18], which applies to two datasets, denoted as *reference* and *experiment*; the former typically corresponds to genes or transcription regulatory regions; the latter consists of multiple, possibly heterogeneous, samples, each constituted by multiple regions. The result is a matrix structure, called *genometric space*, where each row is associated with a reference region, each column refers to a sample, and each matrix entry is computed by means of an aggregate function (computed on a selected attribute of the experiment regions which overlap with the reference regions). GeMSE data exploration consists of three iterative phases:

- *Transition*, where a transformation function is applied on a genometric space resulting in a new genometric space;
- *Analysis*, where a genometric space is analyzed using data analysis functions (e.g., pattern analysis or statistical inference);
- *Visualization*, where a genometric space is visualized (e.g., on heatmaps or graph views).

Tracking multiple transformations of genometric spaces is crucial for data exploration. GeMSE tracks such transitions in a graph data structure called *State-Transition Tree* (STT), whose nodes represent different genometric spaces and whose edges represent the transformations between genometric spaces. From any data exploration state, one can view the related genometric space, visualizing it as a table or a heatmap, and also explore patterns contained into the heatmap. STT visualization facilitates state examination with data exploration and a trial-and-error approach.

3 Future Steps

3.1 Language Interfaces

Several activities are ongoing in GeCo. We fully developed a Python library supporting an interface to the full GMQL language; the library is deployed within the international Python library repository and does not require any installation - as it is customary in Python. It supports a local and a remote execution mode, the former runs in the client desktop, the latter runs on a remote server and as such it connects to the global repository [20]. Similar interfaces are being deployed for R and Galaxy.

3.2 Distributed System

The availability of a core data model as a data interoperability solution and of a high-level data processing language is a strong prerequisite for defining data exchange protocols. We expect that each data repository will be the owner of the data that are locally produced, and that nodes of cooperating organizations will be connected to form a federated database. In such systems, queries move from a requesting node to a remote node; they are locally executed; then results are communicated back to the requesting node. This paradigm allows for distributing the processing to data, transferring only query results which are usually small in size.

Supporting a high-level query interface to a server is already making one big step forward, which is similar to the gigantic step made by SQL in the context of client-server architectures (which dates a couple of decades). Indeed, once a system supports an API for submitting GMQL queries, these have the following properties: they are short texts and produce short answers. In contrast, most of today's implementations are centralized and require first the full data transmission and next the evaluation of server-side programs. This scenario opens up to the design of simple interaction protocols, typically for:

- Requesting information about remote datasets, facilitated by the availability of metadata (for locating data of interest) and of their region schemas (for formalizing queries);
- Transmitting a query in high-level format and obtain data about its compilation, not only limited to correctness, but including also estimates of the data sizes of results;
- Launching query execution and then controlling the transmission of results, so as to be in control of staging resources and of communication load.

3.3 Repository

Very large-scale sequencing projects are emerging; as of today, the most relevant ones include:

- The Encyclopedia of DNA elements (ENCODE) [13], the most general and relevant world-wide repository for basic biology research. It provides public access to more than 4,000 experimental datasets, including the just released data from its Phase 3, which comprises hundreds of epigenetic experiments of processed data in human and mouse;
- The Cancer Genome Atlas (TCGA) [26], a full-scale effort to explore the entire spectrum of genomic changes involved in human cancer;
- The 1000 Genomes Project [1], aiming at establishing an extensive catalogue of human genomic variations from 26 different populations around the globe;
- The Roadmap Epigenomics Project [22], a repository of "normal" (not involved in diseases) human epigenomic data from NGS processing of stem cells and primary ex vivo tissues.

Data collected in these projects are open and public; all the Consortia release both raw and processed data, but biologists in nearly all cases trust the processing, which is of high-quality and well controlled and explained. The use of a high-level model and language, such as GDM and GMQL, is the ideal setting for provisioning next generation services over data collected and integrated from these and other repositories, improving over the current state-of-the-art. We already started to work towards an integrated repository: in [7] we discussed a conceptual representation of metadata, where we presented a minimal conceptual schema that includes data typically found in all platforms, albeit with different names and formats; in [12] we discussed the transformation of TCGA datasets into BED format, which is quite similar to GDM.

We are working towards the development of an integrated repository with the following features:

- Processed datasets available in several sources, including the above ones, will be provided with compatible metadata;
- Processed region datasets of other sources (beyond TCGA) will be translated to GDM;
- It will be possible to choose among a set of custom queries, representing the typical/most needed requests;
- It will be possible to provide user input samples to the services, whose privacy will be protected;
- Deferred result retrieval will be possible, through limited amount of staging at the sites hosting the services.

4 Conclusions

The progress in DNA and RNA sequencing technology has been so far coupled with huge computational efforts in primary and secondary genomic data management. In this paper, we have shown that a new need is emerging: making sense of data produced by these methods, in the so-called tertiary analysis; this is the objective of GeCo, our five-year project. In this paper, we have shown how GeCo is providing a new focus on data extraction, querying, and analysis by raising the level of abstraction of models, languages, and tools.

Acknowledgment. This research is funded by the ERC Advanced Grant project GeCo (Data-Driven Genomic Computing), No. 693174, 2016-2021.

References

1. 1000 Genomes Consortium: An integrated map of genetic variation from 1,092 human genomes. Nature, **491**, 56–65 (2012)
2. Albrecht, F., et al.: DeepBlue epigenomic data server: programmatic data retrieval and analysis of the epigenome. Nucleid Acids Res. **44**(W1), W581–586 (2016)
3. Accelerating bioinformatics research with new software for big data to knowledge (BD2K). Paradigm4 Inc. (2015). http://www.paradigm4.com/)

4. Apache Flink. http://flink.apache.org/
5. Apache Pig. http://pig.apache.org/
6. Apache Spark. http://spark.apache.org/
7. Bernasconi, A., et al.: Conceptual modeling for genomics: building an integrated repository of open data. In: Proceedings of the Entity-Relationship, Valencia, ES (2017)
8. Bertoni, M., et al.: Evaluating cloud frameworks on genomic applications. In: Proceedings of the IEEE Conference on Big Data Management, Santa Clara, CA (2015)
9. Cattani, S., et al.: Evaluating genomic big data operations on SciDB and Spark. In: Cabot, J., De Virgilio, R., Torlone, R. (eds.) ICWE 2017. LNCS, vol. 10360, pp. 482–493. Springer, Cham (2017). https://doi.org/10.1007/978-3-319-60131-1_34
10. Ceri, S., et al.: Data-Driven Genomic Computing (GeCo): Making sense of Signals from the Genome. In: Selected Papers of the XIX International Conference on Data Analytics and Management in Data Intensive Domains (DAMDID/RCDL 2017), CEUR Workshop Proceedings, vol. 2022, pp. 1–2 (2017)
11. Ceri, S., et al.: Data management for heterogeneous genomic datasets. IEEE/ACM Trans. Comput. Biol. Bioinf. **14**(6), 1251–1264 (2016)
12. Cumbo, F., et al.: TCGA2BED: extracting, extending, integrating, and querying the Cancer genome atlas. BMC Bioinf. **18**(6), 1–9 (2017)
13. ENCODE Project Consortium: An integrated encyclopedia of DNA elements in the human genome. Nature **489**(7414), 57–74 (2012)
14. FireCloud. https://software.broadinstitute.org/firecloud
15. Jalili, V., et al.: Indexing next-generation sequencing data. Inf. Sci. **384**, 90–109 (2016). https://doi.org/10.1016/j.ins.2016.08.085
16. Jalili, V., et al.: Explorative visual analytics on interval-based genomic data and their metadata. BMC Bioinf. **18**, 536 (2017)
17. Kaitoua, A., et al.: Framework for supporting genomic operations, IEEE-TC (2016). https://doi.org/10.1109/TC.2016.2603980
18. Masseroli, M., et al.: GenoMetric query language: a novel approach to large-scale genomic data management. Bioinformatics **31**(12), 1881–1888 (2015)
19. Masseroli, M., et al.: Modeling and interoperability of heterogeneous genomic big data for integrative processing and querying. Methods **111**, 3–11 (2016)
20. Nanni, L., et al.: Exploring genomic datasets: from batch to interactive and back. In: Proceedings of the ExploreDB 2018, Co-Located with ACM-Sigmod, June 2018
21. Olston, C., et al.: Pig Latin: a not-so-foreign language for data processing. In: ACM-SIGMOD, pp. 1099–1110 (2008)
22. Romanoski, C.E., et al.: Epigenomics: roadmap for regulation. Nature **518**, 314–316 (2015)
23. SciDB. http://www.scidb.org/
24. Schuster, S.C.: Next-generation sequencing transforms today's biology. Nat. Methods **5**(1), 16–18 (2008)
25. Stephens, Z.D., et al.: Big data: astronomical or genomical? PLoS Biol. **13**(7), e1002195 (2015)
26. Weinstein, J.N., et al.: The Cancer Genome Atlas Pan-Cancer analysis project. Nat. Genet. **45**(10), 1113–1120 (2013)
27. Zaharia, M., et al.: Resilient distributed datasets: a fault-tolerant abstraction for in-memory cluster computing. In: Proceedings of the USENIX, pp. 15–28 (2012)

Novel Approaches to Analyzing and Classifying of Various Astronomical Entities and Events

Data Deluge in Astrophysics: Photometric Redshifts as a Template Use Case

Massimo Brescia[1(✉)], Stefano Cavuoti[1,2,3], Valeria Amaro[2], Giuseppe Riccio[1], Giuseppe Angora[1,2], Civita Vellucci[2], and Giuseppe Longo[2,3]

[1] INAF - Osservatorio Astronomico di Capodimonte, via Moiariello 16, 80131 Napoli, Italy
brescia@oacn.inaf.it
[2] Università degli Studi Federico II - Dipartimento di Fisica "E. Pancini", via Cintia 6, 80135 Napoli, Italy
[3] INFN - Napoli Unit, via Cintia 6, 80135 Napoli, Italy

Abstract. Astronomy has entered the big data era and Machine Learning based methods have found widespread use in a large variety of astronomical applications. This is demonstrated by the recent huge increase in the number of publications making use of this new approach. The usage of machine learning methods, however is still far from trivial and many problems still need to be solved. Using the evaluation of photometric redshifts as a case study, we outline the main problems and some ongoing efforts to solve them.

Keywords: Big data · Astroinformatics · Photometric redshifts

1 The Astronomical Data Deluge

Astronomy has entered the big data era with ground-based and space-borne observing facilities producing data sets in the Tera-byte and Peta-byte domain and, in the near future, instruments such as LSST (Large Synoptic Survey Telescope [1]) and SKA (Square Kilometer Array [2]), to quote just a few, will produce data streams of unprecedented size and complexity [3]. The need to store, reduce, analyze and to extract useful information from this unprecedented deluge of data has triggered the growth of the new discipline of Astroinformatics, placed at the intersection among Information and Communication Technology, Statistics, and Astrophysics, which aims at providing the community with the needed infrastructures and know how. The advent of Astroinformatics is currently driving a true revolution in the methodology of astronomical research. Wordings like "data mining", "machine learning" (or ML), "Bayesian statistics", which rarely appeared in the literature of the last century, have suddenly become very popular in the recent astronomical literature. A trend which will be shortly exemplified in the next section (Sect. 2) where we shall also show the frequency with which these words have been used in the recent literature. In the remaining of this paper we shall instead discuss some problems that are commonly encountered in the application of ML methods to astronomical problems, using the evaluation of photometric redshifts as a case study.

L. Kalinichenko et al. (Eds.): DAMDID/RCDL 2017, CCIS 822, pp. 61–72, 2018.
https://doi.org/10.1007/978-3-319-96553-6_5

This article is the full version of our work, and contains well over the 30% of the extended abstract presented at the DAMDID 2017 Conference and published in the related Proceedings [4].

2 The Fast Uptake of Astroinformatics

In almost all fields, modern technology allows to capture and store huge quantities of heterogeneous and complex data often consisting of hundreds of features for each record and requiring complex metadata structures to understand. This has led to a situation well described by the following sentence: *...while data doubles every year, useful information seems to be decreasing, creating a growing gap between the generation of data and our understanding of it...* [5]. A need has therefore emerged for a new generation of analytics tools, largely automatic, scalable and highly reliable. Strictly speaking, Knowledge Discovery in Databases (KDD) is about algorithms for inferring knowledge from data and ways of validating the obtained results [6], as well as about running them on infrastructures able to match the computational demands. In practice, whenever there is too much data or, more generally, a representation in more than 4 dimensions [7], there are basically three ways to make learning feasible. The first one is straightforward: applying the training scheme to a decimated dataset.

The second method relies on parallelization techniques, the idea being to split the problem into smaller parts, then solving each one by using a separate CPU and finally combining the results together [8]. Sometimes this is feasible due to the intrinsic natural essence of the learning rule (such as genetic algorithms; [9]). However, even after parallelization, the algorithm's asymptotic time complexity cannot be improved. The third and more challenging way to enable a learning paradigm to deal with massive data sets is to develop new algorithms of lower computational complexity, but in many case this is simply unfeasible [10]. Such complexity is one of the main explanations for the gap still existing between the new methodologies and the wide community of potential users who fail to adopt them. In order to be effective, in fact, the KDD methodology requires a good understanding of the mathematics and statistics underlying the methods and learning paradigms, of the computing infrastructures and of the complex workflows that need to be implemented. Most users (even in the scientific community) are usually not willing to make the effort to understand the process and prefer to recur to traditional approaches that are far less powerful, but much more affordable [11]. Astroinformatics copes with these concepts, by exploiting KDD methodologies and by taking into account the fact that background knowledge can make possible to reduce the amount of data to process. It adopts data mining and machine learning paradigms, mostly based on the assumption that in many cases a bunch of the parameter space attributes turn out to be irrelevant when background knowledge is taken into account [8]. Dimensional reduction, classification, highly non-linear regression, prediction, clustering, filtering, are all concrete examples of functionalities belonging to the KDD conceptual domain, in which the various Astroinformatics methodologies can be applied to explore data under peculiar aspects, strictly connected to the associated functionality scope [12].

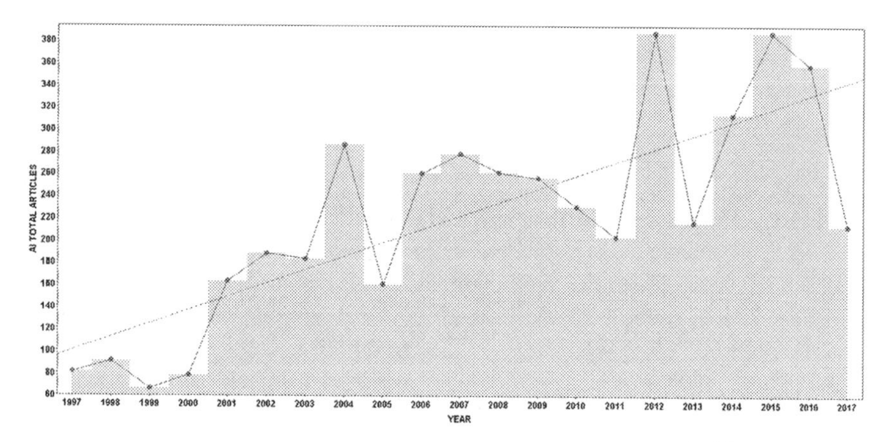

Fig. 1. Quantity of scientific articles focusing Astroinformatics topics, published per year on specialized refereed journals, from January 1997 to July 2017. The dotted red line is the linear regression trend.

In the current Big Data analytics scenario, astronomical observations and related data gathering, reduction and analysis represent one of the clearest examples of *data-driven* science, where the amount of data collected in a single day is enough to keep occupied an entire community of scientists for the rest of their life. Therefore, in the last twenty years an awareness has been reached and consolidated in the scientific community: that the union can make the difference, heterogeneity is not dispersion, multidisciplinary eclecticism is not a presumptuous ambition, but rather they are unavoidable and precious qualities. A characterizing factor of the astrophysical data is that behind them there are specific physical laws supervising complex phenomena. Machine learning methods, by achieving a higher level of complexity, will surely allow disclosing patterns and trends pointing at a higher level of complexity than what has been achieved so far. In fact, in the last decade, in many different fields it has been clearly demonstrated that the emulation of the mechanisms underlying the intelligence in Nature (an expression of the human brain functioning, the sensory organs and the formulation of thoughts), if translated into efficient algorithms and supplied to super computers, is fully and rapidly able to analyze, correlate and extract huge amounts of heterogeneous information. Astroinformatics has grown in the last twenty years precisely as a science identifying the symbiosis between different scientific and technological disciplines, therefore a perfect example of a virtuous resource, through the combination of Astronomy, Astrophysics, Artificial Intelligence, Data Mining and Computer Science. The growing awareness of the need to do science without being submerged by the complexity and quantity of the information to be analyzed is amply demonstrated by the result of our recent bibliographic survey (Fig. 1), in which the positive tendency to increase the study and applicability of these methodologies in the astrophysical field is more than evident. We are hence firmly convinced that the future of Astrophysics, like many other scientific and social disciplines, cannot ignore the awareness that the stereotype of the scientist in the Galilean conception, valid up to twenty years ago, is now undergoing an evolutionary jump. The contemporary scientist must necessarily assume an "open minded" role, must be diversified in professional qualifications, master of all the communicative mechanisms available in the digital age, and with a solid openness and eclectic ability of Darwinian adaptation to the new demands of modern scientific exploitation.

3 A Template Case: Photometric Redshifts

Photometric redshifts (photo-z), by allowing the calculation of the distances for large samples of galaxies, are at the very heart of modern cosmological surveys. The main idea behind photo-z is that there exists a highly nonlinear correlation between broad-band photometry and redshift; a correlation arising from the fact that the stretching introduced by the cosmological redshift causes the main features in a galaxy spectrum to shift through the different filters of a given photometric system [13, 14]. In a rough oversimplification, there are two classes of methods to derive photo-z: the Spectral Energy Distribution (SED) template fitting methods (e.g. [15–18]) and empirical (or interpolative) methods (e.g. [19–23]).

SED methods rely on fitting the multi-wavelength photometric observations of the objects to a library of synthetic or observed template SEDs, which are first convolved with the transmission functions of the given filters and then shifted to create synthetic magnitudes for each galaxy template as a function of the redshift. SED methods are capable to derive photo-z, the spectral type and the Probability Density Function (hereafter PDF) of the photo-z error distribution for each source, all at once. However, they suffer from several shortcomings, such as in particular, the potential mismatch between the templates used for the fitting, and the properties of the selected sample of galaxies [24], color/redshift degeneracies, and template incompleteness.

Among empirical methods, those based on various Machine Learning (ML) algorithms are more commonly used to infer the hidden relation between input (mostly multi-band photometry) and the desired output (redshift). In supervised ML techniques, the learning process is regulated by the spectroscopic redshift available for a subsample of the objects, forming what is known as the Knowledge Base or KB. In the unsupervised approach, the spectroscopic information is not used in the training phase, but only during the validation phase. Over the years many ML algorithms have been used to this purpose; to quote just a few: neural networks [19, 25], boosted decision trees [26], random forests [27], self organized maps [28], quasi Newton algorithm [29], genetic algorithms [30, 31]. Numerous examples (see for instance [32]) show that, when a suitable KB is available, most ML methods outperform the SED fitting ones. ML methods however, suffer from the difficulty of achieving good performances outside the regions of the observed Parameter Space (PS) properly covered by the KB. For instance, they are not effective in deriving the redshifts of objects fainter than the available spectroscopy. A few attempts have been made to overcome this limitation (cf. [33]), but with questionable success. Moreover, there are few examples of combined approaches involving both SED and ML methods (e.g. [34, 35]), which open promising perspectives. Since a detailed analysis of the vast literature is beyond the purpose of this work, we shall focus on some among the main issues still open. Namely: the problem of missing data; the definition of the optimal parameter space and the related problem of feature selection; the characterization of the parameter space in terms of completeness and accuracy of the results; and finally the derivation of reliable PDFs.

3.1 Missing Data

Most ML methods are not robust against missing data (either missing observations or non-detections). In such cases there are two main solution strategies: list-wise deletion and imputation. The list-wise deletion consists basically in the elimination of all the entries with one (at least) missing value; clearly this is the simplest and most commonly used approach [20–22, 36, 37]. However, it has the main drawback of reducing the number of objects for which it is possible to predict the output and this is quite problematic in astronomical applications, especially for objects at high redshift, where a large fraction of objects is not detected in at least one band.

Imputation, instead, consists in replacing the missing data (of any kind) with substituted value. Obviously, the label imputation is just an umbrella under which there exists a plethora of methods. The simplest type of imputation is the usage of the same value for all the entries with a missing data (i.e. −99) that actually creates portions of the parameter space (one for each feature with at least one missing value), in which all such objects will fall. A slightly different approach consists in the substitution of a realistic value. We consider as realistic a value that belongs to the domain of the feature itself, i.e. its minimum, maximum, mean or median. This method can be effective either when a parameter has not been measured for some reason or when, rather than with missing data, the user is dealing with non-detections (e.g. upper limits in the flux). Another approach consists in a sort of hierarchical regression on all features: in practice for each pattern with at least one missing value, the latter is replaced with a prediction based on the other elements of the dataset. Such replacement can be operated either with traditional interpolation or, more effectively, with any ML methods (such as kNN; Cavuoti et al., in preparation). To the best of our knowledge, only a few attempts in this direction have been made on astronomical data sets in spite of the fact that an effective imputation would greatly increase the number of objects for which photo-z could be derived (even though with lower accuracy).

3.2 Biases in the Parameter Space

The performance of any supervised method ultimately depends on how well the training set maps the observed parameter space. From a qualitative point of view, this implies that any ML method will fail or have very low performances either when applied to objects of a missing type in the training set, or when applied to objects falling in regions of the PS under-sampled in the training set.

Focusing on the latter problem, an interesting approach, to robustly map the empirical distribution of galaxies in a multidimensional colour space, has been presented in [38] under the form of unsupervised clustering (via Self Organizing Maps) of the spectroscopic objects in the COSMOS survey [39]. In order to prepare the ground for the future Euclid surveys [40], the photometry was reduced to the Euclid photometric system, and all galaxies in the COSMOS master catalogue were clustered (in a completely data driven approach) to find similar objects. In this way, they found that (cf. Fig. 6 in [38]) more than half of the parameter space was poorly or not covered by training data. Additionally, they found also the first evidence that the colour-redshift relation may be smoother than what is commonly believed on the basis of the photo-z variance estimates derived from template fitting. If confirmed, this result would prove that ML photo-z are intrinsically more robust than those derived with SED fitting techniques. High intrinsic variance in the colour-redshift mapping should in fact have resulted in a large cell-to-cell variation in median spec-z, whereas the actual distribution appeared to be rather smooth overall.

3.3 Optimal Feature Selection

Feature selection, i.e. the choice of the optimal number and combination of parameters to be used in the training of a ML based method, is a crucial step often overlooked on the basis of the false belief that the more are the parameters, the better are the results. This is well known in the data mining literature as "the curse of dimensionality" that can be shortly summarized as it follows: for a fixed number of data points in the training set, the density of training points in a given voxel of the parameter space decreases for increasing number of parameters. Beyond a given threshold, this leads to a decrease in information content and an increase in noise that in turn lowers the performances. Unfortunately, there is not a unique way to find this optimal combination of parameters.

A "Data Driven" approach has been recently explored in [37], by using a data set extracted from the Sloan Digital Sky Survey data release 7. They run two different experiments: in the first one, they explored all possible combinations of four parameters extracted from a subset of 55 parameters provided by the Sloan Digital Sky Survey (SDSS, [41]) archive in order to see whether it was possible to improve the results obtained in [36]. They found that features, selected without relying on the opinion of an expert, improved the results by almost 30%. In a second experiment [42] they used 90 parameters (magnitudes, errors, radii and ellipticities) from the SDSS database and combine them in unusual ways (average, difference, ratio etc.) obtaining a set of $\sim$5000 features. Then they run a forward selection looking for the features that optimized the root mean square error (RMSE). The data driven approach, however, is very demanding in terms of computational time and for most real life applications other approaches need to be explored.

The most surprising result was that, at least at first glance, in both experiments the optimal combination of features makes no sense to the average astronomer. In other words, they proved that the features containing the most information would have never been selected by a human expert.

The coherence between any optimal parameter space found by feature selection and its physical legitimacy is an aspect still poorly explored in the astronomical literature. But for sure it is something to keep in mind while implementing the next generation pipelines for data reduction and analysis.

3.4 The Evaluation of Probability Density Functions

In the context of photo-z and in absence of systematics, the factors affecting the final reliability of the estimates obtained with ML methods are photometric errors, internal errors of the methods and statistical biases in the photometric parameter space or in the KB. All these factors can be summarized by deriving a PDF for both the individual measurements and for large samples of objects (cumulative PDF). We wish to underline that, from a rigorous statistical point of view, a PDF should be an intrinsic property of a certain phenomenon, regardless the adopted method used to quantify the phenomenon itself. However, in the case of photo-z, the PDF turns out to be strictly dependent both on the measurement methods and on the underlying physical assumptions. Such simple statistical consideration raises some doubts on the true meaning of PDFs derived with any photo-z method.

The derivation of PDFs with machine learning method has been subject of many attempts, (cf. [43]) but in most cases, the PDF took into account only the internal errors

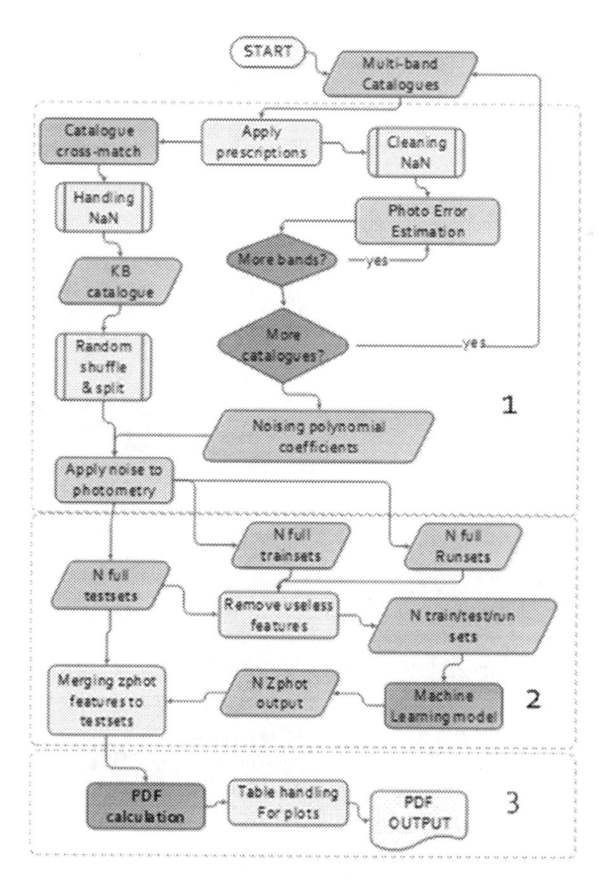

Fig. 2. The processing flow of the METAPHOR pipeline to estimate photometric redshifts by means of machine learning methods and to provide photo-z PDFs.

and biases of the methods (arising from the random initialization of the weights and/or the random extraction of the training sample from the KB).

A method capable to derive a reliable PDF, which takes into account all sources of errors, is METAPHOR (Machine learning Estimation Tool for Accurate PHOtometric Redshifts; [44]).

It was designed as a modular workflow capable to plug-in different ML models (Fig. 2). In order to estimate internal errors, this method runs N trainings on the same training set, or M trainings on M different random extractions from the KB, to derive the initialization errors and the biases induced by the randomness of the training set, respectively. The main advantage of METAPHOR is the possibility to take into account the photometric flux errors: as extensively explained in [44], this is achieved

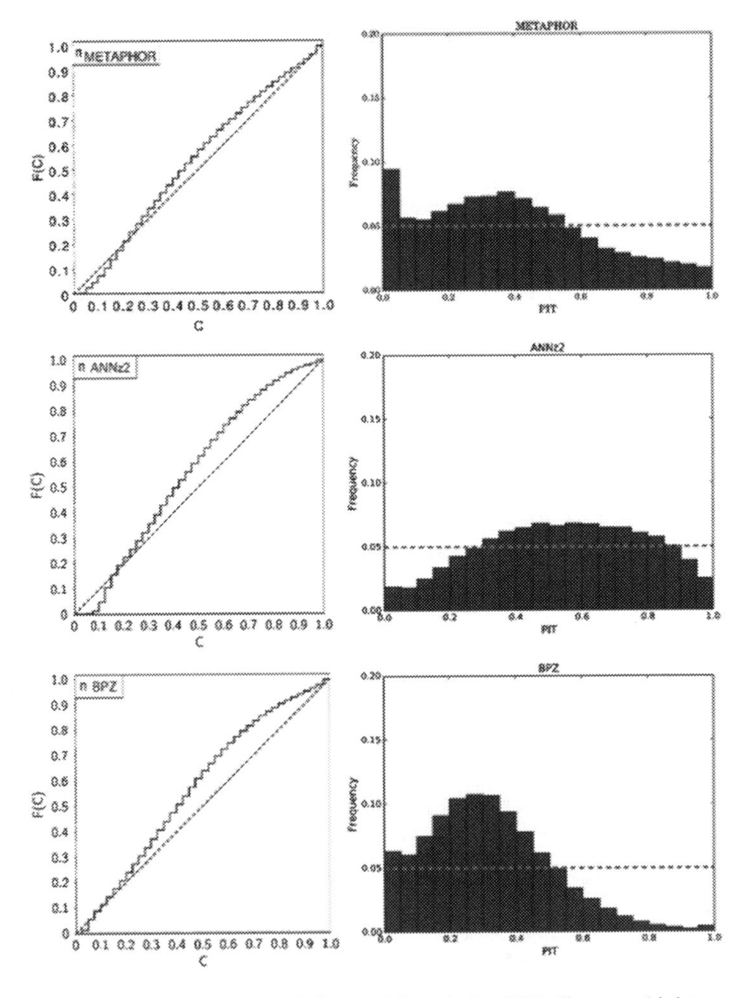

Fig. 3. Wittmann Credibility Curve (left panels) and the PIT diagram (right panels) plots obtained on KiDS-ESO DR3 data for the PDFs provided by METAPHOR (upper panel), ANNz2 (middle), and BPZ (lower).

by introducing a proper perturbation (either measured or parametrized from the error distribution) on the photometric data for the objects in test set. METAPHOR has been successfully tested on SDSS-DR9 [44] and the ESO Kilo-Degree Survey Data Release 3 (KiDS-ESO-DR3 [45]) using the MLPQNA model (Multi Layer Perceptron trained with Quasi Newton Algorithm; [10, 21, 46]) as the internal photo-z estimation ML engine. It is worth stressing, however, that the internal ML model can be easily replaced, thus allowing also METAPHOR as a tool to estimate the effects of different models on the final estimates.

In a recent work [47], three different sets of PDFs obtained with three different methods (METAPHOR; ANNz2 [43] and the SED fitting method BPZ [48]) for the same data set extracted from the KiDS-ESO DR3, were compared using several statistical indicators.

For instance, the PIT (Probability Integral Transform, [49]) histogram and the credibility analysis performed through the Wittmann [50] diagrams show that the three methods fail (even though in different ways) to capture the underlying redshift distribution (Fig. 3). PIT curves show in fact the total under-dispersive character of the reconstructed photometric redshifts distributions; while the Wittmann diagrams confirm an overconfidence of all photo-z estimates. Therefore, these results showed also that much work still remains to be done in order to fix a standard set of always reliable evaluation metrics.

4 Conclusions

Astroinformatics methods have found a widespread use in recent astrophysical contexts and a good knowledge of their methodological background is becoming essential for any observational astronomer.

However, as it always happens, the adoption of a new methodology discloses new problems to solve. By using the derivation of photometric redshifts as template case, we pointed out some general issues that are and will be encountered in most, if not all, astroinformatics applications. However, it needs to be said that at least some of the problems currently encountered derive from the fact that the adoption of machine learning and KDD methodologies is still incomplete and at an early stage. For instance, the results in [37] clearly show that, for a given task, the parameters usually selected by astronomers are not necessarily those carrying most of useful information. This is not only a problem of proper feature selection, but it might require an entirely different and "data driven" approach to the definition and measurement of astronomical parameters. Deep learning can help in some applications but it is not the answer to all problems and it is very likely that in the near future it will become necessary to re-think and re-define the way to measure the parameters.

Furthermore, more approaches to the problems related to missing data need to be explored. For example, techniques based on the co-clustering (i.e. simultaneous clustering of the objects and of the features), but to the best of our knowledge, available methods are too much demanding in terms of computing time to be effectively used on very large data sets.

Finally, further investigation is required about the PDFs obtained with ML methods, in order to assess reproducible and reliable error estimates.

Acknowledgements. MB acknowledges the INAF PRIN-SKA 2017 program 1.05.01.88.04 and the funding from MIUR Premiale 2016: MITIC. MB and GL acknowledge the H2020-MSCA-ITN-2016 SUNDIAL (SUrvey Network for Deep Imaging Analysis and Learning), financed within the Call H2020-EU.1.3.1.

References

1. Ivezic, Z., et al.: LSST: from science drivers to reference design and anticipated data products. arXiv:0805.2366v4 (2008)
2. Blake, C.A., Abdalla, F.B., Bridle, S.L., Rawlings, S.: Cosmology with the SKA. New Astron. Rev. **48**(11–12), 1063–1077 (2004)
3. Allen, M.G., Fernique, P., Boch, T., et al.: An Hierarchical Approach to Big Data. arXiv: 1611.01312 (2016)
4. Longo, G., Brescia, M., Cavuoti, S.: The astronomical data deluge: the template case of photometric redshifts. In: CEUR Workshop Proceedings, vol. 2022, pp. 27–29 (2017)
5. Dunham, M.: Data Mining Introductory and Advanced Topics. Prentice-Hall, Upper Saddle River (2002)
6. Annunziatella, M., et al.: Inside catalogs: a comparison of source extraction software. PASP **125**(923), 68–82 (2013)
7. Odenwald, S.: Cosmology in More Than 4 Dimensions. Astrophysics Workshop, N.R.L. (1987)
8. Paliouras, G.: Scalability of Machine Learning Algorithms. M.Sc. thesis, University of Manchester (1993)
9. Goldberg, D.E., Holland, J.H.: Genetic algorithms and machine learning. Mach. Learn. **3**, 95–99 (1988)
10. Brescia, M., Cavuoti, S., Longo, G., et al.: DAMEWARE: a web cyberinfrastructure for astrophysical data mining. PASP **126**(942), 783–797 (2014)
11. Hey, T., Tansley, S., Tolle, K.: The Fourth Paradigm: Data-Intensive Scientific Discovery. Microsoft Research, Redmond (2009)
12. Brescia, M.: New trends in E-science: machine learning and knowledge discovery in databases. In: Horizons in Computer Science Research, vol. 7, pp. 1–73. Nova Science Publishers (2012)
13. Baum, W.A.: Photometric magnitudes and redshifts. In: McVittie, G.C. (ed.) IAU Symposium, vol. 15, Problems of Extra-Galactic Research, p. 390 (1962)
14. Connolly, A.J., Csabai, I., Szalay, A.S., et al.: Slicing through multicolour space: galaxy redshifts from broadband photometry. AJ **110**, 2655 (1995)
15. Bolzonella, M., Miralles, J.M., Pello, R.: Photometric redshifts based on standard SED fitting procedures. A&A **363**, 476–492 (2000)
16. Arnouts, S., Cristiani, S., Moscardini, L., et al.: Measuring and modelling the redshift evolution of clustering: the Hubble Deep Field North. MNRAS **310**, 540 (1999)
17. Ilbert, O., Arnouts, S., McCracken, H.J., et al.: Accurate photometric redshifts for the CFHT Legacy Survey calibrated using the VIMOS VLT Deep Survey. A&A **457**, 841 (2006)
18. Tanaka, M.: Photometric redshift with Bayesian priors on physical properties of galaxies. AJ **801**, 1, 20 (2015)

19. Tagliaferri, R., Longo, G., Andreon, S., et al.: Neural Networks and Photometric Redshifts, ArXiv e-prints:0203445 (2002)
20. Cavuoti, S., Brescia, M., Tortora, C., et al.: Machine-Learning-based photometric redshifts for the KiDS ESO DR2 galaxies. MNRAS **452**(3), 3100–3105 (2015)
21. Cavuoti, S., Brescia, M., De Stefano, V., Longo, G.: Photometric redshift estimation based on data mining with PhotoRApToR. Exp. Astron. **39**(1), 45–71 (2015)
22. Brescia, M., Cavuoti, S., Longo, G., De Stefano, V.: A catalogue of photometric redshifts for the SDSS-DR9 galaxies (Research Note). Astron. Astrophys. **568**, A126 (2014)
23. Carrasco, K., Brunner, R.J.: Implementing Probabilistic Photometric Redshifts, Astronomical Data Analysis Software and Systems XXII. San Francisco: Astronomical Society of the Pacific, p. 69 (2013)
24. Abdalla, et al.: A comparison of six photometric redshift methods applied to 1.5 million luminous red galaxies. MNRAS **417**, 1891 (2011)
25. Collister, A.A., Lahav, O.: ANNz: estimating photometric redshifts using artificial neural networks. PASP **116**, 345 (2004)
26. Gerdes, et al.: ArborZ: photometric redshifts using boosted decision trees. AJ **715**, 823 (2010)
27. Carrasco, K., Brunner, R.J.: Sparse representation of photometric redshift PDFs: preparing for petascale astronomy. MNRAS **438**(4), 3409–3421 (2014)
28. Carrasco, K., Brunner, R.J.: Exhausting the information: novel bayesian combination of photometric redshift PDFs. MNRAS **442**(4), 3380–3399 (2014)
29. Cavuoti, S., Brescia, M., Longo, G., Mercurio, A.: Photometric redshifts with the quasi Newton algorithm (MLPQNA) results in the PHAT1 contest. A&A **546**, 13 (2012)
30. Cavuoti, S., et al.: Genetic algorithm modeling with GPU parallel computing technology smart innovation. Syst. Technol. **19**, 29–39 (2013)
31. Cavuoti, S., et al.: Astrophysical data mining with GPU. A case study: genetic classification of globular clusters, New Astron. **26**, 12–22 (2014)
32. Hildebrandt, H., et al.: PHAT: PHoto- z Accuracy Testing. A&A **523**, A31 (2010)
33. Hoyle, B., Rau, M.M., Bonnett, C., Seitz, S., Weller, J.: Anomaly detection for machine learning redshifts applied to SDSS galaxies. MNRAS **450**, 305–316 (2015)
34. Cavuoti, S., et al.: A cooperative approach among methods for photometric redshifts estimation: an application to KiDS data. MNRAS **466**(2), 2039–2053 (2017)
35. Duncan, K.J., Jarvis, M.J., Brown, M.J.I., et al.: Photometric redshifts for the next generation of deep radio continuum surveys - II. Gaussian processes and hybrid estimates, arXiv:1712.04476 (2017)
36. Laurino, O., DAbrusco, R., Longo, G., Riccio, G.: Astroinformatics of galaxies and quasars: a new general method for photometric redshifts estimation. MNRAS **418**, 2165 (2011)
37. Polsterer, K.L., Gieseke, F., Igel, C., Goto, T.: Improving the performance of photometric regression models via massive parallel feature selection. In: Manset, N., Forshay, P. (ed.) Data Analysis Software and Systems. ASP Conference Series, vol. 485, p. 425 (2014)
38. Masters, D., Capak, P., Stern, D., et al.: Mapping the galaxy color–redshift relation: optimal photometric redshift calibration strategies for cosmology surveys. ApJ **813**(1), 53 (2015)
39. Laigle, C., et al.: The COSMOS2015 Catalog: Exploring the $1 < z < 6$ Universe with Half a Million Galaxies, ApJ Supp. Ser. **224**(2), 23 (2016). Article id. 24
40. Dubath, P., Apostolakos, N., Bonchi, A., et al.: The euclid data processing challenges. Proc. IAU **12**(S325), 73–82 (2016)
41. Ahn, C.P., Alexandroff, R., Allende Prieto, C., et al.: The ninth data release of the sloan digital sky survey: first spectroscopic data from the SDSS-III baryon oscillation spectroscopic survey. ApJS **203**, 21 (2012)

42. D'Isanto, A., Cavuoti, S., Gieseke, F., Polsterer, K.L.: Return of the features - Efficient feature selection and interpretation for photometric redshifts. Submitted to A&A (2018)
43. Sadeh, I., Abdalla, F.B., Lahav, O.: ANNz2: photometric redshift and probability distribution function estimation using machine learning. PASP **128**, 104502 (2016)
44. Cavuoti, S., Amaro, V., Brescia, M., et al.: METAPHOR: a machine-learning-based method for the probability density estimation of photometric redshifts. MNRAS **465**(2), 1959–1973 (2017)
45. de Jong, J.T.A., Verdoes Kleijn, G.A., Erben, T., Hildebrandt, H., et al.: The third data release of the Kilo-Degree Survey and associated data products. Astron. Astrophys. **604**, A134 (2017)
46. Brescia, M., Cavuoti, S., D'Abrusco, R., Mercurio, A., Longo, G.: Photometric redshifts for quasars in multi-band surveys. ApJ **772**(2), 140 (2013)
47. Amaro, V., Cavuoti, S., Brescia M., Vellucci C., Longo, G., et al.: Statistical analysis of probability density functions for photometric redshifts through the KiDS-ESO-DR3 galaxies. MNRAS submitted (2018)
48. Benitez, N.: Bayesian Photometric Redshift Estimation. ApJ **536**(2), 571–583 (2000)
49. Gneiting, T., Raftery, A.E., Westveld, A.H., Goldman, T.: Calibrated probabilistic forecasting using ensemble model output statistics and minimum CRPS estimation. Mon. Weather Rev. **133**(5), 1098 (2005)
50. Wittman, D., Bhaskar, R., Tobin, R.: Overconfidence in photometric redshift estimation. MNRAS **457**, 4005 (2016)

Fractal Paradigm and IT-Technologies for Processing, Analyzing and Classifying Large Flows of Astronomical Data

Alexei V. Myshev(✉) and Andrei V. Dunin

National Research Nuclear University MEPhI (IATE), Obninsk, Russia
mishev@iate.obninsk.ru

Abstract. In the paper the fractal paradigm of constructing models and logical schemes of algorithms and procedures for information processing, analysis and classification of large flows of astronomical data on the orbits and trajectories of small bodies is considered. The methodology for constructing such models and schemes is based on the construction of estimates of proximity and connectivity criteria for orbits and trajectories in the space of possible states using the corresponding mathematical apparatus of fractal dimensions. The logical, algorithmic and substantial essence of the fractal paradigm is as follows. First, the processing and analysis of the data flow of orbits and trajectories is to determine whether it forms a fractal structure? If so, then one have to determine the centers of fractal connectivity of the flow and obtain estimates of the index of information connectivity of orbits or trajectories. Secondly, isolate the monofractal structures in the flow and classify them according to the attribute of belonging to the classes of a percolating fractal or a fractal aggregate.

Keywords: Connectedness orbits · Fractal measures · Fractal dimension
Percolating fractal · Fractal aggregate

1 Introduction

In the paper new approaches to the construction of models and logical schemes of algorithms and procedures for information processing, analysis and classification of large flows of astronomical data on orbits and trajectories of small bodies is considered. The methodology for constructing such models and schemes is based on the construction of estimates of the proximity and connectivity criteria for orbits and trajectories in the space of possible states using the appropriate mathematical apparatus of fractal dimensions. The logical, algorithmic and substantive essence of the methods and technologies of the fractal paradigm is as follows. First, in the processing and analysis of the data flow of orbits and trajectories in order to determine whether it forms a fractal structure (if so, it is necessary to determine the centers of fractal connectivity of the flow and obtain estimates of the index information connectivity orbits or trajectories). Secondly, in isolating the monofractal structures in the flow and classifying them on the basis of belonging to the classes of a percolating fractal or a fractal aggregate. Fractal paradigm for the development and implementation of information technologies for

L. Kalinichenko et al. (Eds.): DAMDID/RCDL 2017, CCIS 822, pp. 73–85, 2018.
https://doi.org/10.1007/978-3-319-96553-6_6

processing, analyzing and classifying large data flows, in contrast to traditional methods and methods [2–5], allows to take into account the following properties: (1) regularity and irregularity of the state space structure of the information scale of the flow data; (2) dynamic and informational connectivity of the data stream.

The logical schemes of algorithms and procedures for processing, analyzing and classifying large data streams are constructed on the basis of the theory fractal dimensions of spatial and temporal structures. The algorithmic and meaningful meaning of fractal dimensions is as follows. First, the processing of the data stream is to determine whether it forms a fractal structure? If yes, then determine the centers of fractal connectivity data stream and obtain estimates of information connectivity index. Secondly, algorithms and procedures for analyzing and classifying the processed flow allow us to isolate the monofractal structures if the flow forms a multifractal and classify them on the basis of belonging to the classes of the percolating fractal or fractal aggregate, and also to estimate the measure of the discrepancy between the geometric and information fractal dimensions, as indicator of the unity of the quantitative and qualitative characteristics of the flow.

2 Astronomical Data: Objects and Information

2.1 Spatial Location and Dynamics of Astronomical Objects in Space

To determine the location of astronomical objects and modeling their dynamic evolution in space, the following coordinate systems were used (see Fig. 1). Here, *XYZ* is the heliocentric coordinate system with origin *O* in the barycenter of the solar system. The plane *XOY* is the plane of the ecliptic. Axis *OX* – is directed to the point ♈ (the point of the vernal equinox). The *OY* axis is perpendicular to it, the positive direction of the axis coincides with the direction of motion of the planet. The axis *OZ* – is selected so that the system of vectors *OX, OY, OZ* forms a right triple. System $X'Y'Z'$ is a coordinate system associated with an astronomical object, with the origin O' in the center of the object. The axes $O'X'$, $O'Y'$, $O'Z'$ and *OX, OY, OZ* are respectively parallel and co–directed. Ω is the longitude of the ascending node, ω is the perihelion argument, ν is the true anomaly, *i* is the slope angle of the planet's orbit to the ecliptic, N_1N_2 is the node line, and *P* is the perihelion of the astronomical object's orbit.

The orbit of the astronomical object in the heliocentric coordinate system *XYZ* is defined and described by its elements, i.e. the values that characterize its size, shape and spatial orientation (see Fig. 1). The size of the orbit is determined by the semimajor axis a, which can take an imaginary and real value, and its shape is eccentricity *e*, whose values can vary from 0 to ∞. Three types of orbits are considered in celestial mechanics: a circle ($e = 0$), an ellipse ($0 < e < 1$), a parabola ($e = 1$) and a hyperbola ($e > 1$). The eccentricity of the orbit e is a numerical characteristic of the conic section. The semimajor axis of the orbit a for a closed orbit (circle and ellipse) determines its size, and for a body in orbit determines its energy. The spatial orientation of the orbit in the heliocentric coordinate system *XYZ* is determined by the longitude of the ascending node Ω, the perihelion argument ω and the inclination angle *i* between the orbital plane and the *XOY* plane (in our case this is the ecliptic plane). The position of the celestial body in orbit is given by the angle v. The relationship between the true anomaly (note that Eq. 1 does not contain true anomaly) and the evolution parameter or the time t within the two-body problem is determined from the following relation, which is called the Kepler equation:

$$E = M + e\,sin(E) \tag{1}$$

where M is the mean anomaly, which is determined from the following relation:

$$M = n(t - \tau) \tag{2}$$

where n is the average angular velocity of the object in the orbit, τ is the perihelion transit time, E is the eccentric anomaly, and e is the eccentricity of the orbit [1].

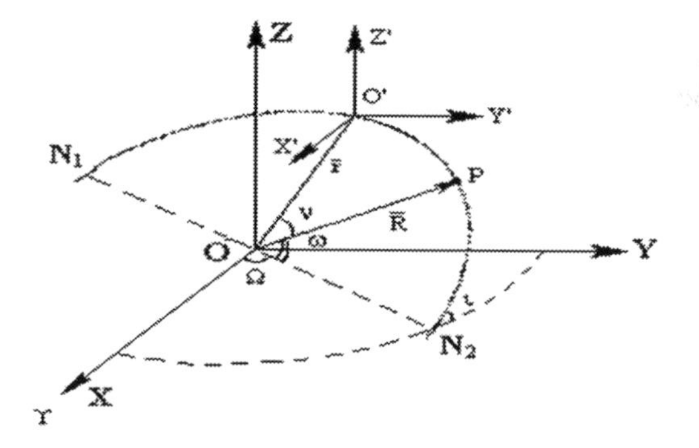

Fig. 1. Orientation of the orbit in space and coordinate systems

The relationship between the elements of the Keplerian orbit and the spatial coordinates of the radius vector and the velocity vector is determined by the following formulas that allow us to calculate the spatial coordinates of the object in the orbit and the projection of its velocity on the axis of the heliocentric coordinate system XYZ [1]:

$$\begin{cases} x = r(\cos(\omega + v)\cos\Omega - \sin(\omega + v)\sin\Omega\cos i) \\ y = r(\cos(\omega + v)\sin\Omega + \sin(\omega + v)\cos\Omega\cos i) \\ z = r\sin(\omega + v)\sin i \end{cases} \tag{3}$$

$$\begin{cases} \dot{x} = \frac{x}{r}V_r + V_n(-\sin(\omega + v)\cos\Omega - \cos(\omega + v)\sin\Omega\cos i) \\ \dot{y} = \frac{y}{r}V_r + V_n(-\sin(\omega + v)\sin\Omega + \cos(\omega + v)\cos\Omega\cos i) \\ \dot{z} = \frac{z}{r}V_r + V_n\cos(\omega + v)\sin i \end{cases} \tag{4}$$

where

$$r = \frac{a(1 - e^2)}{1 + e\cos v}$$

$$V_r = \sqrt{\frac{k^2(m_0 + m)}{a(1 - e^2)}}e\sin v$$

$$V_n = \sqrt{\frac{k^2(m_0 + m)}{a(1 - e^2)}}(1 + e \cos v) \tag{5}$$

Here r is the radius vector of the astronomical object on the orbit, m_0 is the mass of the central body (in this case, the mass of the Sun), m is the mass of the object, V_r and V_n are the radial and transversal components of the object's speed in orbits.

If Keplerian elements of the orbit of the astronomical object are known and its true anomaly, then using formulas 1–5, it is easy to calculate the values of the space coordinates of the radius vector and the velocity vector of the astronomical object in orbit at any time t. To simulate the dynamic evolution of astronomical objects within the N–body problem and the more accurate determination of the elements of its Keplerian orbit at the time t, the $X'Y'Z'$ coordinate system associated with the planets of the Solar System was used. It was used in order to take into account close encounters with planets when modeling the dynamic evolution of an object. In this case, the geometry of the 6-body problem can be illustrated in the following form (see Fig. 2).

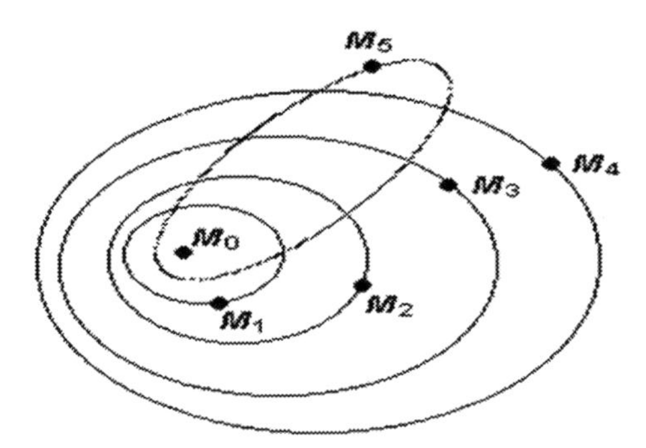

Fig. 2. Geometry of 6–body problem

On Fig. 2 M_0 denotes the Sun, M_1 – Earth, M_2 – Mars, M_3 – Jupiter, M_4 – Saturn, M_5 – the studied astronomical object.

3 Astronomical Data: Processing, Analyzing and Classification

3.1 Astronomical Data and Fractal Paradigm

The solution of many problems of space exploration is in one way or another connected with the development and implementation of intelligent information technologies for the processing and analyzing of large data flows and the interpretation of the results and conclusions obtained. The main goal is to obtain new knowledge, facts, detection and description of new regularities. For the construction of mathematical and logical schemes of algorithms and procedures of designated technologies, statistical and probabilistic approaches to the processing and analysis of data streams of observations or measurements are widely used now [2–5]. The features of the application of the indicated classical approaches in the processing and analysis of data flows of astronomical observations consist in the fact that they give only average estimates and averaged description of the space of possible states of the object under study. In this case, the properties of regularity and irregularity of the structure of the state space, as well as dynamic and information connectivity, are not taken into account. Such an approach does not allow more fully decoding information that is hidden in the streams of astronomical observations or measurements. In this case, an incomplete and deformed content and semantic information picture of the object of investigation is obtained, which does not allow, on the one hand, to give a more complete mathematical and logical description of the space of possible states of the observed objects on the trajectories of their evolution under conditions of closure, limitations, exchange and uncertainty. And on the other hand, to give a mathematical description and reflection of the distribution of the observed objects in real three–dimensional space both at local instants of time and in evolutionary development. To overcome these difficulties, new approaches and methods for developing information technologies for processing and analyzing large data streams of astronomical observations and modeling results have been developed that allow them to be reflected, described and interpreted as genetic information that is the bearer of dominant and recessive traits. These attributes are defined as dominant for technologies of decoding information contained in observational data and modeling results. Such an approach of mathematical description and representation, logical structuring and organization of streams of observational data and modeling results allows us to reveal hidden regularities that can not be obtained within the framework of traditional analytical methods. Logical schemes of algorithms and procedures for processing, analysis and classification of data flows are constructed on the basis of the theory of fractal dimensions of spatial and temporal structures, the algorithmic and substantial essence of which consists in the following. First, the processing of the data stream is to determine whether it forms a fractal structure? If yes, then determine the centers of fractal connectivity of the data stream and obtain estimates of the information connectivity index. Secondly, algorithms and procedures for analyzing and classifying the processed flow allow us to isolate the monofractal structures if the flow forms a multifractal and classify them on the basis of belonging to the classes of the percolating fractal or fractal aggregate, and also to estimate the measure of the discrepancy between the geometric and information fractal dimensions, as indicator of the unity of the quantitative and qualitative characteristics of the flow.

3.2 Formulation of Problem

To process astronomical data streams about the orbits and trajectories of space objects, new criteria for their proximity have been obtained. As a criterion for the closeness of two orbits, a quantitative estimate of the fractal measure on the set of distances between the corresponding pairs of points of orbits is introduced. Such an estimate of a fractal measure is a criterion for the connection of two orbits, through which the degree of their geometric and information proximity is reflected. The basic premise and meaning of the entered entity is explained and explained by the following logical scheme and algorithm. First, in the orbits, reference (or reference) points are defined, which may be the perihelion or aphelion of the orbit. Secondly, N_0 of discrete points with a sampling rate for the true anomaly is chosen with respect to the reference points $\Delta v = \frac{360^0}{N_0}$, where the value of N_0 is determined from the following conditions: (1) the statistical significance and representativeness of the sample; (2) the level of reliability of estimating the fractal measure. Third, at each *i*–th step of discretization with respect to Δv, the distance ri between the corresponding points on orbits is calculated. Fourthly, after the completion of the previous step, r_{min} and r_{max} are determined over all discrete points of orbits, $\Delta = |r_{max} - r_{min}|$ is calculated, which is divided into *K* intervals. Fifthly, for a given level of geometric proximity of two orbits $r_{\partial o \textit{в}}$ (radius of the confidence circle), an estimate of the confidence level pd is calculated, which reflects the fraction of point $r_i \epsilon \Delta$ for which the following condition is true: $r_i \leq r_{\partial o \textit{в}}$. The evaluation of $p_{\partial o \textit{в}}$ is a criterion and a quantitative measure of the fractal nature of the proximity of two orbits, i.e. for a given $r_{\partial o \textit{в}}$, and with which $p_{\partial o \textit{в}}$ value, two orbits can be considered to be connected. To obtain a quantitative estimate of the fractal measure on the *K* intervals of the set of points $r_i \epsilon \Delta$ taking into account simultaneously the capacitive and information dimensions of the fractal, we use the universal formula that was obtained by one of authors [6] and is defined by the following expression:

$$d_b = \lim_{\varepsilon \to 0} \frac{B(\varepsilon)}{\log(1/\varepsilon)} = \lim_{\varepsilon \to 0} \frac{\sum_{i=1}^{K} p_i \log \sum_{j=1}^{K} (1 - \rho_{ij}) p_j}{\log \varepsilon}, \tag{6}$$

where p_i is the probability of the value ri falling into the *i*–th sub–interval $\Delta = |r_{max} - r_{min}|$; ε is the length of the subinterval for a given partition of the interval Δ; ρ_{ij} is a randomized metric between the centers of the *j*–th and *i*–th subintervals; $B(\varepsilon)$ is *B*–entropy. The randomized metric ρ_{ij} is defined by the following formula:

$$\rho_{ij} = \frac{|r_i - r_j|}{|r|} \tag{7}$$

where $|r_i - r_j|$ is the distance (geometric or informational) between the *i*–th and *j*–th intervals; $|\, r \,|$ – length of interval Δ. To calculate the estimate of the capacitive fractal dimension of set $\{r_i\}$, which reflects the properties of the fractal geometry of the "holey" set of points $r_i \epsilon \Delta$, we used the well known formula [6]:

$$d_f = - \lim_{\varepsilon \to 0} \frac{\log N(\varepsilon)}{\log \varepsilon} \tag{8}$$

where $N(\varepsilon)$ is the number of coverings of the set of points $r_i \epsilon \Delta$; ε is the radius of the coating sphere. The capacitive fractal dimension d_f is introduced and is defined as an estimate of the irregularity measure for the topological and geometric structure of the set of points r_i. Formulas (6) and (8) were used to obtain estimates of the connectivity criteria for two orbits that describe and reflect the degree of spatial proximity of two orbits in a certain spatial neighborhood of *r* with a given level of confidence *p*. The logical scheme for obtaining estimates of the fractal measure of the orbital flow connectivity with respect to the optimal reference orbit and finding such an orbit in the flow is expressed as the following algorithm. The first step is arbitrarily chosen the first fixed orbit of the flow O_j ($j = 1 \div L$, where L is the number of flow orbits) and the pairwise indices of the I_k relationships between it and all other orbits O_k ($k = 1 \div L$, $j \neq k$, k is a variable index) of the flow by formulas (6) and (7). The result of this step is the set of connectedness indices $K_j = \{I_k\}$ for all orbits of the flow relative to the chosen reference orbit O_j. At this step, the indices of the connections I_m are determined, which are calculated from formula (8) and reflect the properties of the fractal geometry of the proximity of two orbits, forming the set $G_j = \{I_m\}$. The second step calculates the connectedness of the orbit flow $\{O_k\}$ relative to the reference orbit O_j by formula (6), i.e. the fractal dimension R_j of the set K_j is computed. The first and second step procedures are performed for the entire set of orbits $\{O_j\}$, where j is the index of the reference orbit. At the last step of the algorithm, we get the set $S_{opб} = \{R_j\}$, whose elements show how much an orbit can be the "center" of the orbit flow, with respect to which the orbits are most densely and compactly grouped. The condition for choosing such a "center" of the flow of orbits is determined by the following expression:

$$R_j \rightarrow \min_j \tag{9}$$

The basic premise and meaning of the indicated logic scheme in the information processing and analyzing technologies is to give a mathematical and logical description of pace possible states of observed small bodies, taking into account the geometric, dynamic and information aspects of their evolution. To represent the location of bodies at the time of their observations and further evolution, a mathematical–logical scheme for the transition from the orbital description of the positions of the bodies to their geometric arrangement in real space has been developed, which is implemented as the following procedure. First, a region of space (a cube or a rectangular parallelepiped) is localized, in which the observable objects of the body are. Secondly, an elementary volume of partitioning of this region is distinguished, that is, the partitioning region is represented as a three-dimensional lattice whose node is the identifier of an elementary volume. Thirdly, the problem of taxonomy and classification of the spatial distribution of small bodies on the nodes of the three-dimensional lattice is solved. The problem consists in the following: does the distribution of bodies in this volume form a regular or irregular spatial structure? For this, the apparatus of the theory of fractal dimensions and fractal geometry was used [6–8]. The solution of the problem is the allocation of volume, the definition of spatial geometry and the distribution of bodies at the lattice sites, i.e., are the corresponding subsets of lattice sites fractal objects or regular ones? Fourthly, the problem of spatial clustering of the observable objects at the lattice sites was solved: the separation of fractal clusters is a percolation fractal or a fractal aggregate.

3.3 Fragments of Results of Processing and Analyzing of Observational Data

For processing and analyzing, astronomical observations for small bodies of the solar system were grouped, based on the experience of previous studies [10–13].

The classification of small bodies by groups, based on the restrictions on the values of the elements of their Keplerian orbits, is shown in Table 1.

Table 1. Classification of small bodies by elements Keplerian orbits

Group 1	Main belt of asteroids: e < 1/3; i < 20°; 2.1 < a < 3.5 a. e
Group 2	Short–period comets and meteoric bodies (including asteroids of the Apollo–Amur group): 1/3 < e < 0.95; i < 30°; a < 15 a. e
Group 3	Long–period comets and meteoric bodies: e > 0.95; i – random; a > 15 a. e
Group 4	Trojans (captured by Jupiter and oscillating about its front and back Lagrangian libration points): a ≈ 5.2 a. e
Group 5	Asteroids of Guild group: e ≈ 0.2; i ≈ 10°; a ≈ 3.95 a. e
Group 6	Asteroids of Hungarian group: e ≈ 0.1; i ≈ 25°; a ≈ 1.9 a. e
Group 7	Small bodies whose orbits cross the Earth's orbit in the neighborhood of the radius of its sphere of influence

Illustration and reflection of the results of processing and analysis of the flow of astronomical data on the example of group 7 objects in the form of fragment are given in Table 2. Columns (subsets) for the sets G_j and K_j $(j = 2)$ are given in columns 2, 3, 4 and the values of *B*–entropy corresponding to their elements. The values of the elements of these sets were calculated relative to the second orbit. As can be seen from Table 2, the values of all the elements of the set G_j are greater than one (the condition of geometric regularity is the equality of all elements to unity), reflecting thereby the fractal properties of the spatial structure of the orbit flow of group 7. Logical analogies for the elements of the set K_j (the condition for the homogeneity of the flow geometry is equality of all elements to zero) indicate that the "leaky" spatial geometry of the trajectory flow has an inhomogeneous density. Based on the elements of the fifth column, the orbit was determined, which is the "center of attraction" of the stream, i.e. with respect to its most closely grouped orbits of the bodies of the group. Elements of this orbit at the time of observation have the following values: $a = 2.211545$ AU, $i = 7.506260^0$, $e = 0.5894668$, $\omega = 123.263920^0$, $\Omega = 313.243320^0$. This stage of processing and analyzing of data in the logical chain of information technologies makes it possible to obtain estimates of the quantitative measure of the fractal nature and the fractal geometry of the orbit flow of the group of bodies under consideration, and also to determine the center of the flow. The second stage in the logical chain of information processing and analysis technologies is the spatial representation and description of the distribution of small bodies of the investigated group in the local region of the heliocentric rectangular coordinate system. Let us illustrate the results obtained at this stage on the example of the objects-bodies of group 1. The partition of the volume in which the bodies are localized at the time of observation is determined by the bounded three–dimensional lattice Z^3 of the following scale: along X axis – 15 grills, along Y axis – 14 grills, Z axis – 12 grills. The procedure for spatial clustering on the lattice Z^3 allowed the following results. First, 31 clusters were identified: one cluster of percolation fractal type and thirty types of fractal aggregate. The size of the percolation fractal was 473 grills, and the sizes of fractal aggregates ranged from one to several knots. Secondly, percolation is detected in the *XOY* plane. Thirdly, the degree of filling of the lattice sites Z^3 was within twenty–one percent: in the vicinity of eighteen percent, a percolation fractal occupies the rest, and the rest is fractal aggregates. Similar calculations for bodies of group 7 with the same dimensions of the Z^3 lattice showed the following results: no percolation fractal was detected, but only clusters of the fractal aggregate type in an amount of 31 (sizes from 1 to 9 grills), i.e. objects of this group do not form compact spatially extended entities. The processes of percolation of small bodies at the lattice sites of Z^3 sufficiently reflect and describe the spatial and evolutionary connection of these objects in their orbits. These processes are most characteristic for the objects of group 7, indicating that the space-time geometry of these objects in the space of possible states is fractal in nature, thereby determining the type of their fractal dynamics. This type of dynamics is characterized by the most irregular spatiotemporal fractal geometry. Such a geometry with small space–time scales is possessed by small bodies such as meteoric bodies and a number of others. Fractal aggregation in the space of possible states is most clearly manifested in other groups of small bodies. Such fractal structures reflect a different type of fractal dynamics, manifested in different spatial–temporal scales of fractal geometry in different ways: for large ones it is more regular, and for smaller ones it is less regular type of dynamics.

Table 2. The values for the elements of the sets G_j and K_j $(j = 2)$ and the values of *B*–entropy corresponding to these elements.

Number of orbit	B - entropy	Elements of set G_j	Elements of set K_j	Elements of set $S_{орб}$
1	0.4750731	1.2494696	0.20632162	0.026309
2	0	1	0	0.036631
3	0.4301281	1.2460074	0.18680227	0.033831
4	0.4807579	1.2096293	0.20879053	0.025871
5	0.5052235	1.1657263	0.21941579	0.028186
6	0.4058040	1.2416164	0.17623847	0.036397
7	0.4985102	1.1521853	0.21650024	0.026056
8	0.4528685	1.2466675	0.19667829	0.022999
9	0.4474228	1.1198819	0.19431325	0.030783
10	0.4823662	1.2204674	0.20948901	0.026293
11	0.4808495	1.2688664	0.20883028	0.022580
12	0.5008954	1.2177231	0.21753612	0.031327

4 Some Fragments of Model Algorithms and Procedures of Technological Chains of Data Processing

The streams of astronomical observational data about the orbital elements of space objects, as illustrated and noted above, are unrelated and non-substantive lists. The model of algorithms and procedures of technological chains of primary processing of astronomical data streams at the first stage includes the software component of converting the MPCORB.dat information file into a programming environment for further

	w (град.)	l (град.)	i (град.)	e	a (a.e.)	M (град.)
1	72,14554	80,3505	10,5878	0,0777895	2,768817	242,44438
2	248,22602	169,90317	12,97943	0,2552218	2,6712512	167,42159
3	150,08873	103,89537	7,13426	0,0882196	2,3619105	110,5317
4	358,80654	141,60955	5,36719	0,1905003	2,5742862	21,74034
5	239,33102	138,72579	14,74869	0,2021124	2,4251184	128,40079
6	145,31524	259,64046	5,52358	0,2304981	2,3865682	164,94757
7	285,25189	110,93171	5,8875	0,156383	2,2014558	130,35494
8	6,16241	68,94772	5,57549	0,1228462	2,3859879	302,03019
9	312,92153	283,41394	3,84159	0,1162722	3,1380714	50,8372
10	195,121	125,59676	4,62597	0,0988868	2,4526593	332,28621
11	69,70161	235,49007	8,36923	0,2210903	2,3335634	134,43513
12	79,74785	43,26903	16,54551	0,0848434	2,578151	288,86212
13	97,74444	86,16449	9,11723	0,1656877	2,5873537	268,8823
14	97,59162	293,19807	11,73852	0,1883149	2,6428544	48,09277
15	227,04681	150,29921	3,09902	0,1372278	2,9215039	137,49956
16	135,50409	125,59141	5,58816	0,1342074	2,470264	4,30551
17	227,95195	150,50098	10,12696	0,2181288	2,296033	250,40827
18	182,29749	211,23385	1,57339	0,1578252	2,4424587	245,49551

Fig. 3. Fragment of the converted information file MPCORB.dat

	x (a.e.)	y (a.e.)	z (a.e.)	T (r.)	Vx. (a.e./сутки)	Vy. (a.e./сутки)	Vz. (a.e./сутки)
1	2,51912386463626	1,33235702167991	-0,422477839297213	4,59725224508105	-0,00503553470122269	0,00847218229155991	0,00119340128321885
2	-2,11787883827869	-2,50212298344277	0,653361548392502	4,35641304458273	0,00604683921648515	-0,00536375480642689	0,000972806555561912
3	2,363288915581	0,571285043811837	-0,304313793222421	3,62203055511926	-0,00169482079310829	0,0106470993646522	-0,000114108542591181
4	-2,11440851623031	0,284490574903165	0,102415109704696	4,12137561575123	-0,00272399670433857	-0,0123909211467882	0,00107172089217733
5	-2,61935694141764	0,888530678218003	0,279077459892076	3,76839461482056	-0,00400144413896922	-0,00832879333850445	0,002342747261568
6	-2,37850259216205	-1,68954715411886	-0,196881307955361	3,67889784599285	0,00471408881849874	-0,00747304648599337	0,000578396946019049
7	-2,43933030032897	0,0655571441432973	0,232527228782375	3,25928178720595	-0,00135283349534262	-0,0102437433031122	0,000507676064570449
8	2,24367307783448	0,171368851936596	-0,198396492193554	3,67755612780497	-0,00213906333003335	0,0115342923471592	0,000599344614937726
9	1,39787821852388	-2,58144699674901	0,0510928687517182	5,5469108452808	0,00951871057226173	0,00402498506180874	0,000684434755283269
10	0,661713920740265	-2,1433380469014	0,0574107919629003	3,83277028584789	0,0112108448252901	0,00406890865939729	-0,00092924684187679
11	-0,204861225211406	2,72194351920198	-0,251708766701636	3,55702044416192	-0,00935234754111943	0,000495692629714201	-0,00117509838348988
12	1,86338599636898	1,70343775338045	-0,0109515385719715	4,13066028812472	-0,00770498384097817	0,00711444635937834	0,00310788551911437
13	0,718026612598478	2,56509828887496	-0,0874361735950925	4,15279657647443	-0,0101822286098897	0,00114857383178962	0,00164273949689435
14	-0,324843521610121	2,34700397472171	0,130063019706179	4,28713155896932	-0,0115144529724196	0,00041152537431152	-0,00216547372876621
15	-3,11628917134597	0,88115267579433	0,0421562152738244	4,98272203039838	-0,00317043570847075	-0,00843582363617454	0,000481768686791696
16	-0,114830146839894	-2,13290644158049	0,130593182890157	3,87411054051073	0,0124546025215714	-0,00087558571887195	-0,00094107999185283

Fig. 4. Rectangular coordinates, the period of revolution small body around the Sun and the projection of velocities in the heliocentric rectangular coordinate system *XYZ*

analyzing and processing. Fragment of program interface and visualization of the conversion results to the programming environment is shown in Fig. 3. The first column of Fig. 3 shows the orbital numbers, and in the other columns – the values of elements of the corresponding Keplerian orbits. At the second stage of the technological chain of processing, the procedure for the transition from the values of the Keplerian orbital elements to the spatial rectangular coordinates and the projections of the velocities of astronomical objects in the heliocentric rectangular *XYZ* coordinate system is realized. Visualization of result of program component implementing this procedure is shown in Fig. 4.

Logical schemes of algorithms and procedures of technological chains for the implementation of solutions to the main task formulated in Sect. 3.2, include the construction of russified interface and visualization of the results of processing.

A fragment of the visualization of the operation of the software component for calculating the orbital boundness indexes, the orbital boundness coefficients relative to the reference orbit, the determination of the "center" of the orbital constraints, and the output of the Keplerian "center" orbit elements is shown on Fig. 5.

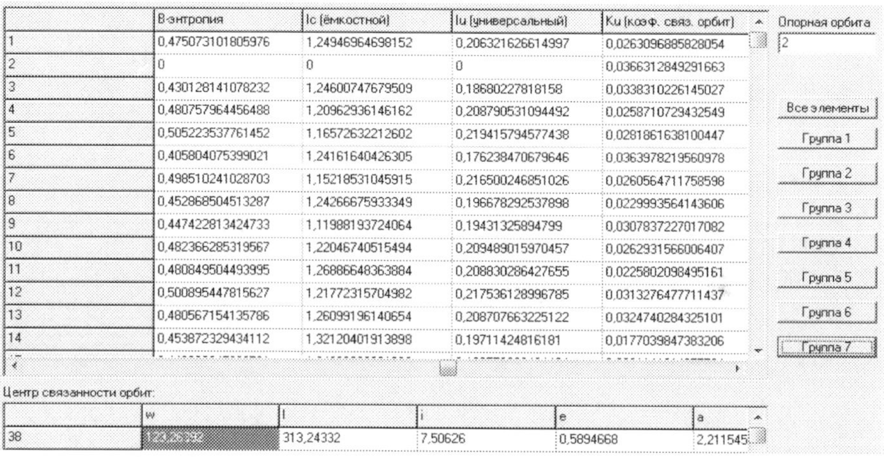

	В-энтропия	Ic (ёмкостной)	Iu (универсальный)	Ku (коэф. связ. орбит)
1	0,475073101805976	1,24946964698152	0,206321626614997	0,0263096885828054
2	0	0	0	0,0366312849291663
3	0,430128141078232	1,24600747679509	0,18680227818158	0,0338310226145027
4	0,480757964456488	1,20962936146162	0,208790531094492	0,0258710729432549
5	0,505223537761452	1,16572632212602	0,219415794577438	0,0281861638100447
6	0,405804075399021	1,24161640426305	0,176238470679646	0,0363978219560978
7	0,498510241028703	1,15218531045915	0,216500246851026	0,0260564711758598
8	0,452868504513287	1,24266675933349	0,196678292537898	0,0229993564143606
9	0,447422813424733	1,11988193724064	0,19431325894799	0,0307837227017082
10	0,482366285319567	1,22046740515494	0,209489015970457	0,0262931566006407
11	0,480849504493995	1,26886648363884	0,208830286427655	0,0225802098495161
12	0,500895447815627	1,21772315704982	0,217536128996785	0,0313276477711437
13	0,480567154135786	1,26099196140654	0,208707663225122	0,0324740284325101
14	0,453872329434112	1,32120401913898	0,19711424816181	0,0177039847383206

Опорная орбита: 2

Все элементы | Группа 1 | Группа 2 | Группа 3 | Группа 4 | Группа 5 | Группа 6 | Группа 7

Центр связанности орбит:

	w	l	i	e	a
38	[illegible]	313,24332	7,50626	0,5894668	2,211545

Fig. 5. The values of the indices cohesion, coupling coefficients of the orbits relative to the reference orbit, and Keplerian elements of the orbit coupling center

The technological chain of transition from the orbital description of positions bodies to their geometric arrangement in real space involves following procedures, namely: (1) filling of spatial volume lattice of given dimension; (2) the implementation of spatial clustering for observable small bodies of different groups – each cluster is assigned a unique label, the cluster size is calculated; 3) determination of the type of fractal cluster: percolation fractal or fractal aggregate; 4) visualization of the data clustering display – filling one of the given layers of the volume grating; 5) calculation of the degree of filling of a given grating and the number of clusters; 6) identification of percolation processes at the nodes of the volume lattice – whether there are percolation processes for a given lattice scale parameter and a group of small bodies. A fragment of the visualization of the execution of the stages of the indicated technological chain is shown on Fig. 6.

The fragments of model algorithms and procedures for processing large flows of astronomical data described in this section [13] reflect the mathematical, logical and algorithmic nature of the fractal paradigm in the development and implementation of IT– technologies of a new generation. The IT – technology for processing large flows of astronomical data presented in this paper is part of the technological system for processing large data flows developed on the basis of fractal paradigm in the form of synergy of cognitive technologies and scientific visualization systems.

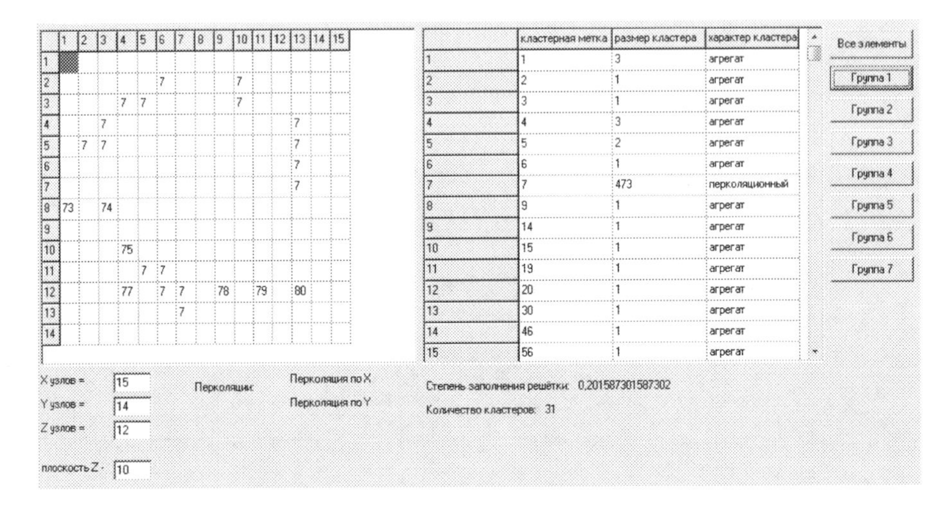

Fig. 6. Classification of the data stream orbits on the basis of fractal connectivity criteria for the example of group 1

5 Conclusions and Some Generalizations

The results of processing and analyzing observational data for groups of small bodies, indicated above, make it possible to draw a number of conclusions and generalizations of the following character. First, fractal methods in information processing and analysis of large flows of astronomical data on the basis of logical schemes of cognitive analytics of decoding hidden information in them are a promising and unique paradigm in the development of new-generation information technologies for a wide class of problems not only of modern astronomy. Secondly, the streams of astronomical observations can be processed using various processes and methods of the theory of fractals and genetic data both for obtaining aggregates and populations of sample data and for their analysis. These methods and processes reflect and determine the characteristics of the received estimates of fractal measures and dimensions, as well as the scope of the conclusions that can be drawn from these data. In this case, two types of sampling are used: genetic and statistical. Statistical selectivity is associated with the definition of the spatial scales of the lattice Z^3, and genetic – with the distribution of information and objects at the nodes of this lattice. In a broad aspect of fundamental astronomical studies of the problem of the formation and evolution of planetary systems, the results of this paper for the first time made it possible to show how and in what way the synergy of the geometry of the spatial structure and the dynamic evolution of the objects of the designated systems is manifested and how this can be described and explained within the framework of the fractal paradigm. Is it possible to conduct such analogies within the framework of traditional models, algorithms, schemes, etc.? If yes, then show the results of the indicated analogies and formulate the trends of their theoretical development and practical continuation. Applied aspects of the results of the work are closely related to the solution of the problems of asteroid–cometary–meteoric safety and the problem of space debris. On the one hand, methods of fractal theory for solving complex nonlinear problems of processing, analyzing and interpreting the results of the dynamic evolution of space objects with irregular spatially-temporal fractal geometry are proposed. And on the other hand, a new IT technology was developed and implemented in the DAMDID trend of processing, analyzing and classifying the orbital data of small bodies of the Solar System (for software implementation of IT technologies, data from the MPC portal – www.cfa.harvard.edu and NASA – www.portal.nasa.gov).

References

1. Roy, A.E.: Orbital motion. CRC Press, USA (2004)
2. Guseva, I.S., Lih, Y.S.: Statistical analysis of the orbits comets. News Main Astron. Obs. RAS **220**, 219–224 (2012)
3. Kochetova, O.M., Kuznetsov, V.B., Medvedev, Y.D., Shor, V.A.: Catalog of elements orbits of numbered asteroids of institute applied astronomy RAS. News Main Astron. Obs. RAS **220**, 255–258 (2012)
4. Malkin, Z.M.: Some results of statistical analysis of the definitions of the galactic distance of the sun. News Main Astron. Obs. RAS **220**, 401–406 (2012)
5. Bruno, A.D., Varin, V.P.: On asteroid distribution. Sol. Syst. Res. **45**(4), 323–330 (2011)
6. Myshev, A.V.: Metrological theory of the dynamics interacting objects in the information field of a neural network and a neuron. Inf. Technol. **4**, 52–63 (2012)
7. Pavlov, A.N., Anishchenko, V.S.: Multifractal analysis of complex signals. Adv. Phys. Sci. **7**(8), 819–834 (2007)
8. Feder, J.: Fractals. Plenum Press, New York (1988)
9. Emel'yanenko, V.V., Naroenkov, S.A., Shustov, B.M.: On asteroid distribution. Sol. Syst. Res. **45**(6), 498–504 (2011)
10. Gaftonyuk, N.M., Gorkavyi, N.N.: Asteroids with satellites: analysis of observational data. Sol. Syst. Res. **47**(6), 196–203 (2013)
11. Naroenkov, S.A.: Storing and processing astrometric and photometric data on NEA: current state and future in Russia. Cosm. Res. **48**(5), 455–459 (2010)
12. Alfven, H., Arrhenius, G.: Evolution of the Solar System. NASA SP, Washington (1976)
13. Myshev, A.V., Dunin, A.V.: Fractal methods in information technologies for processing, analyzing and classifying large flows of astronomical data. Selected Papers of XIX International Conference on Data Analytics and Management in Data Intensive Domains (DAMDID/RCDL 2017), Moscow, Russia. vol. 2022, pp. 172–176. CEUR Workshop Proceedings (2017). http://ceur-ws.org/Vol-2022/

Neural Gas Based Classification of Globular Clusters

Giuseppe Angora[1(✉)], Massimo Brescia[2], Stefano Cavuoti[1,2,3], Giuseppe Riccio[2], Maurizio Paolillo[1], and Thomas H. Puzia[4]

[1] University of Naples Federico II - Dept. of Physics “E. Pancini”, via Cintia 6, 80135 Napoli, Italy
gius.angora@gmail.com

[2] INAF - Astronomical Observatory of Capodimonte, via Moiariello 16, 80131 Napoli, Italy

[3] INFN - Napoli Unit, via Cintia 6, 80135 Napoli, Italy

[4] Institute of Astrophysics, Pontificia Universidad Catolica de Chile, Av. Vicuña Mackenna 4860, Macul, Santiago, Chile

Abstract. Within scientific and real life problems, classification is a typical case of extremely complex tasks in data-driven scenarios, especially if approached with traditional techniques. Machine Learning supervised and unsupervised paradigms, providing self-adaptive and semi-automatic methods, are able to navigate into large volumes of data characterized by a multi-dimensional parameter space, thus representing an ideal method to disentangle classes of objects in a reliable and efficient way. In Astrophysics, the identification of candidate Globular Clusters through deep, wide-field, single band images, is one of such cases where self-adaptive methods demonstrated a high performance and reliability. Here we experimented some variants of the known Neural Gas model, exploring both supervised and unsupervised paradigms of Machine Learning for the classification of Globular Clusters. Main scope of this work was to verify the possibility to improve the computational efficiency of the methods to solve complex data-driven problems, by exploiting the parallel programming with GPU framework. By using the astrophysical playground, the goal was to scientifically validate such kind of models for further applications extended to other contexts.

Keywords: Data analytics · Astroinformatics · Globular Clusters Machine learning · Neural Gas

1 Introduction

The current and incoming astronomical synoptic surveys require efficient and automatic data analytics solutions to cope with the explosion of scientific data amounts to be processed and analyzed. This scenario, quite similar to other scientific and social contexts, pushed all communities involved in data-driven disciplines to explore data mining techniques and methodologies, most of which

L. Kalinichenko et al. (Eds.): DAMDID/RCDL 2017, CCIS 822, pp. 86–101, 2018.
https://doi.org/10.1007/978-3-319-96553-6_7

connected to the Machine Learning (hereafter ML) paradigms, i.e. supervised and unsupervised self-adaptive learning and parameter space optimization [6,8]. Following this premise, this paper is focused on the investigation about the use of a particular kind of ML methods, known as Neural Gas (NG) models [21], to solve classification problems within the astrophysical context, characterized by a complex multi-dimensional parameter space. In order to scientifically validate such models, we decided to approach a typical astrophysical playground, already solved with ML methods [7,11,12] and to use in parallel other two ML techniques, chosen among the most standard, respectively, Random Forest [5] and Multi Layer Perceptron Neural Network [23], as comparison baseline. The astrophysical case is related to the identification of Globular Clusters (GCs) in the galaxy NGC1399 using single band photometric data obtained through observations with the Hubble Space Telescope (HST) [7,25,27]. The physical identification and characterization of a Globular Cluster (GC) in external galaxies is considered important for a variety of astrophysical problems, from the dynamical evolution of binary systems, to the analysis of star clusters, galaxies and cosmological phenomena [27]. Here, the capability of ML methods to learn and recognize peculiar classes of objects, in a complex and noising parameter space and by learning the hidden correlation among objects parameters, has been demonstrated particularly suitable in the problem of GC classification [7]. In fact, multi-band wide-field photometric data (colours and luminosities) are usually required to recognize GCs within external galaxies, due to the high risk of contamination of background galaxies, which appear indistinguishable from galaxies located few Mpc away, when observed by ground-based instruments. Furthermore, in order to minimize the contamination, high-resolution space-borne data are also required, since they are able to provide particular physical and structural features (such as concentration, core radius, etc.), thus improving the GC classification performance [25]. In [7] we demonstrated the capability of ML methods to classify GCs using only single band images from Hubble Space Telescope with a classification accuracy of 98.3%, a completeness of 97.8% and only 1.6% of residual contamination. Thus confirming that ML methods may yield low contamination by minimizing the observing requirements and extending the investigation to the outskirts of nearby galaxies. These results gave us an optimal playground where to train NG models and to validate their potential to solve classification problems characterized by complex data with a noising parameter space. The paper is structured as follows: in Sect. 2 we describe the data used to test various methods. In Sect. 3 we provide a short methodological and technical description of the models. In Sect. 4 we describe the experiments and results about the parameter space analysis and classification experiments, while in Sect. 5.2 we discuss the results and draw our conclusions. This paper is an extended version of the work presented at the DAMDID 2017 Conference and published in the related Proceedings [2]. It includes three new sections, respectively, Sects. 3.6, 4.3, 5.2, plus a new figure (Fig. 3) and the re-phrased abstract section, whose total amount is more than 30% of the previous work [2].

2 The Astrophysical Playground

The HST single band data are very suitable to investigate the classification of GCs. They, in fact, represent deep and complete in terms of wide field coverage i.e. able to sample the GC population, to ensure a high S/N ratio required to measure structural parameters [10]. Furthermore, they provide the possibility to study the overall properties of the GC populations, which usually may differ from those of the central region of a galaxy. With such data we intend to verify that Neural Gas based models could be able to identify GCs with low contamination even with single band photometric information. Throughout the confirmation of such behavior, we are confident that these models could solve other astrophysical problems as well as other data-driven problem contexts.

2.1 The Data

The data used in the described experiment consist of wide field single band HST observations of the giant elliptical NGC1399 galaxy, located in the core of the Fornax cluster [27]. Due to its distance (D = 20.130 Mpc, see [14]), it is considered an optimal case where to cover a large fraction of its GC system with a restricted number of observations. This dataset was used by [25] to study the GC-LMXB connection and the structural properties of the GC population. The optical data were taken with the HST Advanced Camera for Surveys, in the broad V band filter, with 2108 seconds of integration time for each field. The detection of GCs relies basically upon two aspects: the shape of the image (which differs from the instrumental PSF) and the colours (i.e. the ratio of observed fluxes at different wavelengths). The shape allows to disentangle large systems from stars (which are PSF-like), while the colours are needed to disentangle GCs from other extended systems, such as background galaxies.

The source catalog was generated using Sextractor [3,4], by imposing a minimum area of 20 pixels: it contains 12915 sources and reaches 7σ detection at m_V = 27.5, i.e. 4 mag below the GC luminosity function, thus allowing to sample the entire GC population (see [7] for details). The source subsample used to build our Knowledge Base (KB) to train the ML models, is composed by 2100 sources with 11 features (7 photometric and 4 morphological parameters). Such parameter space includes three aperture magnitudes within 2, 6 and 20 pixels (*mag_aper1, mag_aper2, mag_aper3*), isophotal magnitude (*mag_iso*), kron radius (*kron_rad*), central surface brightness (*mu0*), FWHM (*fwhm_im*), and the four structural parameters, respectively, *ellipticity*, King's tidal, effective and core radii (*calr_t, calr_h, calr_c*). The target values of the KB required as ground truth for training and validation, i.e. the binary column indicating the source as GC or not GC, is provided through the typical selection based on multi-band magnitude and colour cuts. The original 2100 sources having a target assigned have been randomly shuffled and split into a training (70%) and a blind test set (30%).

3 The Machine Learning Models

In our work we tested three different variants of the Neural Gas model, using two additional machine learning methods, respectively feed-forward neural network and Random Forest, as comparison benchmarks. In the following all main features of these models are described.

3.1 Growing Neural Gas

Growing Neural Gas (GNG) is presented by [16] as a variant of the Neural Gas algorithm (introduced by [21]), which combines the Competitive Hebbian Learning (CHL, [22]) with a vector quantization technique to achieve a learning that retains the topology of the dataset. Vector quantization techniques [22] encode a data manifold, e.g. $V \subseteq R^m$, using a finite set of reference vectors $w = w_1 \dots w_N, w_i \in R^m, i = 1 \dots N$. Every data vector $v \in V$ is described by the best matching reference vector $w_{i(v)}$ for which the distortion error $d(v, w_{i(v)})$ is minimal. This procedure divides the manifold V into a number of subregions: $V_i = \{v \in V : ||v - w_i|| \leq ||v - w_j|| \forall j\}$, called Voronoi polyhedra [24], within which each data vector v is described by the corresponding reference vector w_i. The Neural Gas network is a vector quantization model characterized by N neural units, each one associated to a reference vector, connected to each other. When an input is extracted, it induces a synaptic excitation detected by all the neurons in the graph and causes its adaptation. As shown in [21], the adaptation rule can be described as a "winner-takes-most" instead of "winner-takes-all" rule:

$$\Delta w_i = \epsilon h_\lambda(v, w_i) \cdot (v - w_i), \quad i = 1 \dots N. \tag{1}$$

The step size ε describes the overall extent of the adaptation. While $h_\lambda(v, w_i) = h_\lambda(k_i(v, w))$ is a function in which k_i is the "neighborhood-ranking" of the reference vectors. Simultaneously, the first and second Best Matching Units (BMUs) develop connections between each other [21]. Each connection has an "age"; when the age of a connection exceeds a pre-specified lifetime T, it is removed [21]. Martinez's reasoning is interesting [22]: they demonstrate how the dynamics of neural units can be compared to a gaseous system. Let's define the density of vector reference at location u through $\rho(u) = F^{-1}_{BMU(u)}$, where $F_{BMU(u)}$ is the volume of Voronoi polyhedra. Hence, $\rho(u)$ is a step function on each Voronoi polyhedra, but we can still imagine that their volumes change slowly from one polyhedra to the next, with $\rho(u)$ continuous. In this way, it is possible to derive an expression for the average change:

$$\langle \Delta w_i \rangle \propto \frac{1}{\rho^{1+2/m}} \Big(\partial_u P(u) - \frac{2+m}{m} \frac{P}{\rho} \partial_u \rho(u) \Big) \tag{2}$$

where $P(u)$ is the data point distribution. The equation suggests the name Neural Gas: the average change of the reference vectors corresponds to a motion of particles in a potential $V(u) = -P(u)$. Superimposed on the gradient of this potential there is a force proportional to $-\partial_u \rho(u)$, which points toward the direction of the space where the particle density is low. Main idea behind the GNG network is to successively add new units to an initially small network, by evaluating local statistical measures collected during previous adaptation steps [16]. Therefore, each neural unit in the graph has associated a local reconstruction error, updated for the BMU at each iteration (i.e. each time an input is extracted): $\Delta error_{BMU} = ||w_{BMU} - v||$. Unlike the Neural Gas network, in the GNG the synaptic excitation is limited to the receptive fields related to the Best Matching Unit and its topological neighbors: $\Delta w_i = \epsilon_i(v - w_i)$, $i \in (BMU, n), \forall n \in neighbours(BMU)$. It is no longer necessary to calculate the ranking for all neural units, but it is sufficient to determine the first and the second BMU. The increment of the number of units is performed periodically: during the adaptation steps the error accumulation allows to identify the regions in the input space where the signal mapping causes major errors. Therefore, to reduce this error, new units are inserted in such regions [16]. An elimination mechanism is also provided: once the connections, whose age is greater than a certain threshold, have been removed, if their connected units remain isolated (i.e. without emanating edges), those units are removed [16].

3.2 GNG with Radial Basis Function

Fritzke describes an incremental Radial Basis Function (RBF) network suitable for classification and regression problems [16]. The network can be figured out as a standard RBF network [9], with a GNG algorithm as embedded clustering method, used to handle the hidden layer. Each unit of this hybrid model (hereafter GNGRBF) is a single perceptron with an associated reference vector and a standard deviation. For a given input-output pair $(v, y), v \in R^n, y \in R^m$, the activation of the i-th unit is described by $D_i(v) = e^{-||v - w_i||/\sigma_i^2}$. Each of the single perceptron computes a weighted sum of the activations: $O_i = \sum_j w_{ij} D_j(v)$, $i = 1 \ldots m$. The adaptation rule applies to both reference vectors forming the hidden layer and the RBF weights. For the first, the adaptation rule is the same of the updating rule for the GNG network, while for the weights:

$$\Delta w_{ij} = \eta D_j(y_i - o_i), i = 1 \ldots m, j \in N. \tag{3}$$

Similarly to the GNG network, new units are inserted where the prediction error is high, updating only the Best Matching Unit at each iteration: $\Delta error_{BMU} = \sum_{i=1}^{m} y_i - O_i$.

3.3 Supervised Growing Neural Gas

The Supervised Growing Neural Gas (SGNG) algorithm is a modification of the GNG algorithm that uses class labels of data to guide the partitioning of data into optimal clusters [15,20]. Each of the initial neurons is labelled with a unique class label. To reduce the class impurity inside the cluster, the original learning rule 1 is reformulated by considering the case where the BMU belongs or not to the same class of the neuron whose reference vector is the closest to the current input. Depending on such situation the SGNG learning rule is expressed alternatively as:

$$\begin{cases} \Delta w_n = -\epsilon \frac{v-w_n}{||v-w_n||} & or \\ \Delta w_n = +\epsilon \frac{v-w_n}{||v-w_n||} + repulsion(sn, n) \end{cases} \quad (4)$$

where sn is the nearest class neuron and $repulsion()$ is a function specifically introduced to maintain neurons sufficiently distant one each other. For the neuron that is topologically close to the neuron sn, the rule intends to increase the clustering accuracy [20]. The insertion mechanism has to reduce not only the intra-distances between data in a cluster, but also the impurity of the cluster. Each unit has associated two kinds of error: an aggregated and a class error. A new neuron is inserted close to the neuron having a highest class error accumulated, while the label is the same as the neuron label with the greater aggregated error.

3.4 Multi Layer Perceptron

The Multi Layer Perceptron (MLP) architecture is one of the most typical feed-forward neural networks [23]. The term feed-forward is used to identify basic behavior of such neural models, in which the impulse is propagated always in the same direction, e.g. from neuron input layer towards output layer, through one or more hidden layers (the network brain), by combining the sum of weights associated to all neurons. As easy to understand, the neurons are organized in layers, with proper own role. The input signal, simply propagated throughout the neurons of the input layer, is used to stimulate next hidden and output neuron layers. The output of each neuron is obtained by means of an activation function, applied to the weighted sum of its inputs. The weights adaptation is obtained by the Logistic Regression rule [18], by estimating the gradient of the cost function, the latter being equal to the logarithm of the likelihood function between the target and the prediction of the model. In this work, our implementation of the MLP is based on the public library Theano [1].

3.5 Random Forest

Random Forest (RF) is one of the most widely known machine learning ensemble methods [5], since it uses a random subset of candidate data features to build an ensemble of decision trees. Our implementation makes use of the public library scikit-learn [26]. This method has been chosen mainly because it provides for each input feature a score of importance (rank) measured in terms of its informative contribution percentage to the classification results. From the architectural point of view, a RF is a collection (forest) of tree-structured classifiers $h(x, \theta_k)$, where the θ_k are independent, identically distributed random vectors and each tree casts a unit vote for the most popular class at input. Moreover, a fundamental property of the RF is the intrinsic absence of training overfitting [5].

3.6 Implementation Details

In this work we compare the classification results and computational performance referred to three different implementations: besides the original networks (GNG_{old}, $GNGRBF_{old}$), we developed different versions, including one exclusively using the numpy methods [29] in the case of GNG, and another completely written using theano library [1] in all cases of GNG, GNGRBF and MLP. The MLP model was selected as test bench to compare NG-based models with the most general-purpose feedforward neural network. Concerning the *old* version of NG methods, main modification in the new versions is the random extraction of a sample batch, whose size varies between 1 (corresponding to the *old* case) and the total number of samples. The advantages are: (i) the method allows to use the matrices algebra; and (ii) during the learning phase the adaptation of the weights and the reference vectors reflect the complexity of the data set. The use of the theano library allows the possibility to compute the gradient of a cost function depending on the parameters available. For the GNG the cost function could be represented by the quantization error, which is a measure of similarity among the samples allocated by the same BMU. Given a data set $v \in V$ composed by $|V|$ records, distributed among p neurons, the quantization error is given by:

$$\mathcal{QE} = \frac{1}{2|V|} \sum_{i=1}^{p} \sum_{n \in BMU_i} ||\mathbf{v}_n - \mathbf{w}_i||^2.$$

In addition, the reference vectors are adapted with:

$$\Delta \mathbf{W} = -\eta \nabla_{\mathbf{W}}(\mathcal{QE}),$$

where $\mathbf{W}$ is the matrix of the reference vectors, η is the learning rate and $\nabla_{\mathbf{W}}$ is the gradient derived from reference vectors. For the supervised networks ($GNGRBF_{theano}$ and MLP) the cost function is computed as the negative logarithm of the likelihood function between the target and the output:

$$\begin{aligned} \mathcal{C} &= \frac{1}{|V|}\mathcal{L}(\theta = \{\mathbf{W}, \mathbf{b}\}, V) \\ &= \frac{1}{|V|}\sum_{i=0}^{|V|} \log(P(y = y_i|\mathbf{X}_i, \mathbf{W}, \mathbf{b})) \end{aligned} \tag{5}$$

4 The Experiments

The five models previously introduced have been applied to the dataset described in Sect. 2.1 and their performances have been compared to verify the capability of NG models to solve particularly complex classification problems, like the astrophysical identification of GCs from single-band observed data.

4.1 The Classification Statistical Estimators

In order to evaluate the performances of the selected classifiers, we decided to use three among the classical and widely used statistical estimators, respectively, average efficiency, purity, completeness and F1-score, which can be directly derived from the confusion matrix [28]. The average efficiency (also known as accuracy, hereafter AE), is the ratio between the sum of correctly classified objects on both classes (true positives for both classes, hereafter tp) and the total amount of objects in the test set. The purity (also known as precision, hereafter pur) of a class measures the ratio between the correctly classified objects and the sum of all objects assigned to that class (i.e. tp/[tp+fp], where fp indicates the false positives). While the completeness (also known as recall, hereafter comp) of a class is the ratio tp/[tp+fn], where fn is the number of false negatives of that class. The quantity tp+fn corresponds to the total amount of objects belonging to that class. The F1-score is a statistical test that considers both the purity and completeness of the test to compute the score (i.e. 2[pur·comp]/[pur+comp]). By definition, the dual quantity of the purity is the contamination, another important measure which indicates the amount of misclassified objects for each class. In statistical terms, it is well known the classical tradeoff between purity and completeness in any classification problem, particularly accentuated in astrophysical problems [13]. In the specific case of the GC identification, from the astrophysical point of view, we were mostly interested to the purity, i.e. to ensure the highest level of true GCs correctly identified by the classifiers [7]. However, within the comparison experiments described in this work, our main goal was to evaluate the performances of the classifiers mostly related to the best tradeoff between purity and completeness.

4.2 Analysis of the Data Parameter Space

Before to perform the classification experiments, we preliminarily investigated the parameter space, defined by the 11 features defined in Sect. 2.1, identifying each object within the KB dataset of 2100 objects. Main goal of this phase was to measure the importance of any feature, i.e. its relevance in terms of informative contribution to the solution of the problem. In the ML context, this analysis is usually called feature selection [17]. Its main role is to identify the most relevant features of the parameter space, trying to minimize the impact of the well known problem of the curse of dimensionality, i.e. the fact that ML models exhibit a decrease of performance accuracy when the number of features is significantly higher than optimal [19]. This problem is mainly addressed to cases with a huge amount of data and dimensions. However, its effects may also impact contexts with a limited amount of data and parameter space dimension. The Random Forest model resulted particularly suitable for such analysis, since it is intrinsically able to provide a feature importance ranking during the training phase. The feature importance of the parameter space, representing the dataset used in this work, is shown in Fig. 1. From the astrophysical point of view, this ranking is in accordance with the physics of the problem. In fact, as expected, among the five most important features there are the four magnitudes, i.e. the photometric log-scale measures of the observed object's photonic flux through different apertures of the detector. Furthermore, almost all photometric features resulted as the most relevant. Finally, by looking at the Fig. 1, there is an interesting gap between the first six and the last five features, whose cumulative contribution is just ~11% of the total. Finally, a very weak joined contribution (3%) is carried by the two worst features (kron_rad and calr_c), which can be considered as the most noising/redundant features for the problem domain.

Table 1. Statistical analysis of the classification performances obtained by the five ML models on the blind test set for the four selected experiments. All quantities are expressed in percentage and related to average efficiency (AE), purity for each class (purGC, purNotGC), completeness for each class (compGC, compNotGC) and the F1-score for GC class. The contamination is the dual value of the purity. The numerical index of used features corresponds to the list in Fig. 1.

ID	Features	Estimator	RF%	MLP%	SGNG%	GNGRBF%	GNG%
E1	1, 2, 3, 5	AE	88.9	84.4	88.1	88.1	88.4
		purGC	85.9	80.1	89.7	85.4	83.7
		compGC	87.3	82.6	80.3	85.7	89.2
		F1-scoreGC	86.6	81.3	84.7	85.5	86.4
		purNotGC	91.0	87.6	87.2	90.0	92.1
		compNotGC	89.7	85.6	93.0	89.6	88.1
E2	1, 2, 3, 4, 5, 6	AE	89.0	85.1	87.3	88.3	83.2
		purGC	84.9	77.0	81.0	82.9	74.0
		compGC	89.2	90.7	90.3	90.0	91.1
		F1-scoreGC	87.0	83.3	85.4	86.3	81.7
		purNotGC	92.2	92.6	92.7	92.6	92.6
		compNotGC	89.0	85.6	85.7	87.4	80.0
E3	1, 2, 3, 4, 5, 6, 10	AE	89.0	83.2	85.1	89.2	86.8
		purGC	85.2	77.2	80.0	86.0	84.1
		compGC	88.8	83.8	84.9	88.0	83.8
		F1-scoreGC	87.0	80.4	82.4	87.0	83.9
		purNotGC	91.9	88.0	89.0	91.5	88.7
		compNotGC	89.9	83.2	85.1	89.8	88.4
E4	1, 2, 3, 4, 5, 6, 7, 8, 9	AE	89.5	86.0	88.1	88.7	83.8
		purGC	85.3	82.5	84.1	83.8	78.3
		compGC	90.0	83.8	87.6	90.0	83.8
		F1-scoreGC	87.6	83.1	85.8	86.8	81.0
		purNotGC	92.7	88.6	91.1	92.6	88.1
		compNotGC	89.1	87.5	88.1	88.2	84.1

Based on such considerations, the analysis of the parameter space provides a list of most interesting classification experiments to be performed with the selected five ML models. This list is reported in Table 1. The experiment E1 is useful to verify the efficiency by considering the four magnitudes. The experiment E2 is based on the direct evaluation of the best group of features as derived from the importance results. The classification efficiency of the full photometric subset of features is evaluated through the experiment E3. Finally, the experiment E4 is performed to verify the results by removing only the two worst features.

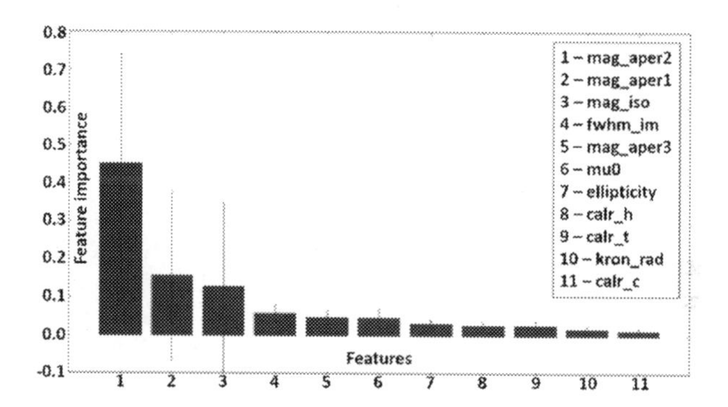

Fig. 1. The feature importance ranking obtained by the Random Forest on the 11-feature domain of the input dataset during training (see Sect. 2.1 for details). The vertical lines report the importance estimation error bars.

4.3 The Classification of Globular Clusters

By applying the feature space analysis, the original parameter domain has been simplified, reducing the number of features used. The classification experiments have been performed on the data, presented in Sect. 2.1, composed by 2100 objects and represented by a parameter space with up to a maximum of 9 features (Table 1). The dataset has been randomly shuffled and split into a training set of 1470 objects (70% of the whole KB) and a blind test set of 630 objects (the residual 30% of the KB). These datasets have been used to train and test the selected five ML classifiers. The analysis of results, reported in Table 1, has been performed on the blind test set, in terms of the statistical estimators defined in Sect. 4.2. Besides the classification quality evaluation, the three models and their variants have been also investigated in terms of computing efficiency. In fact, NG-based models are not intrinsically scalable with the data volume, but they require a careful optimization. The *old* versions of GNG and GNGRBF have the strong limitation of the absence of any batch system, being forced to evaluate one by one the data entries during the learning phase. Therefore, a direct comparison of computing time vs data size among *old* and evolved batch-based variants is intrinsically in favor of the latter at least by one order of magnitude. However, an equally interesting comparison involving *old* and evolved versions of the models is shown in the two top panels of Fig. 3, where are evaluated, respectively, the three variants of the GNG model in terms of execution time as function of the quantization error along the training process and the two versions of GNGRBF vs MLP in terms of execution time as function of the Root Mean Square (RMS) of the learning error along the training process. While a comparison of the computing efficiency, in terms of execution time as function of training data increasing size is reported in the two bottom panels of Fig. 3, showing, respectively, a direct evaluation among all NG-based models and a direct comparison among the *theano*-based version for all three types of ML models. The incremental dataset was obtained as multiple repetitions of the original dataset with the addition of white noise.

5 Discussion

As already underlined, main goal of this work was the validation of NG models as efficient classifiers in noising and multi-dimensional problems, with performances at least comparable to other ML methods, considered traditional in terms of their use in such kind of problems.

5.1 Analysis of Classification Performance

By looking at Table 1 and focusing on the statistics for the three NG models, it is evident that their result is able to identify GCs from other background objects, reaching a satisfying tradeoff between purity and completeness in all experiments and for both classes. The occurrence of statistical fluctuations is mostly due to the different parameter space used in the four experiments. Nevertheless, none of the three NG models overcome the others in terms of the measured statistics. If we compare the NG models with the two additional ML methods (Random Forest and MLP neural network), their performances appears almost the same. This implies that NG methods show classification capabilities fully comparable to other ML methods. Another interesting aspect is the analysis of the degree of coherence among the NG models in terms of commonalities within classified objects. Table 2 reports the percentages of common predictions for the objects correctly classified by considering, respectively both and single classes. On average, the three NG models are in agreement among them for about 80% of the objects correctly classified.

Table 2. Statistics for the three NG models related to the common predictions of the correctly classified objects. Second column is referred to both classes, while the third and fourth columns report, respectively, the statistics for single classes.

EXP ID	GC+notGC%	GC%	notGC%
E1	86.0	85.4	86.9
E2	79.8	79.8	79.8
E3	81.1	82.5	79.2
E4	77.8	77.4	78.4

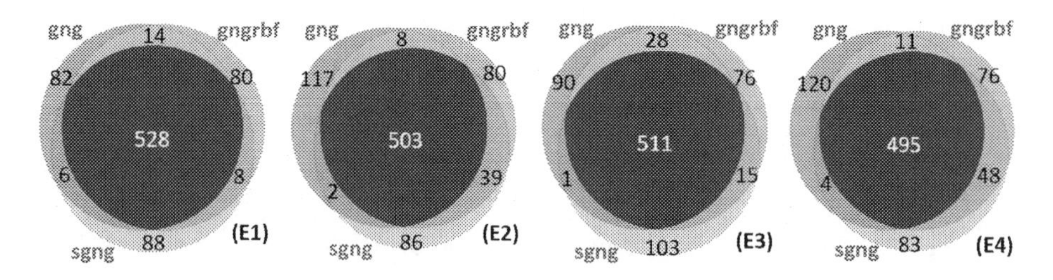

Fig. 2. The Venn diagram related to the prediction of all (both GCs and not GCs) correctly classified objects performed by the three Neural Gas based models (GNG, GNGRBF and SGNG) for the experiments, respectively, E1 (a), E2 (b), E3 (c) and E4 (d). The intersection areas (dark grey in the middle) show the objects classified in the same way by different models. Internal numbers indicate the amount of objects correctly classified for each sub-region.

This is also confirmed by looking at the Fig. 2, where the tabular results of Table 2 are showed through the Venn diagrams, reporting also more details about their classification commonalities. Finally, from the computational efficiency point of view, the NG models have theoretically a higher complexity than Random Forest and neural networks. But, since they are based on a dynamic evolution of the internal structure, their complexity strongly depends on the nature of the problem and its parameter space. Nevertheless, all the presented ML models have a variable architectural attitude to be compliant with the parallel computing paradigms. Besides the embarrassingly parallel architecture of the Random Forest, the use of optimized libraries, like Theano [1], make also models like MLP highly efficient. From this point of view NG models have a high potentiality to be parallelized. By optimizing GNG, the GNGRBF would automatically benefit, since both share the same search space, except for the RBF training additional cost.

In practice, the hidden layer of the supervised network behaves just like a GNG network whose neurons act as inputs for the RBF network. Consequently, with the same number of iterations, the GNGRBF network performs a major number of operations. On the other hand, the SGNG network is similar to the GNG network, although characterized by a neural insertion mechanism over a long period, thus avoiding too rapid changes in the number of neurons and excessive oscillations of reference vectors. Therefore, on average, the SGNG network computational costs are higher than the models based on the standard Neural Gas mechanism.

5.2 Analysis of Computing Performance

From the computing efficiency point of view, all NG-based models have a complexity of the order $O(D \times N \times M)$, with D dimensions of the parameter space, N amount of data and M number of neurons. But, since they are based on a dynamic evolution of the internal structure, their complexity strongly depends on the nature of the problem and its parameter space. In Fig. 3 (top left panel) we can see how the different architectures influence the computing time of the same model. In fact, different ways to store the memory induce different *loading times*. The *old* implementations having a simpler data structure are, in fact, faster at the beginning of the training than the *theano* implementations. During the training such initial gap is filled by means of the theano optimization, while at the end of the process we obtain a final quantization error which is similar

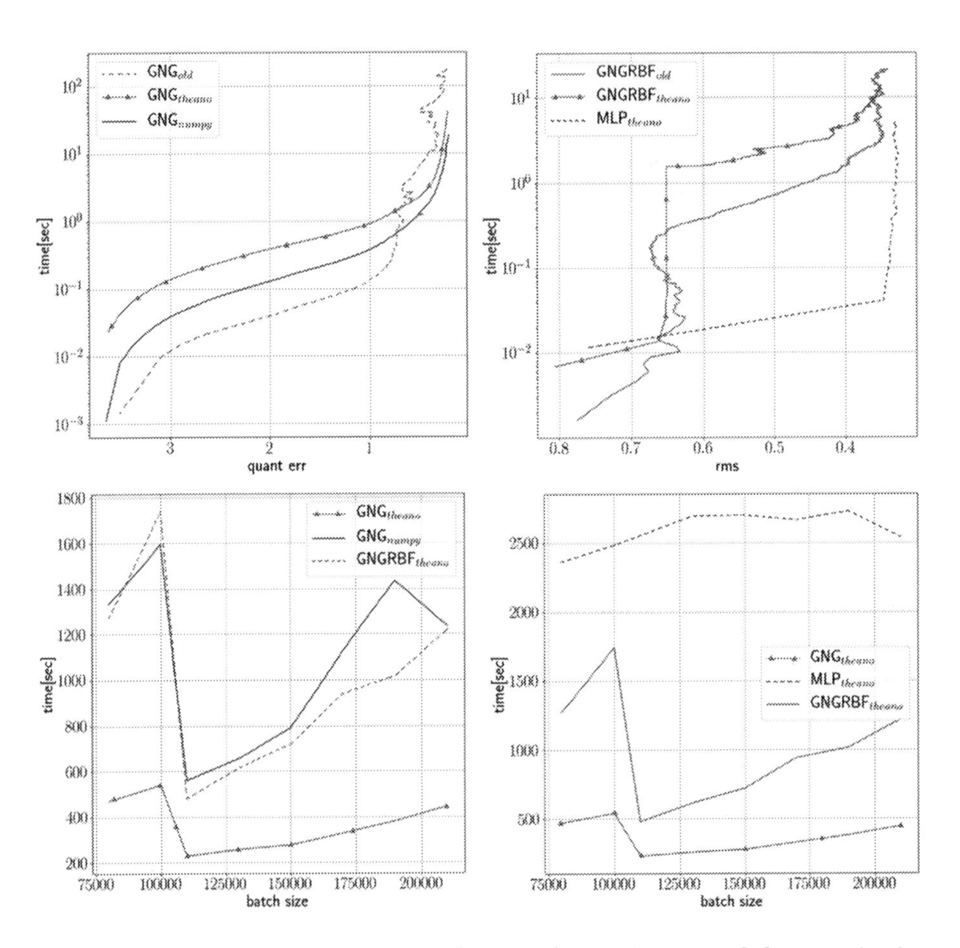

Fig. 3. Top Left panel: execution time (log scale) as a function of the quantization training error for the three GNG networks: GNG_{old} (dotted line), GNG_{theano} (triangle line), GNG_{numpy} (whole line). Top Right panel: Execution time (log scale) as a function of the root mean square of the training error for the three supervised networks: $GNGRBF_{old}$ (whole line), $GNGRBF_{theano}$ (triangle line), MLP (dotted line). Bottom panels: execution time vs increasing data size for, respectively, the three NG-based models (left) and the three *theano*-based models (right). $GNGRBF_{theano}$ (dotted line in left panel, whole line in right panel), GNG_{theano} (triangle line in both panels), GNG_{numpy} (whole line in left panel) and MLP (dotted line in right panel).

but reached at different times: the *old* implementation is the slower while *numpy* and *theano* are comparable. This should be even more remarkable with a wider dataset and we plan to perform such kind of analysis in the next future. By looking at the top right panel of Fig. 3 we can see the trend of time as function of RMS for the supervised models (i.e. GNGRBF_{theano}, GNGRBF_{old} and MLP_{theano}). Between the two versions of GNGRBF it appears an initial gap. After that, the GNGRBF implementation needs a sufficient number of neurons to codify the input data in order to reduce the RMS error. This is clearly connected to the use of the *batch system* in the implementation. Once out from the local minimum the *theano* implementation is close to the computing performance of the *old* implementation, reaching the final solution in less time. As in the previous case we expect that the improvement should be more evident with a larger dataset. We noticed that MLP has a completely different behavior, by immediately reaching a good solution, spending much of the computing time to slightly improve the learning quality. From what computing time vs data batch size concerns, the two bottom panels of Fig. 3 report the execution time required by different models as function of the growing batch size of input data. As evident in the bottom left panel of Fig. 3, the *theano* version of GNG is outperforming its *numpy* version. The almost same trend between GNG_{numpy} and GNGRBF_{theano} can be easily motivated by taking into account the higher computational cost of RBF network as well as an increased number of calculations imposed by the hybrid model, which shades the improvements carried by the *theano* optimization. The bottom right panel of Fig. 3 poses a direct comparison among the three *theano* versions of the three ML models. As in the previous case, the GNG model outperforms the others. It is also remarkable the behavior of MLP, which shows an almost uniform trend, motivated by considering its intrinsic learning mode, less sensible to the variation of data batch size. While a usual sharp decay is again present in the behavior of GNGRBF model.

6 Conclusions

In conclusion, although a more intensive test campaign on these models is still ongoing, we can assert that Neural Gas based models are very promising as problem-solving methods, also in presence of complex and multi-dimensional classification and clustering problems, especially if preceded by an accurate analysis and optimization of the parameter space within the problem domain.

Acknowledgements. MB acknowledges the *INAF PRIN-SKA 2017 program 1.05.01.88.04* and the funding from *MIUR Premiale 2016: MITIC*.

References

1. Al-Rfou, R., Alain, G., Almahairi, A., et al.: Theano: A Python framework for fast computation of mathematical expressions. ArXiv e-prints, May 2016
2. Angora, G., Brescia, M., Riccio, G., Cavuoti, S., Paolillo, M., Puzia, T.H.: Astrophysical data analytics based on neural gas models, using the classification of globular clusters as playground. In: CEUR Workshop Proceedings, vol. 2022, pp. 381–388 (2017)
3. Annunziatella, M., Mercurio, A., Brescia, M., Cavuoti, S., Longo, G.: Inside catalogs: a comparison of source extraction software. Publ. Astron. Soc. Pac. **125**(923), 68–82 (2013). https://doi.org/10.1086/669333
4. Bertin, E., Arnouts, S.: SExtractor: software for source extraction. Astron. Astrophys. Suppl. **117**, 393–404 (1996). https://doi.org/10.1051/aas:1996164
5. Breiman, L.: Random forests. Mach. Learn. **45**(1), 5–32 (2001). https://doi.org/10.1023/A:1010933404324
6. Brescia, M., Cavuoti, S., Longo, G., et al.: Dameware: a web cyberinfrastructure for astrophysical data mining. Publ. Astron. Soc. Pac. **126**(942), 783–797 (2014)
7. Brescia, M., Cavuoti, S., Paolillo, M., Longo, G., Puzia, T.: The detection of globular clusters in galaxies as a data mining problem. Mon. Not. R. Astron. Soc. **421**(2), 1155–1165 (2012). https://doi.org/10.1111/j.1365-2966.2011.20375.x
8. Brescia, M., Longo, G.: Astroinformatics, data mining and the future of astronomical research. Nuclear Instrum. Methods Phys. Res. A **720**, 92–94 (2013). https://doi.org/10.1016/j.nima.2012.12.027
9. Broomhead, D., Lowe, D.: Radial basis functions, multi-variable functional interpolation and adaptive networks RSRE-MEMO-4148, March 1988
10. Carlson, M., Holtzman, J.: Measuring sizes of marginally resolved young globular clusters with the hubble space telescope. Publ. Astron. Soc. Pac. **113**(790), 1522–1540 (2001). https://doi.org/10.1086/324417
11. Cavuoti, S., Garofalo, M., Brescia, M., Paolillo, M., Pescape', A., Longo, G., Ventre, G.: Astrophysical data mining with gpu. a case study: genetic classification of globular clusters. New Astron. **26**, 12–22 (2014). https://doi.org/10.1016/j.newast.2013.04.004
12. Cavuoti, S., Garofalo, M., Brescia, M., Pescape, A., Longo, G., Ventre, G.: Genetic algorithm modeling with GPU parallel computing technology. Smart Innov. Syst. Technol. **19**, 29–39 (2013). https://doi.org/10.1007/978-3-642-35467-0_4
13. D'Isanto, A., Cavuoti, S., Brescia, M., Donalek, C., Longo, G., Riccio, G., Djorgovski, S.: An analysis of feature relevance in the classification of astronomical transients with machine learning methods. Mon. Not. R. Astron. Soc. **457**(3), 3119–3132 (2016). https://doi.org/10.1093/mnras/stw157
14. Dunn, L., Jerjen, H.: First results from sapac: toward a three-dimensional picture of the fornax cluster core. Astron. J. **132**(3), 1384–1395 (2006). https://doi.org/10.1086/506562
15. Fritzke, B.: Supervised learning with growing cell structures. In: Proceedings of the 6th International Conference on Neural Information Processing Systems, NIPS 1993, pp. 255–262. Morgan Kaufmann Publishers Inc., San Francisco (1993). http://dl.acm.org/citation.cfm?id=2987189.2987222
16. Fritzke, B.: A growing neural gas network learns topologies. In: Proceedings of the 7th International Conference on Neural Information Processing Systems, NIPS 1994, pp. 625–632. MIT Press, Cambridge (1994). http://dl.acm.org/citation.cfm?id=2998687.2998765

17. Guyon, I., Elisseeff, A.: An introduction to variable and feature selection. J. Mach. Learn. Res. **3**, 1157–1182 (2003). http://dl.acm.org/citation.cfm?id=944919.944968
18. Harrell Jr., F.E.: Regression Modeling Strategies. Springer-Verlag New York Inc., Secaucus (2006)
19. Hughes, G.: On the mean accuracy of statistical pattern recognizers. IEEE Trans. Inf. Theor. **14**(1), 55–63 (2006). https://doi.org/10.1109/TIT.1968.1054102
20. Jirayusakul, A., Auwatanamongkol, S.: A supervised growing neural gas algorithm for cluster analysis. Int. J. Hybrid Intell. Syst. **4**(2), 129–141 (2007). http://dl.acm.org/citation.cfm?id=1367006.1367011
21. Martinetz, T., Schulten, K.: A "neural-gas" network learns topologies. Artif. Neural Networks **1**, 397–402 (1991)
22. Martinetz, T., Berkovich, S., Schulten, K.: "neural-gas" network for vector quantization and its application to time-series prediction. IEEE Trans. Neural Networks **4**(4), 558–569 (1993). https://doi.org/10.1109/72.238311
23. McCulloch, W., Pitts, W.: A logical calculus of the ideas immanent in nervous activity. Bull. Math. Biophys. **5**(4), 115–133 (1943). https://doi.org/10.1007/BF02478259
24. Montoro, J.C.G., Abascal, J.L.F.: The voronoi polyhedra as tools for structure determination in simple disordered systems. J. Phys. Chem. **97**(16), 4211–4215 (1993). https://doi.org/10.1021/j100118a044
25. Paolillo, M., Puzia, T.H., Goudfrooij, P., et al.: Probing the GC-LMXB Connection in NGC 1399: a wide-field study with the hubble space telescope and Chandra. Astrophys. J. **736**, 90 (2011). https://doi.org/10.1088/0004-637X/736/2/90
26. Pedregosa, F., Varoquaux, G., Gramfort, et al.: Scikit-learn: machine learning in python. J. Mach. Learn. Res. **12**, 2825–2830 (2011)
27. Puzia, T.H., Paolillo, M., Goudfrooij, P., et al.: Wide-field hubble space telescope observations of the globular cluster system in NGC 1399, ApJ **786**, 78 (2014). https://doi.org/10.1088/0004-637X/786/2/78
28. Stehman, S.V.: Selecting and interpreting measures of thematic classification accuracy. Remote Sens. Environ. **62**(1), 77–89 (1997). http://www.sciencedirect.com/science/article/pii/S0034425797000837
29. van der Walt, S., Colbert, S.C., Varoquaux, G.: The numpy array: a structure for efficient numerical computation. Comput. Sci. Eng. **13**(2), 22–30 (2011). https://doi.org/10.1109/MCSE.2011.37

Matching and Verification of Multiple Stellar Systems in the Identification List of Binaries

Nikolay A. Skvortsov[1(✉)], Leonid A. Kalinichenko[1], Alexey V. Karchevsky[2], Dana A. Kovaleva[2], and Oleg Yu. Malkov[2]

[1] Federal Research Center "Computer Science and Control" of Russian Academy of Sciences, Moscow, Russia
nskv@mail.ru, leonidandk@gmail.com

[2] Institute of Astronomy of Russian Academy of Sciences, Moscow, Russia
geisterkirche@gmail.com, {dana,malkov}@inasan.ru

Abstract. Binary and multiple stellar systems have been observed using various methods and tools. Catalogs of binaries of different observational types are independent and use inherent star identification systems. Catalog rows describing components of stellar systems refer to identifiers of surveys and catalogs of single stars. The problem of cross-identification of stellar objects contained in sky surveys and catalogs of binaries of different observational types requires not only combining lists of existing identifiers of binary stars, but rather matching components and of multiple systems and pairs of components by their astrometric and astrophysical parameters. Existing identifiers are verified for belonging to matched components, pairs and systems. After that, they may be matched to one another. The framework of multiple system cross-matching presented in the paper uses domain knowledge of binaries of different observational types to form sets of matching criteria. The Identification List of Binaries (ILB) has been created after accurate matching of systems, their components and pairs of all observational types. This work continues research of binary and multiple system identification methods.

Keywords: Binary stars · Cross-matching methods · Entity resolution
Identification systems · Identification List of Binaries (ILB)

1 Introduction

Binary stars are numerous, and they make up much of stellar population of the Galaxy (between 20 and 90% of estimated binaries in different selections). In fact, a significant part of binary stars belongs to systems of greater multiplicity. The largest catalog of visual binaries WDS [1] contains more than 100,000 pairs, 25,000 of which are in systems of multiplicity 3 or higher.

There are strong grounds to think that components of a binary star are formed simultaneously and evolve in parallel within the system. Initial mass distribution between the components is a factor defining further evolution of stars. Therefore, to determine belonging of stellar objects to a system it is necessary to evaluate commonality of star evolution.

L. Kalinichenko et al. (Eds.): DAMDID/RCDL 2017, CCIS 822, pp. 102–112, 2018.
https://doi.org/10.1007/978-3-319-96553-6_8

Binaries are divided into several types depending on the observation method. There are catalogs specialized in certain observational types and contained their own observed parameters. Observation types of binaries are visual, astrometric, orbital, interferometric, eclipsing, spectral, and others.

Among visual pairs, optical and physical ones are distinguished. Optical pairs consist of quite distant and distinct stars in the space, which are projected closely to one another in the direction of observation on the celestial sphere. Physical pairs represent close components in the space bound by gravity, revolving around common center of mass according to the Kepler's laws. If observation continues long enough, then orbital movement of them may be detected. Observation of physical binaries includes angular separations and position angles of components. Major catalogs of visual binaries are WDS, CCDM [2], and Tycho [3].

If one component in a pair is not visible for some reason, duality may be discovered from periodical change of visible component position on the celestial sphere. In this case, they say about astrometric binary stars.

Eclipsing variable binary stars are pairs with a separation of components comparable to sizes of the stars and the line-of-sight of an observer lying in the orbit plane of the components. These stars change in brightness periodically due to the eclipse phenomenon of components in a binary system. Observation of these stars includes determination of light curve parameters reflecting patterns of star brightness change over time. Major catalogs of eclipsing binaries are GCVS [4] and CEV2 [5].

Interferometric binaries are discovered using Fourier analysis of photometric images, which increases resolution up to the diffraction limit. Binaries discovered with the interferometric method are listed in the INT4 catalog [6].

Spectroscopic binary stars are pairs with planes of orbits slightly inclined from the line of sight of the observer. They are detected using spectroscopic method of measurement of radial velocity. Spectral lines of such stars regularly shift or split due to the Doppler effect. Curves, amplitudes, and periods of radial velocity are determined from observations of spectroscopic binaries. Major data source for them is the SB9 catalog [7]. There are several other types of binaries and specialized catalogs.

Communities of researchers are usually specialized in particular observational types of binaries. Thus, specialized catalogs of binaries of different types have been developed in independent ways. Substantial heterogeneity of catalogs leads to conflicts during catalog integration. Most of the catalogs use their own identification systems of binaries. Some of the catalogs contain references to identifiers of other types including identifiers of surveys of single stars. In particular, HIP2 [8] and HD [9] surveys are referred in some catalogs of binaries.

The Binary Star DataBase (BDB) [10] developed by the authors of this paper includes data on binary and multiple stellar systems of all observational types collected from different catalogs. It contains some common parameters used for search and references to the original catalog records by identifiers. It was necessary to development generalized identification system of binary and multiple stars (BSDB) [11] defining rules for identifier generation for multiple systems of any type, components, and pairs of components within systems. However, generalized identification system itself does not solve initial heterogeneity of identification systems and requires accurate matching of BSDB identifiers to ones of other types.

For that purpose, the Identification List of Binaries (ILB) has been developed, which joins identifiers of a number of identification systems of binaries and of surveys. To create this list, it is not enough just tabulating necessary identifiers. Identification of binaries based on existing cross-references between catalogs faces many conflicts. Resolution of these conflicts is based on astrometric and astrophysical approaches to cross-matching of components, pairs and systems of stars.

An approach to multi-component entity resolution and its implementation as a framework used for correct and verified cross-matching of binary and multiple stars from various catalogs of binaries have been developed. The results of cross-matching are used to solve conflicts of heterogeneous identifiers of binaries.

The tool has been developed for populating ILB from original catalogs of binaries, as well as for keeping it up to date and possible extension, since some catalogs and surveys are periodically updated, and new catalogs may be integrated into ILB creation tools.

The objective of this paper is description of the framework development for multiple system cross-matching and creation of the Identification List of Binaries (ILB) which brings together information on identifiers of multiple stellar systems of all observational types. This issue has been initially discussed in [12], and enhanced here with details of implementation and principles of the result verification briefly outlined in [13].

The framework is described in Sects. 2 and 3. Section 4 focuses on software tools implementation issues. Preliminary results of ILB creation have been presented in Sect. 5.

2 The Framework for Cross-Matching of Binary and Multiple Stellar Systems

Common approach to single and multi-component entity matching includes construction of a set of candidates for identification for every entity or its component and application of sets of criteria constraining such sets of candidates. Criteria are formed from the domain knowledge limiting the interpretation of objects. Interfaces of criteria are unified, they take both matched objects as arguments but may use just one or several parameter values from these objects to match entities.

If a criterion is applicable but not satisfied for a couple of objects, then a matched object is excluded from the set of candidates. If data on an entity required for a criterion are not available, then this criterion cannot influence the set of candidates. Any accessible data on parameters of entities make possible application of criteria using these parameters for matching objects of a certain type. In contrast, absence of data on some parameters precludes only application of related criteria, but not general ability to match objects with the other criteria.

The approaches used for the entity resolution include various similarity criteria for subsets of parameter values, which allow estimating identity multi-component entities. Criteria based on domain knowledge may include expression and comparison of secondary attributes from observed parameters, limit parameter values or their combinations for a single entity, or declare restrictions on variability of parameter values.

Behavior of objects beyond the limits of the subject domain means that matched objects describe different entities.

There are graph-based criteria, which include matching rules restricting structure of multi-component entities, or making conclusions about entities or their components identity based on previously established identification of other components.

Ordering of criteria application is important in cases where a criterion uses results of another object identification process. Thus, special priority may be assigned to graph-based criteria. Criteria restricting sets of candidates by common parameters of objects have equal importance, and matching results are not dependent on the order of such criteria application. They are applied in any order when there are parameters of matched objects sufficient for their application. Their priorities may only have an impact on the effectiveness of candidate set constraining. To generate initial sets of candidates, most effective criteria should be applied first.

2.1 Conceptual Approach to Data Unification

The approach to multi-component entity resolution being described is driven by conceptual specifications of a research domain. Ontologies are used to define what types of entities are there in the domain, which of them should be recognized and resolved by their parameters, which observed and estimated parameters may be used, and how this parameters are restricted and dependent on one another.

To define knowledges related to binary and multiple stellar systems we use ontologies of most general astronomical domains like astrometry, photometry, spectroscopy, astronomical objects, stellar objects as well as more specific domains like binary stars, eclipsing stars, variable objects, orbital movement, and others [14].

Domain ontologies are the basis for conceptual scheme development for binaries. All astronomical objects and their relationships in the domain of binary and multiple stellar systems are considered in terms of general entity types: systems, their components and pairs of components [14]. Identifiers are specific parameters, which refers to one of entity types: component, pair, or system.

The relational scheme (see Fig. 1) has been developed for binaries. Using an object-relational approach it has been also applied for internal information representation in the framework of binary and multiple stellar system matching. The scheme refers to ontological specifications to obtain sets of entity types and their parameters. On the other hand, ontological knowledges are used to develop object matching criteria as behavior specifications.

The scheme contains tables allowing flexible definition of component linkage:

- an entity description (EntitySource) identifying an object form a source catalog;
- a parameter description (Parameters) with unified name from the ontology and a value from the source catalog;
- linking entity within the system (EntityGraph) allowing to define pairs of components or components themselves as star considered as single in a survey;
- results of matching criteria application (MatchingCriteria);
- and additional structure (Fields) for allocation of fields near certain stellar systems.

A multiple system may be represented by a graph with components as vertices, and pairs as arcs from primary component to secondary one. Every vertex, arc and graph as a whole system obtained from different sources should be correctly identified to one another. A vertex does not always identified to another vertex but to an arc too, because a single component in one catalog may be recognized as a compound object in another catalog.

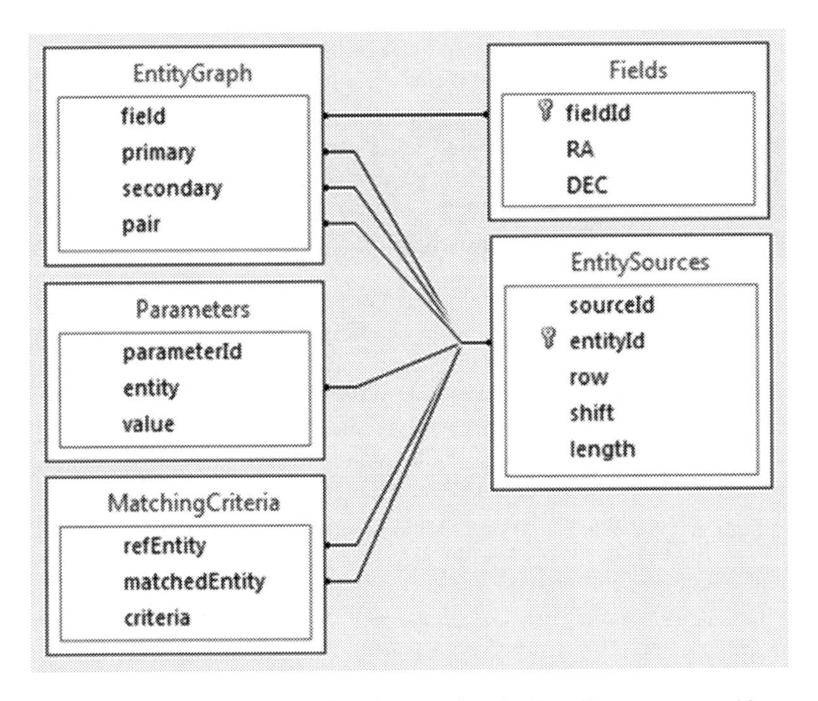

Fig. 1. The database scheme for binary and multiple stellar system matching

Formally, for each entity type X a tuple of type $T_X[a_1, \ldots, a_i, \ldots]$ with unified attributes a_i necessary for description of entity parameter is defined in the conceptual scheme. There is a set of data sources D_j (surveys and catalogs), which store data on certain types of entities structured as tuples of type $X_j[a_{1j}, \ldots, a_{ij}, \ldots]$ containing attributes a_{ij} relevant to some parameters of the entity X in T_X. Representations may be very different in various catalogs. For instance, WDS records describe pairs of stellar objects while CCDM considers separate visual components as records. So gathering data from multiple sources needs mapping M of a source data representation X_j into unified representation T_X in terms of the conceptual schema:

$$M_{jX} : X_j \rightarrow T_X$$

Mapping function is a set of transformation rules of subsets of source attributes into conceptual schema type attributes. To unify representation of data and to improve its quality, mapping also includes standardization functions as cleaning and unification rules applied to attributes of X_j.

For example, the right ascension parameter is represented in WDS as HHMMSS.ss, in SB9 it is HHMMSSSSS, and in TDSC it is in the degree scale, so these attributes need to be transformed to common representation. Records of INT4 have non-trivial structure in general; they need sophisticated transformation functions providing unified presentation of data in the conceptual scheme.

Any data processing and analysis is performed in unified representation. So analytical algorithms are developed in terms of the conceptual scheme of the domain.

2.2 Management of Criteria of Entity Matching

After unification of data representation in the conceptual scheme, the process of object matching begins. The sets of candidates for identification $C : \{T_X\}$ is generated for each object. Every candidate has a set of associated flags (F) as well (stored together with the results of criteria application).

A set of candidates for identification for a reference object is constructed using a set of criteria in accordance with the domain knowledge constraints related to entity type X. Matching criteria are applied with functions:

$$T_X \times \{T_X\} \rightarrow \{T_X\}$$

An object (single or multiple one) of type T_X is considered in this function. It also operates with a set of possible candidates for identification $C : \{T_X\}$ being matched to that object. The function defines one of the criteria of object matching, and it matches every candidate from the set C with respect to the reference object using certain criterion. The function returns a reduced set of candidates that comply with the matching criterion.

A criterion k reducing a set of candidates for identification C to an object of type X, is defined as a R_{kX} predicate containing certain restriction of the domain over referencing (x) и matched (c) objects:

$$\{c \in C | R_{kX}(x, c)\}$$

A predicate R_{kX} may use attribute values of X as well as attributes or flags of components of X if X is multiple one. Particularly, to match data on pairs of stellar components, values of own parameters of pairs may be compared, but components should be matched too. For that purpose, matching algorithm may access attributes of components related to a pair and data on component matching results. For example, identical pairs should have close values of component proper motion. The function defining this matching criterion compares proper motion parameter values of a reference pair and a candidate pair, or if there is no data on proper motion of a pair, it gets those values from attributes of its components.

3 Multiple Stellar System Matching Criteria

The process of multi-component entity resolution for multiple stellar systems is split into several interacting phases by entity types and by hierarchical structure of systems. The process begins from matching of system components as separate entities. It uses data from catalogs of visual binaries, and surveys of single stars, at least those ones that contain identifiers used in catalogs of binaries. This phase also includes splitting of sources data to fields of systems by coordinates of components. Preliminary sets of candidates for component identification has been created. After that, wide visual pairs are matched using results of the first phase. In the next phase close pairs of other observational types are taken to analysis. Identification of stellar systems as a whole is a consequence of identification of all their components and pairs. Finally, existing identifiers of systems, pairs, and components are matched using the results of previous phases.

For matching of components and pairs of multiple stellar systems, it is mostly sufficient using positional information including coordinates of components corrected with observation epochs, precession and proper motion. Matching of pairs is mostly based on results of component identification and restriction of difference of relative position components in a pair [15]. Close pairs are matched excluding relative position since it can change fast.

Thus, the following restrictions of the subject domain of single star have been chosen as basic criteria for matching components of stellar systems:

- similarity of component coordinates;
- similarity of coordinates corrected for axial precession and proper motion in different epochs;
- similarity of proper motions.

If both components have been visually recognized in pairs, the sets of candidates for pair identification are formed on the basis of previously defined sets of candidates for component identification. The criteria for pair matching include:

- similarity of position angles and separations;
- limiting difference of position angles on the basis of certain values of angular separation to evaluate possible orbital movement of components in a pair
- taking into account known observational type of a pair from different catalogs.

After identification of systems on the basis of cross-matching of all components and pairs, other criteria of subjects domain could be applied for verification of source data and established identifications. They are structured in the same way as matching criteria, and they can be used as matching criteria too. In cases of difference of positional accuracy or magnitude limits of catalogs as well as in case of matching component in high-density stellar fields, some criteria used only for verification become more effective for cross-matching of binaries.

The list of verification criteria for components include:

- similarity of trigonometric parallax of components defined by distance from observer;
- similarity of brightness or color indices if photometric passbands, sensitivity to low and high limits of magnitude, as well as possible variability of stars in case of different values of brightness are known;
- compliance of spectral classification of stars.

To verify pair source data or identifications the following criteria can be used:

- similarity of periods (low estimate of period available for any visual pair);
- similarity of other characteristics of orbital movement;
- similarity of proper motions of a pair as a whole and ability of different proper motions of optical pairs;
- similarity of differences of component brightness in pairs;
- taking into account effective angular resolution (separation of the pair cannot be less than minimal limit of object distinguishing in the catalog);
- similarity of chemical composition and evolution statuses of components in pairs.

Unique identification of objects is a special case, in which a set of candidates after application of all possible criteria contains exactly one object from particular source as a candidate for identification. In most cases, it means that most probable candidate has been found. Such object is marked with special flag. It creates conditions for applying criteria using this flag.

Graph criteria are used for matching multi-component entities or their components in the basis of previously identified components. They use unique identifications of components or pairs to propagate candidate set restriction for other elements of graphs.

There are such criteria as:

- identification of pairs inverse to uniquely identified pairs;
- identification of components of uniquely identified pair;
- identification of pairs with both uniquely identified components, and other like these ones.

Identification of multi-component objects depends on the identification of their components. Unique identification of all components means identification of the entire object. Identification of stellar systems as a whole is equivalent to construction of connected graphs from identified components and pairs.

Identifiers are matched on the basis of belonging to identical entities: components, pairs or systems. Often the same identifier should be related to a component in one catalog while in another catalog it should be related to a pair since angular resolution allows distinguishing both visual components in it, or it turns a pair of observational type different from visual. Such conflicts are resolved by quality matching and identification of components, pairs and systems. The quality of ILB depends on quality of system matching results.

4 Implementation Issues

The cross-matching tools for binary and multiple stars is implemented using Java programming language. To store and analyze results of cross-matching, a PostgreSQL repository is used. Spring Data JPA and Hibernate object-relational mapping are used as an interface to the database. All results of criterion application for a couple of objects are accessible in the repository for analysis and reuse. Data retrieved from various catalogs for a certain field of the celestial sphere and multiple stellar systems inside it are processed in operating memory.

Tools include methods necessary for multicomponent entity matching and supporting methods for access to catalogs, caching of objects in a field of the celestial sphere, data cleaning, statistics gathering. The tools are not developed for single use to generate ILB. The possibility to integrate additional catalogs and to analyze for binaries new more precise data was taken into account in the architecture of tools. The tools may be reused to develop multi-component entity resolution instruments for other subject domains.

Sets of entity types and of their parameters used in the tools are formed directly from the domain ontology of binaries. A set of matching and verification criteria for various types of entities may be completed with new implementations of the interface common for all criteria. Involvement of additional catalogs demands integrating them in accordance with domain specifications of binaries and with internal representation of data in the tools.

The implementation is suitable for large catalog data processing. In particular, a large release of Gaia survey is expected in 2018. Such surveys and catalogs are accessed to retrieve data for fields near known binary and multiple systems. The Kubernetes [16] tool may be used to run cross-matching in a cluster. Parallelization is performed at fields of the celestial sphere.

5 The Identification System BSDB and Considered Identifier Types

The main identification system of ILB is BSDB. It allows to describe multiple systems, their pairs and components, and to extend multiple systems with possible newly discovered components. It is suitable to refer to binaries of any observational type. ILB formation tools generate BSDB identifiers after the cross-matching process.

ILB is a basic catalog for BDB, and it contains all the data for cross-identification of more than 130,000 binary and multiple stellar systems in BDB. ILB includes coordinates and cross-identifications for the following identifier types: Bayer/Flamsteed, DM (BD/CD/CPD), HD, HIP, ADS, WDS, CCDM, TDSC, GCVS, SBC9. Preliminary statistics of created ILB are:

- 136885 multiple systems;
- 313811 pairs;
- 627460 components;
- 36560 HIP identifiers, 20701 unique ones;

- 29882 HD identifiers, 27917 unique ones;
- 121067 DM identifiers, 105569 unique ones;
- 49067 ADS identifiers, 15389 unique ones;
- 1622 FLAMSTEED identifiers, 1529 unique ones;
- 600 BAYER identifiers, 548 unique ones.

6 Conclusion

The Identification List of Binaries (ILB) has been created to resolve the problem of matching identifier used to refer to binary and multiple stars of various observational types.

The paper describes the problem of multiple stellar system cross-matching, a framework for multi-component entity resolution applied for this problem resolution. Matching of pairs and components of multiple systems involves all their known astrometric and astrophysical parameters. The framework with a set of matching criteria for the domain of binary and multiple stellar systems has been implemented for the ILB creation.

Identifications contained in catalogs of binaries are not sufficient for unambiguous relating of components and pairs of systems. Accurate matching of systems by parameters allowed to resolving most of existing conflicts. The ILB catalog contains cross-identifications of binary and multiple stellar systems compiled from data of the catalogs of binary stars of all observational types.

ILB provides all necessary identifications for the Binary Star Database (BDB) and contains BSDB identifiers generated for every component, pair and system in BDB.

Acknowledgements. This work has been supported by the Russian Foundation for Basic Research (grants 16-07-01162, 16-07-01028, 18-07-01434).

References

1. Mason, B.D., et al: The Washington Visual Double Star Catalog. VizieR on-line data catalog: B/wds (2016). http://cdsarc.u-strasbg.fr/viz-bin/Cat?B/wds
2. Dommanget, J., Nys, O.: Catalogue of the Components of Double and Multiple Stars (CCDM). Observations et Travaux **54**(5) (2002). http://cdsarc.u-strasbg.fr/viz-bin/Cat?I/274
3. Fabricius, C., Hog, E., Makarov, V.V., et al.: The Tycho double star catalogue. Astron. Astrophys. **384**(1), 180–189 (2002)
4. Samus, N.N., Durlevich, O.V., Kazarovets, E.V., et al: General Catalogue of Variable Stars. VizieR On-line Data Catalog: B/gcvs (2013). http://cdsarc.u-strasbg.fr/viz-bin/Cat?B/gcvs
5. Avvakumova, E.A., Malkov, O.Y., Kniazev, A.Y.: Eclipsing variables: catalogue and classification. Astron. Nachr. **334**(8), 860–865 (2013)
6. Hartkopf, W.I., Mason, B.D., Wycoff, G.L., McAlister, H.A.: Fourth Catalog of Interferometric Measurements of Binary Stars (2001). http://www.usno.navy.mil/USNO/astrometry/optical-IR-prod/wds/int4

7. Pourbaix, D., Tokovinin, A.A., Batten, A.H., et al.: SB9: 9th catalogue of spectroscopic binary orbits. Astron. Astrophys. **424**, 727–732 (2004). http://cdsarc.u-strasbg.fr/viz-bin/Cat?B/sb9
8. van Leeuwen, F.: Validation of the new Hipparcos reduction. Astron. Astrophys. **474**(2), 653–664 (2007). https://doi.org/10.1051/0004-6361:20078357. http://cdsarc.u-strasbg.fr/viz-bin/Cat?I/311
9. Cannon, A.J., Pickering, E.C.: Henry Draper Catalogue and Extension. VizieR Online Data Catalog. CDS/ADC Collection of Electronic Catalogues, 3135, 0 (1993). http://cdsarc.u-strasbg.fr/viz-bin/Cat?III/135A
10. Malkov, O., Kaygorodov, P., Kovaleva, D., et al.: Binary star database BDB: datasets and services. Astron. Astrophys. Trans. (AApTr) **28**(3), 235–244 (2014)
11. Kovaleva, D.A., Malkov, O.Yu., Kaygorodov, P.V., et al: BSDB: a new consistent designation scheme for identifying objects in binary and multiple stars. Open Astron. **24**(2), 185–193. (2015). https://doi.org/10.1515/astro-2017-0218
12. Skvortsov, N.A., Kalinichenko, L.A., Karchevsky, A.V., Kovaleva, D.A., Malkov, O.Y.: Development of Identification List of Binaries ILB. In: The XIX International Conference on Data Analytics and Management in Data Intensive Domains (DAMDID/RCDL 2017), Moscow, Russia, vol. 2022, pp. 43–49. CEUR WS (2017). http://ceur-ws.org/Vol-2022/. (In Russian)
13. Skvortsov, N.A., Kalinichenko, L.A., Kovaleva, D.A., Malkov, O.Y.: Hierarchical multiple stellar systems. In: Kalinichenko, Leonid, Kuznetsov, Sergei O., Manolopoulos, Yannis (eds.) DAMDID/RCDL 2016. CCIS, vol. 706, pp. 119–129. Springer, Cham (2017). https://doi.org/10.1007/978-3-319-57135-5_9
14. Skvortsov, N.A., et al.: Conceptual approach to astronomical problems. Astrophys. Bull. **71** (1), 114–124 (2016)
15. Isaeva, A.A., Kovaleva, D.A., Malkov, O.Y.: Visual binaries: cross-matching and compiling of a comprehensive list. Open Astron. **24**, 157–165 (2015)
16. Kubernetes: Production-Grade Container Orchestration (2018). https://kubernetes.io/

Aggregation of Knowledge on Star Cluster Structure and Kinematics in Data Intensive Astronomy

Sergei V. Vereshchagin(✉) and Ekaterina S. Postnikova

Institute of Astronomy, Russian Academy of Sciences,
Pyatnitskaya str., 48, 119017 Moscow, Russia
svvs@ya.ru, es_p@list.ru

Abstract. A technique for studying the star motions inside the open star clusters is proposed. It allows for revealing the details of the spatial and kinematic cluster structure on the basis of precise measurements of the astrometric parameters of the stars. By successively scanning a lot of clusters and applying a uniform technique, a processing pipeline is built. Analysis of such mass processing results reveals the patterns and relationships of the internal arrangement of clusters with their parameters and position in the Galaxy. Actually, the problem has become more interesting by the recently launched space telescope Gaia, in particular, for the membership of star clusters.

Keywords: Star catalogue · Star group · Open star cluster
Star cluster structure · AD-diagram · Star apex

1 Introduction

Open star clusters (OSC) are important representatives of the Milky Way population. An open cluster is a gravitationally bound group of stars having common origin, chemical composition and some co-directionality of movement. The OSC contain from tens to several thousand member stars. This multiplicity of samples makes a strong advantage of clusters over individual stars as it leads to increased accuracy and reliability of measurements of various astrophysical parameters, like masses or ages.

With the help of OSC, both the star and planet formation [59, 62] and the dynamic, photometric and chemical evolution of the Galaxy can be studied. But the internal structure of the clusters is poorly understood. The promising direction is the search for previously unknown substructures inside OSC. The task of detecting and cataloging such substructures is of interest and is discussed in this article.

The primary sources of information used in our research are the mass star surveys performed by ground based and space telescopes. The rate of data produced by these tools increases yearly (Sect. 4). The number of OSC in our Galaxy is expected to be about one hundred thousand. With so many objects and data about them, it becomes clear that dealing with Big Data requires intensive use of reduction and analysis technologies. For that purpose the processing conveyor is being developed.

L. Kalinichenko et al. (Eds.): DAMDID/RCDL 2017, CCIS 822, pp. 113–127, 2018.
https://doi.org/10.1007/978-3-319-96553-6_9

The method of apex point diagrams (AD diagrams, Sect. 5) is principal in our study [13]. It yields such a representation of the data that helps in seeking patterns and various irregularities in the spatial motions of stars within cluster allowing studying its internal structure.

The structure of the paper is as follows. Section 2 explains the object of research, Sect. 3 discusses the public data archives. Section 4 is devoted to the dynamics of data replenishment. Section 5 gives the overview of an AD method and its application to various OSC. The architecture of this system is given in Sect. 6.

This work is an extension of [79]. Comparing with [79], Sect. 7 devoted Pleiades OSC and the Ursa Major (UMa) star flow is added, Sect. 8 with discussion and conclusions is extended.

2 OSCs as Objects of Our Attention

2.1 The Subgroups of Stars Within OSC

Star clusters are traditionally studied either as a subsystem of Galactic objects, or as a gravitating star systems. Detailed investigation of OSCs internal structure, such as subgroups of stars, has begun not so long ago and still remains promising. The concept of a "subclusterings" of stars appeared at least in 1969 [81], historical overview can be found in [56].

For the first time van Albada showed in 1968 [72] the possibility of forming wide double and multiple stars with characteristic dimensions of about 10 astronomical units (AU) and more. Such configurations can occur as a result of the decay of small stellar groups with sizes from 10^2 to 10^5 AU (the maximum spatial size is approximately 0.5 parsec). In the review by Larson [38] it was shown that most stars in the vicinity of the Sun (single, double and stable multiples) could be formed as a result of the decay of non-hierarchical small groups of stars used to be composed from several pieces to several dozen objects. The process of formation of multiple systems in clusters was considered in [73].

In [22] it is shown that there are groups of higher level visible in the wavelet space, which correspond to the flow of Hercules, the Pleiades and Hyades, and the Sirius group. The dynamic models indicate their resonant origin at Lagrange points of the Galaxy bulge or from spiral arms or a combination of these effects.

2.2 Detection of Previously Unknown OSC

We easily find the Pleiades on the sky, but do not see many thousands of similar clusters that are either small or located far enough from the Sun and blended with the star background, or field. Figure 1 shows the two OSC in the Perseus constellation. Their names are NGC 869 and NGC 884 (NGC stands for the New General Catalog of Nebulae and Clusters of Stars), also known as the famous double cluster h and χ Perseus. For the observer they look like a simple stellar density fluctuations. The search for such fluctuations is the basis for the discovery of previously unknown clusters.

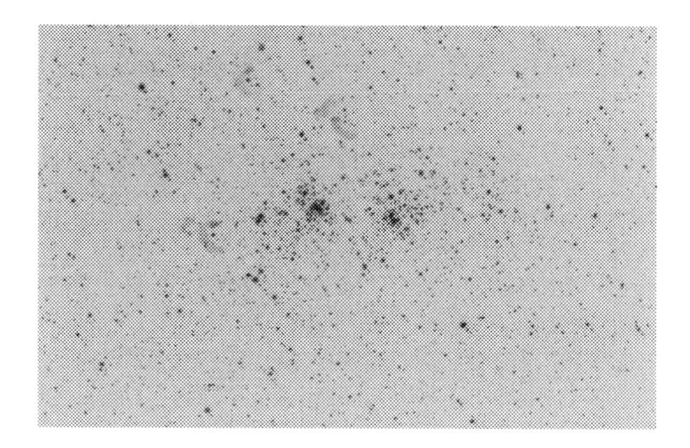

Fig. 1. A fragment of a photographic plate with the image of double OSC h and χ Perseus. The photographic plate was obtained on a 40 cm astrograph telescope of the Zvenigorod Observatory by V. P. Osipenko.

Fluctuations can be of a genetic nature or have a random character. The art of the science is not only to be able to find a fluctuation, but also to determine that it is a physically connected star system. For this reason, the criteria based on the community of the stars spatial motions vectors are applied. Clusters can also be discriminated by their AD diagrams, in particular, by the presence or absence of internal substructures, or star subgroups. Additional criteria are based on the ages and chemical compositions of stars, as well as on their photometric characteristics. The whole set of criteria mentioned above may indicate the physically connected star cluster.

Various methods have been developed for OSC examination. For example, the "moving cluster" method [26, 44] works well for distances up to 100 parsecs (pc) from the Sun. The method of trigonometric parallaxes is also applicable to sufficiently close clusters [46]. Methods based on the chemical composition of the cluster stars are also known [39]. Methods of an automatic and supervised search for fluctuations in stellar density from astrometric and photometric data were developed [17, 24]. The method of AD diagrams is used to identify internal substructures [13].

The visible image of a cluster in the sky is determined by two factors - the distance from the Sun and its physical dimensions. Often it is difficult to distinguish clusters from background field without additional tests. Thus a large cluster located close to the observer can occupy such a large area of the sky that it is indistinguishable from the background too (for instance, UMa star stream in Sect. 7).

3 Data Archives

How to understand that a new, previously unknown cluster is found? It's already said about the selection criteria, but that's not enough. One also needs to prove a novelty of the discovery by a search within already known clusters [58, 59]. For that purpose we need the data archives on OSCs.

We list some of the most famous archives below. VizieR [80], SIMBAD (Set of Identifications, Measurements and Bibliography for Astronomical Data) [63] contain data about stars and OSC. ADS NASA (Astrophysics Data System) [70] contains research papers. ScienceDirect (data from Elsevier Publishing) [60], IOPscience [34], Wiley Online Library [83] and Astronomy and Astrophysics International Journal [4] contains papers, thus requiring additional processing.

ADS NASA is the largest archive containing information on more than 7 million documents. WEBDA [82], Lynga [42], Alter and Ruprecht [2], Barkhatova [6], Piskunov [52] archives contain data on OSCs and are of mostly historical interest. Nevertheless information from them is relevant for comparing results.

Dias [16] continues to be updated with new OSC data. It is also valuable due to use of active links to WEBDA [82], Lynga [42], Kharchenko et al. [35]. Additionally they also included a library publication codes (Bibcod).

A team of authors (see the list of participants in [35]) created the most important modern Global Survey of Milky Way Star Clusters (MWSC). It includes ten directories providing data on all currently known clusters. The MWSC catalog system is the most complete and cited in modern cluster studies. The main result given in [35] is a directory containing all the metadata. Also there are actual additional data like star charts and color-magnitude diagrams. These diagrams reveal the stellar composition and age of the cluster. Currently Gaia data is being actively connected to MWSC [25].

4 Increasing of the Data on OSCs with Time

Recently, in the astrophysics and other scientific disciplines, the direction of "data science" (the extraction of scientific facts from large data sets) becomes more popular. The variety of composition and growing volume of information (measured in the number of publications) about clusters is shown in Fig. 2. A sharp jump in the number of publications happened around 2000. Such ascents normally happen following the launches of new large telescopes.

Statistical data of the large telescopes launching are the following. 11 refractor telescopes with lens diameter more than 70 cm were built in the period from 1880 till 1917. 14 reflector telescopes with the mirror diameter 6 m and more were constructed and became operational in 1975–2005. The famous Palomar 200-inch (5.1-m) Hale Telescope was commissioned in 1948. It marks the noticeable growth of information (the upper curve in Fig. 2).

The recent biggest contribution to the overall astrometric data has been made by Global Astrometric Interferometer for Astrophysics, the Gaia project [25]. The Gaia survey [25] includes approximately 1 billion stars, which is already comparable with the population of the Galaxy (approximately 1% of its stellar population). The limiting stellar magnitude in wavelength range from 400 to 1000 nm (G-band mean magnitude) is 20. The microsecond measurement accuracy allows researchers to derive new information about the motions of stars inside clusters. The first list of interesting objects studied with the help of the Gaia mission data were published in [45].

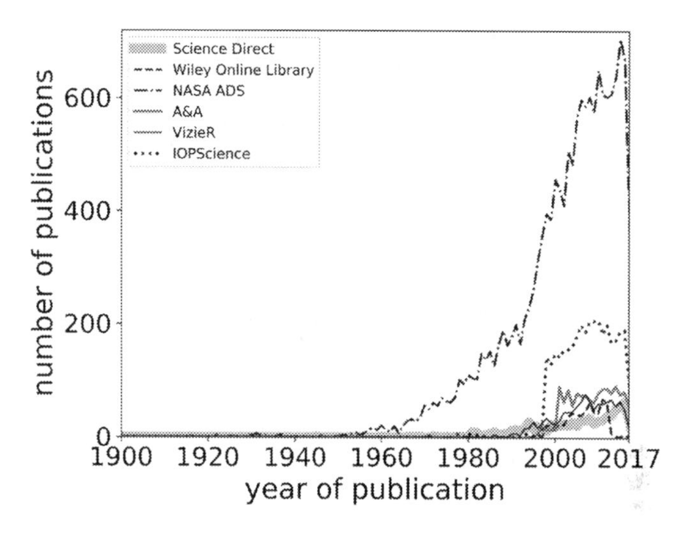

Fig. 2. The growth of the number of the new publications over time as the result of keyword search with "open" and "cluster" in their abstracts.

MWSC [35] survey contain 3754 OSCs which are only 0.3% of the galactic disk OSC population. Vereshchagin [78] estimated that the Gaia can give about one million clusters located within the Milky Way.

5 Application of the AD Diagrams

There are many patterns of stellar motions inside the Galaxy. An example is the involvement of stars in the galactic disk rotation. There are stars that form streams, and there are "escaping" stars that move inexplicably fast relatively to the Sun.

The coordinates of the point in space, in the direction of which the motion of the star is observed, is called its *apex*. The AD diagram is the distribution of individual star apexes in the equatorial coordinate system. The coordinates of stellar apexes are obtained from the solution of a geometric problem, in which there are intersections of the vectors of spatial velocities of stars with the celestial sphere, while the vectors begin to move to the observation point [13].

The AD method was successfully used for the study of many OSCs (Hyades [75], Praesepe [76], NGC188 [20], M67 [77] and so on). Star groups in Orion and some other clusters were detected using the method [12]. A map representing the distribution of apexes helps to investigate the kinematic features of any star fields. Thus, Fig. 3 shows the distribution of 249603 stars with the most accurately measured velocities and distances on AD diagram [21, 37]. The concentration of the stars of one of the streams is clearly visible on Fig. 3. Along with the classical methods of investigation, our approach allows us to pay special attention to the internal kinematic structure for the two specific star systems – Pleiades OSC and UMa star stream.

Fig. 3. A map representing the distribution of apexes with the most accurately measured velocities and distances. The concentration of stars of one of the streams is clearly visible as a high-density spot.

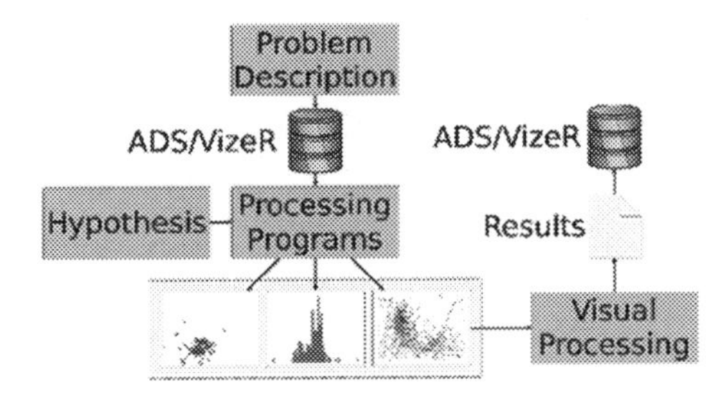

Fig. 4. The information processing pipeline.

6 An Overview of the Data Accumulation Conveyor

Despite the OSC variability a single unified technique to examine the internal structure can be used for the most of them. Figure 4 shows the pipeline of data processing for OSCs within the proposed paradigm. This pipeline starts with the statement of a problem on the basis of large data analysis and the hypothesis is being verified. The next step is applying synthesized algorithms which result in series of visual representations. These visualizations are analyzed by a researcher, and new results and deductions are produced. Successful results are published and eventually added to the public dataset for reuse and independent verification.

Further, using the gathered information, it is possible to do a comparative analysis and make the conclusions about the relationship between an internal structure and other metadata with main OSC parameters in the galactic disk. This might provide the new knowledge and lead to discovery of new regularities concerning the OSC population in the disk of the Galaxy.

Accumulation of results about groups in clusters is carried out both with publishing articles (e.g., in ADS [70] and VizieR [80]) and placement of the obtained data in SIMBAD [63].

7 Data Accumulation for the Pleiades and UMa Stream

Though rather universal method of investigation is presented, the study of OSC implies consideration of various peculiarities. As OSCs differ in many quantitative and qualitative parameters, methods of their investigation also vary. The observation instrument, used model of Galaxy and mathematical apparatus also matter. An overview of history on investigation of two contrast OSCs presented below exposes of the impact of mentioned factors on the subject.

7.1 An Overview of Pleiades and UMa OSC

Famously called as "the Seven Sisters", Pleiades open cluster (M45, Melotte 22, or NGC 1432) has attracted observers since antiquity. Its stellar content has been always studied extensively. The Pleiades is located in the constellation of Taurus and contains about two thousand member stars. The age of the cluster varies over a wide range from 77 million years (Myr) [47] to 141 Myr [35]. The metallicity [Fe/H] ~ -0.03 is reported by Friel and Boesgaard [23] and Stauffer [67], Takeda [69]. The distance to the Pleiades has puzzled the astronomers. There are several estimates of the heliocentric distance to the Pleiades obtained by different methods. Using the moving cluster method, Galli et al. [26] gave the distance value as $134.4^{+2.9}_{-2.8}$ pc.

The Ursa Major Moving Group, also known as Collinder 285 or Ursa Major association, is a nearby stellar moving group – a set of stars with common velocities in space and thought to have a common origin (formed about 300 million years ago [36]) in space and time. The UMa stream is a possibly disintegrated OSC inside which the Sun is located. This location determines the peculiarities of studying the UMa stream "from inside".

7.2 The History of Data Accumulation for the Pleiades

The main papers whose authors studied the stellar composition of the cluster are presented in Table 1. In the columns of Table 1 we list the reference and the estimated number of membership stars.

Bouy et al. [11] applied a probabilistic method based on multivariate data analysis [57] to select high probability members of the Pleiades on the wide field data taken in multi-epoch data (Bouy et al. [10]) and Tycho-2 (Høg et al. [33]) catalogs. They identified 2107 high probability member stars in the Pleiades region and produced the fullest member list.

Figure 5 presents the history of the accumulation of stars in the Pleiades. John Mitchell in 1767 stated that the probability of an accidental concentration of stars in the Pleiades area is extremely small. Only in the early 20th century it was finally established that the Pleiades is a star cluster (van Leeuwen [74]) due to the study of the proper motions of the stars in the cluster. At about the same moment a purposeful search for star-members of this cluster began.

The first precise positions of stars were given by Le Monnier in 1755 [40]. Bessel observed 27 Pleiades stars in 1829 [9], Wolf [84] in 1874 observed 625 stars in the Pleiades field up to the 13th magnitude and measured their proper motions. A key role

Table 1. Pleiades membership surveys.

Reference	Number of stars
Trumpler [71]	246
Hertzsprung [32]	247
Artyukhina [3]	~200
Haro et al. [30]	519
van Leeuwen et al. [74]	193
Stauffer et al. [66]	225
Hambly et al. [29]	440
Pinfield et al. [51]	339
Adams et al. [1]	1200
Deacon and Hambly [15]	916
Bouy et al. [11]	2107

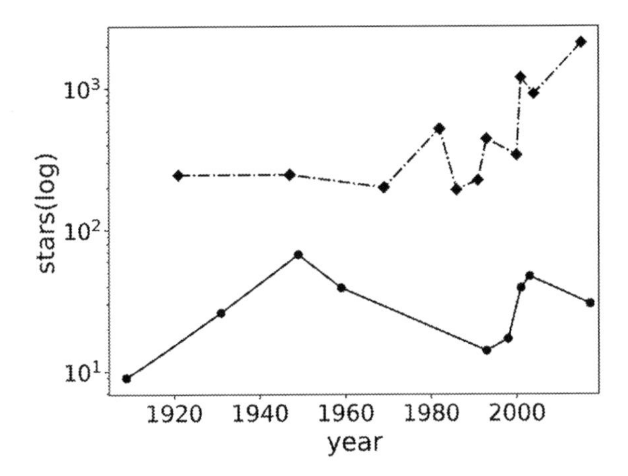

Fig. 5. Comparison of the growth in the number of data on the membership stars in the OSC Pleiades (dot-dashed polyline) and the UMa stream (solid polyline).

in these studies was played by R. Trumpler, who determined the 246 members (up to 9 magnitude within 1° from the center and up to 13 magnitude in the region of 3.5°) [71]. Hertzsprung studied the cluster by the measurements of photographic plates, carefully examined this area and obtained the catalog of 3259 stars and its proper motions for 2920 of them [32]. This examination resulted in a list of 247 stars belonging to the Pleiades – almost the same as Trumpler result 20 years earlier.

The modern era brings an accelerating data growth due to mass sky surveys. Lodieu et al. [43] discovered about a thousand new candidates of the Pleiades cluster thanks to the new sky survey in the infrared (IR) spectrum. The IR survey [67] was conducted to search for low-mass members [1]. Such members were hard to discover due to their tarnish, but the improvement of observational technology allows researchers to discover new objects of this type.

7.3 Accumulation of Data About the UMa Stream

The UMa stream is a possibly disintegrated OSC. Observations provide a rich material, since the stream is located all over the celestial sphere. It is difficult to choose a new membership of stars because of the impossibility of distinguishing them from the field stars. Therefore, it is important to know the characteristics of the stars as accurately as possible for qualitative data filtering, which allows identification of stars related to the UMa stream.

The UMa stream was discovered almost 150 years ago. Table 2 presents the main papers whose authors studied the membership of stars to the stream. In the columns of Table 2 we present the reference and the number of stars in the cluster. The search for members of this stream and the study of its motion was carried out by Eggen [18].

The number of stream stars continues to be refined due to the appearance of new data. In the works of Bannister and Jameson [5] and Shkolnik et al. [61] were discovered some substellar objects possibly belonging to UMa. This discovery was previously impossible due to the lack of the necessary tools. The Fig. 5 shows the data growth for UMa.

Table 2. UMa membership surveys.

Reference	Number of stars
Ludendorff [41]	5
Hertzsprung [31]	15
Plummer [53]	6
Rasmuson [54]	28
Haas [28]	26
Bertaud [8]	74
Nassau and Henyey [50]	126
Smart [64]	42
Bartholeyns [7]	add 17 new member
Gliese [27]	80
Roman [55]	69
Mohr [48]	96
Eggen [18, 19]	50, 67
Soderblom and Mayor [65]	add 14 new memeber
Montes et al. [49]	39
King et al. [36]	47
Shkolnik et al. [61]	add 8 new member
Tabernero et al. [68]	30
Bannister and Jameson [5]	add 4 new member

8 Discussion and Conclusions

We have examined two extreme types of OSC in the sense of their location on the sky sphere. The first is the Pleiades occupying a limited area in the sky. The second is the UMa star stream combining stars scattered all over the sky. Below we discuss some features of replenishing knowledge about these OCSs and their difference in terms of data usage.

High growth of interest to Pleiades OC rather than to UMa stream can be spotted at Fig. 5, which outlines a number of stellar formations members. This can be explained with the ease of observation of the Pleiades cluster: single cluster observations are usually feasible with single observation instrument. The stream is scattered across the sky sphere, thus more resources are required and more complexities have to be overcame in order to conduct observations.

On the other hand, wide-sky surveys like Hipparcos or Gaia do not take into account stellar density fluctuations which look like OSCs. This gives an advantage to star streams of the UMa type.

Figure 5 shows the data growth for both types of OSCs over the years. It is interesting that the data on Pleiades are "growing" more rapidly. This is explained by the permanent attention of researchers to this object and the relative ease of observation mentioned above. Observations of the UMa stream stars scattered all over the sky are much more complicated and expensive. For this reason, in Fig. 5 we do not see the growth of the UMa stream polyline.

This work is aimed at further expanding and improving of data on subsystems or "star groups" in star clusters of the Galaxy. The ultimate goal of the project is the accumulation of information on the structure of clusters in a unified format by pipelining the data. Such information will serve a basis for identifying the star substructures inside clusters. The approach can be also extended to clusters of other galaxies. Thus, in the near future large amount of new data will be obtained, and the proposed methodology will allow them to be efficiently processed in order to extract the new knowledge.

A prototype implementation of the conveyor for processing the data on clusters has been created and is being successfully applied. The conveyor includes the selection of suitable clusters, the determination of necessary parameters, and the promotion of hypotheses. The results produced by these automated methods can contain new knowledge suitable for publishing. Within the paradigm AD diagrams application an example of the discovery of substructures in the corona of the UMa stream is shown [14]. Examples of application of the conveyor are NGC 188, M67 objects [20, 77]. The methodological results of the work include the development of the individual clusters research process in order to create a catalog of clusters apexes and unique scientific web applications.

Acknowledgements. S. V. Vereshchagin and E. S. Postnikova are partly supported by the Russian Foundation for Basic Research (RFBR, grant number is 16-52-12027).

References

1. Adams, J., Stauffer, J., Monet, D., Skrutskie, M., Beichman, C.: The mass and structure of the Pleiades star cluster from 2MASS. Astron. J. **121**, 2053 (2001). https://doi.org/10.1086/319965
2. Alter, G., Israel, B.Y., Ruprecht, J.: Catalogue of star clusters and associations. Astronomical Institute Czechoslovakia, Prague (1964)
3. Artyukhina, N.: The structure of the Pleiades. Sov. Astron. **12**, 987 (1969)
4. Astronomy & Astrophysics. Worldwide astronomical and astrophysical research. https://www.aanda.org
5. Bannister, N.P., Jameson, R.F.: L and T dwarfs in the Hyades and Ursa Major moving groups. Mon. Not. R. Astron. Soc. **378**, L24–L28 (2007). https://doi.org/10.1111/j.1745-3933.2007.00312.x
6. Barkhatova, K.A.: Atlas of color-luminosity diagrams of the of open star clusters. Izd. Academy of Sciences of the USSR, Moscow, 127 pp. (1958). (in Russian)
7. Bartholeyns, R.: Le Courant de la Grande Ourse. Astron. Nachr. **269**(3), 149 (1939). https://doi.org/10.1002/asna.19392690304
8. Bertaud, Ch.: Les Étoiles A dans le courant de la Grande Ourse. B.A., vol. 8, p. 109 (1932)
9. Bessel, F.W.: Beobachtungen verschiedener Sterne der Plejaden. Astronomische Untersuchungen, Ersterband. V, pp. 209– 238 (1841)
10. Bouy, H., Bertin, E., Moraux, E., Cuillandre, J.-C., Bouvier, J., Barrado, D., Solano, E., Bayo, A.: Dynamical analysis of nearby clusters. Automated astrometry from the ground: precision proper motions over a wide field. Astron. Astrophys. **554**, 22 (2013). https://doi.org/10.1051/0004-6361/201220748. A101
11. Bouy, H., Bertin, E., Sarro, L.M., Barrado, D., Moraux, E., Bouvier, J., Cuillandre, J.-C., Berihuete, A., Olivares, J., Beletsky, Y.: The Seven Sisters DANCe. I. Empirical isochrones, luminosity, and mass functions of the Pleiades cluster. Astron. Astrophys. **577** (2015). https://doi.org/10.1051/0004-6361/201425019. A148
12. Chupina, N.V., Vereshchagin, S.V.: Star clusters in the Sword region in Orion. Star formation from the small to the large scale. In: Favata, F., Kaas, A., Wilson, A. (eds.) ESLAB symposium (33: 1999: Noordwijk, The Netherlands), Proceedings of the 33rd ESLAB Symposium on Star Formation From the Small to the Large Scale, ESTEC, Noordwijk, The Netherlands, 2–5 November 1999, 445, p. 347. ESA SP, Noordwijk (2000)
13. Chupina, N.V., Reva, V.G., Vereshchagin, S.V.: The geometry of stellar motions in the nucleus region of the Ursa Major kinematic group. Astron. Astrophys. **371**, 115–122 (2001). https://doi.org/10.1051/0004-6361:20010337
14. Chupina, N.V., Reva, V.G., Vereshchagin, S.V.: Kinematic structure of the corona of the Ursa Major flow found using proper motions and radial velocities of single stars. Astron. Astrophys. **451**, 909–916 (2006). https://doi.org/10.1051/0004-6361:20054009
15. Deacon, N., Hambly, N.: Proper motion surveys of the young open clusters Alpha Persei and the Pleiades. Astron. Astrophys. **416**, 125–136 (2004). https://doi.org/10.1051/0004-6361:20034238
16. Dias, W.S., Alessi, B.S., Moitinho, A., Lepine, J.R.D.: New catalog of optically visible open clusters and candidates. Astron. Astrophys. **389**, 871 (2002). https://doi.org/10.1051/0004-6361:20020668
17. Drake, A.J.: Cluster candidates from the USNO-A2.0 catalogue. Astron. Astrophys. **435**(2), 545–550 (2005). https://doi.org/10.1051/0004-6361:20041568
18. Eggen, O.J.: Stellar groups. I. The Hyades and Sirius groups. Mon. Not. R. Astron. Soc. **118**, 65 (1958). https://doi.org/10.1093/mnras/118.1.65

19. Eggen, O.J.: The Sirius supercluster and missing mass near the Sun. Astron. J. **116**(2), 782–788 (1998). https://doi.org/10.1086/300465
20. Elsanhoury, W.H., Haroon, A.A., Chupina, N.V., Vereshchagin, S.V., Sariya, D.P., Yadav, R.K.S., Jiang, I.-G.: 2MASS photometry and kinematical studies of open cluster NGC 188. New Astron. **49**, 32–37 (2016). https://doi.org/10.1016/j.newast.2016.06.002
21. Famaey, B., Jorissen, A., Luri, X., Mayor, M., Udry, S., Dejonghe, H., Turon, C.: Local kinematics of K and M giants from CORAVEL/Hipparcos/Tycho-2 data. Revisiting the concept of superclusters. Astron. Astrophys. **430**, 165–186 (2005). https://doi.org/10.1051/0004-6361:20041272
22. Famaey, B., Siebert, A., Jorissen, A.: On the age heterogeneity of the Pleiades, Hyades, and Sirius moving groups. Astron. Astrophys. **483**(2), 453–459 (2008). https://doi.org/10.1051/0004-6361:20078979
23. Friel, E.D., Boesgaard, A.M.: Chemical composition of open clusters. II. C/H and C/Fe in F dwarfs from high-resolution spectroscopy. Astrophys. J. **351**, 480 (1981). https://doi.org/10.1086/168485
24. Froebrich, D., Scholz, A., Raftery, C.L.: A systematic survey for infrared star clusters with |b| < 20° using 2MASS. Mon. Not. R. Astron. Soc. **374**, 399 (2007). https://doi.org/10.1111/j.1365-2966.2006.11148.x
25. GAIA DR1 (Gaia Collaboration 2016) Brown, A.G.A., et al.: Gaia Data Release 1 Summary of the astrometric, photometric, and survey properties. Astron. Astrophys. **595**, id. A2, 23 pp. (I/337/tgas) (2016). https://doi.org/10.1051/0004-6361/201629512
26. Galli, P.A.B., Moraux, E., Bouy, H., Bouvier, J., Olivares, J., Teixeira, R.: A revised moving cluster distance to the Pleiades open cluster. Astron. Astrophys. **598**, 22 (2017). https://doi.org/10.1051/0004-6361/201629239. A48
27. Gliese, W.: Die Untersuchung der Raumgeschwindigkeiten des FK3. Astron. Nachri. **272**, 97–126 (1941). https://doi.org/10.1002/asna.19412720302
28. Haas, J.: Über die Ursa-Gruppe. Astron. Nachr. **241**(14), 233 (1931). https://doi.org/10.1002/asna.19312411402
29. Hambly, N., Hawkins, M.R.S., Jameson, R.: Very low mass proper motion members in the Pleiades. Astron. Astrophys. Suppl. Ser. **100**(3), 607 (1993). (ISSN 0365-0138)
30. Haro, G., Chavira, E., Gonzalez, G.: A catalog and identification charts of the Pleiades flare stars. Instituto de Tonantzintla, Boletin **3**, 3–68 (1982)
31. Hertzsprung, E.: On new members of the system of the stars beta, gamma, delta, epsilon, zeta Ursae Majoris. Astrophys. J. **30**, 135–143 (1909)
32. Hertzsprung, E.: Catalogue de 3259 étoiles dans les Pléiades. Ann. Leiden Obs. **19**, Part 1A (1947)
33. Høg, E., Fabricius, C., Makarov, V.V., Urban, S., Corbin, T., Wycoff, G., Bastian, U., Schwekendiek, P., Wicenec, A.: The Tycho-2 catalogue of the 2.5 million brightest stars. Astron. Astrophys. **355**, L27–L30 (2000)
34. IOPscience. http://iopscience.iop.org
35. Kharchenko, N.V., Piskunov, A.E., Schilbach, E., Röser, S., Scholz, R.-D.: Global survey of star clusters in the Milky Way. II. The catalogue of basic parameters. Astron. Astrophys. **558**, 1–8 (2013). https://doi.org/10.1051/0004-6361/201322302. A53
36. King, J.R., Villarreal, A.R., Soderblom, D.R., Gulliver, A.F., Adelman, S.J.: Stellar kinematic groups. II. A reexamination of the membership, activity, and age of the Ursa Major group. Astron. J. **125**(4), 1980–2017 (2003)
37. Kunder, A., et al.: The Radial Velocity Experiment (RAVE): fifth data release. Astron. J. **153**(2), article id 75, 30 pp. (2017). https://doi.org/10.3847/1538-3881/153/2/75

38. Larson, R.B.: Implications of binary properties for theories of star formation. The formation of binary stars. In: Zinnecker, H., Mathieu, R.D. (eds.) Proceedings of IAU Symposium 200, held, Potsdam, Germany, 10–15 April 2000, p. 93 (2001)
39. Liu, F., Yong, D., Asplund, M., Ramírez, I., Meléndez, J.: The Hyades open cluster is chemically inhomogeneous. Mon. Not. R. Astron. Soc. **457**(4), 3934–3948 (2016). https://doi.org/10.1093/mnras/stw247
40. l'abbé Outhitr, M.: Tome II des Mémoires présente's & 1'Academie royale des Sciences par dives Savants, p. 607 (1755)
41. Ludendorff, H.: Über die Radialgeschwindigkeiten von β, ε, ζ Ursae majoris und über die Bewegung und Parallaxe der sieben Hauptsterne des Großen Bären. Astron. Nachr. **180**(17), 265
42. Lynga, G.: VizieR On-line Data Catalog. Open Cluster Data. VII/92A. 5th Edition (1987). Originally published in: Lund Observatory (1995)
43. Lodieu, N., Deacon, N.R., Hambly, N.C.: Astrometric and photometric initial mass functions from the UKIDSS Galactic Clusters Survey - I. The Pleiades. Mon. Not. R. Astron. Soc. **422** (2), 1495–1511 (2012). https://doi.org/10.1111/j.1365-2966.2012.20723.x
44. Mamajek, E.E.: A moving cluster distance to the exoplanet 2M1207b in the TW Hydrae association. Astrophys. J. **634**(2), 1385–1394 (2005). https://doi.org/10.1086/468181
45. Mamajek, E.E.: A Pre-Gaia census of nearby stellar groups. In: Young Stars & Planets Near the Sun, Proceedings of the International Astronomical Union, IAU Symposium, vol. 314, pp. 21–26 (2016). https://doi.org/10.1017/s1743921315006250
46. Melis, C., Reid, M.J., Mioduszewski, A.J., Stauffer, J.R., Bower, G.C.: A VLBI resolution of the Pleiades distance controversy. Science **345**(6200), 1029–1032 (2014). https://doi.org/10.1126/science.1256101
47. Mermilliod, J.C.: Comparative studies of young open clusters. III - Empirical isochronous curves and the zero age main sequence. Astron. Astrophys. **97**(2), 235–244 (1981)
48. Mohr, J.M.: Sur le courant d'étoiles Ursa Major. Bull. Astrono. **6**, 147–153 (1930)
49. Montes, D., López-Santiago, J., Gálvez, M.C., Fernández-Figueroa, M.J., De Castro, E., Cornide, M.: Late-type members of young stellar kinematic groups - I. Single stars. Mon. Not. R. Astron. Soc. **328**(1), 45–63 (2001). https://doi.org/10.1046/j.1365-8711.2001.04781.x
50. Nassau, J.J., Henyey, L.G.: The Ursa Major group. Astrophys. J. **80**, 282 (1934). https://doi.org/10.1086/143604
51. Pinfield, D., Hodgkin, S., Jameson, R., Cossburn, M., Hambly, N., Devereux, N.: A six-square-degree survey for Pleiades low-mass stars and brown dwarfs. Mon. Not. R. Astron. Soc. **313**(2), 347–363 (2000). https://doi.org/10.1046/j.1365-8711.2000.03238.x
52. Piskunov, A.Eh.: Catalogue of masses and ages of 68 open clusters. Astronomical Council of the USSR Academy of Sciences, Moskva, 17 + 294 p. (1977). (in Russian)
53. Plummer, H.C.: On the motion of the brighter stars of class A in relation in the Milky Way, LicOB, N. 212, V. VII, p. 30 (1912)
54. Rasmuson, N.H.: A research on moving clusters. Meddelanden fran Lunds Astronomiska Observatorium Series II **26**, 3–74 (1921)
55. Roman, N.G.: The Ursa Major group. Astrophys. J. **110**, 205 (1949). https://doi.org/10.1086/145199
56. Rubinov, A.V., Orlov, V.V.: The N-body problem in stellar dynamics. Textbook of St. Petersburg, Izd. St. Petersburg University, 97 pp. (2008). (in Russian)
57. Sarro, L.M., Bouy, H., Berihuete, A., Bertin, E., Moraux, E., Bouvier, J., Cuillandre, J.-C., Barrado, D., Solano, E.: Cluster membership probabilities from proper motions and multi-wavelength photometric catalogues I. Method and application to the Pleiades cluster. Astron. Astrophys. **563**, 14 (2014). https://doi.org/10.1051/0004-6361/201322413. A45

58. Schmeja, S., Kharchenko, N.V., Piskunov, A.E., Röser, S., Schilbach, E., Froebrich, D., Scholz, R.-D.: Global survey of star clusters in the Milky Way. III. 139 new open clusters at high Galactic latitudes. Astron. Astrophys. **568**, 1–9 (2014). https://doi.org/10.1051/0004-6361/201322720. A51
59. Scholz, R.-D., Kharchenko, N.V., Piskunov, A.E., Röser, S., Schilbach, E.: Global survey of star clusters in the Milky Way. IV. 63 new open clusters detected by proper motions. Astron. Astrophys. **581**, 1–15 (2015). https://doi.org/10.1051/0004-6361/201526312. A39
60. Science direct: http://www.sciencedirect.com/
61. Shkolnik, E.L., Anglada-Escudé, G., Liu, M.C., Bowler, B.P., Weinberger, A.J., Boss, A.P., Reid, I.N., Tamura, M.: Identifying the young low-mass stars within 25 pc. II. Distances, kinematics, and group membership. Astrophys. J. **758**(1), 56 (2012). https://doi.org/10.1088/0004-637x/758/1/56. 23 pp.
62. Siegler, N., Muzerolle, J., Young, E.T., Rieke, G.H., Mamajek, E.E., Trilling, D.E., Gorlova, N., Su, K.Y.L.: Spitzer 24 μm observations of open cluster IC 2391 and debris disk evolution of FGK stars. Astrophys. J. **654**, 580–594 (2007). https://doi.org/10.1086/509042
63. SIMBAD Astronomical Database. http://simbad.u-strasbg.fr/simbad/
64. Smart, W.M.: The Ursa Major cluster. Mon. Not. R. Astron. Soc. **99**, 441S (1939). https://doi.org/10.1093/mnras/99.5.441
65. Soderblom, D.R., Mayor, M.: Stellar kinematic groups. I - The Ursa Major group. Astron. J. **105**(1), 226–249 (1993). https://doi.org/10.1086/116422. (ISSN 0004-6256)
66. Stauffer, J., Klemola, A., Prosser, C., Probst, R.: The search for faint members of the Pleiades. I - A proper motion membership study of the Pleiades to M(V) of about 12.5. Astron. J. **101**, 980–1005 (1991). https://doi.org/10.1086/115741
67. Stauffer, J.R., Hartmann, L.W., Fazio, G.G., Allen, L.E., Patten, B.M., Lowrance, P.J., Hurt, R.L., Rebull, L.M., Cutri, R.M., Ramirez, S.V., Young, E.T., Rieke, G.H., Gorlova, N.I., Muzerolle, J.C., Slesnick, C.L., Skrutskie, M.F.: Near- and mid-infrared photometry of the Pleiades and a new list of substellar candidate members. Astrophys. J. Suppl. Ser. **172**(2), 663–685 (2007). https://doi.org/10.1086/518961
68. Tabernero, H.M., Montes, D., González Hernández, J.I., Ammler-von Eiff, M.: Chemical tagging of the Ursa Major moving group. A northern selection of FGK stars. Astron. Astrophys. **597**, 25 (2017). https://doi.org/10.1051/0004-6361/201322526. A33
69. Takeda, Y., Hashimoto, O., Honda, S.: Photospheric carbon and oxygen abundances of F-G type stars in the Pleiades cluster. Publ. Astron. Soc. Jpn. **69**(1), 17 (2017). https://doi.org/10.1093/pasj/psw105. id. 1
70. The SAO/NASA Astrophysics Data System. http://www.adsabs.harvard.edu
71. Trumpler, R.J.: The physical members of the Pleiades group. Lick Obs. Bull. **10**, 110 (1921)
72. van Albada, T.S.: The evolution of small stellar systems and its implications for the formation of double stars. Bull. Astron. Inst. Netherlands. **20**, 57 (1968)
73. van den Berk, J., Portegies Zwart, S.F., McMillan, S.L.W.: The formation of higher order hierarchical systems in star clusters. Mon. Not. R. Astron. Soc. **379**(1), 111–122 (2007). https://doi.org/10.1111/j.1365-2966.2007.11913.x
74. van Leeuwen, F., Alphenaar, P., Brand, J.: A VBLUW photometric survey of the Pleiades cluster. Astron. Astrophys. Suppl. Ser. **65**(2), 309–347 (1986). (ISSN 0365-0138)
75. Vereshchagin, S.V., Reva, V.G., Chupina, N.V.: The structure of the AD diagram for the Hyades cluster. Astron. Rep. **52**(2), 94–98 (2008). https://doi.org/10.1134/s1063772908020029
76. Vereshchagin, S.V., Chupina, N.V.: Apex and internal structure of the Praesepe open cluster. Astron. Nachr. **334**(8), 892 (2013). https://doi.org/10.1002/asna.201311938

77. Vereshchagin, S.V., Chupina, N.V., Sariya, D.P., Yadav, R.K.S., Kumar, B.: Apex determination and detection of stellar clumps in the open cluster M 67. New Astron. **31**, 43–50 (2014). https://doi.org/10.1016/j.newast.2014.02.008
78. Vereshchagin, S.V., Chupina, N.V., Fionov, A.S.: Star clusters: the growth of knowledge based on data intensive research. In: Kalinichenco, L., Manolopoulos, Y., Kuznetsov, S. (eds.) XVIII International Conference on Data Analytics and Management in Data Intensive Domains, DAMDID/RCDL 2016, Ershovo, Moscow Region, Russia, 11–14 October 2016, pp. 323–327. FRC CSC RAS, Moscow (2016)
79. Vereshchagin, S.V., Postnikova, E.S.: Accumulation of new knowledge about the internal structure of an open star clusters on the basis of intensive use of data. In: Kalinichenco, L., Manolopoulos, Y., Skvortsov, N.A., Sukhomlin, V.A. (eds.) XIX International Conference on Data Analytics and Management in Data Intensive Domains, DAMDID/RCDL 2017 and CEUR Workshop Proceedings, 10–13 October 2017. Moscow State University, Moscow, Russia, vol. 2022, pp. 30–36 (2017). http://ceur-ws.org/Vol-2022/paper08.pdf. (in Russian)
80. VizieR Catalogue Service. http://vizier.u-strasbg.fr
81. Walker, M.F.: Studies of extremely young clusters. V. Stars in the vicinity of the Orion nebula. Astrophys. J. **155**, 447 (1969). https://doi.org/10.1086/149881
82. WEBDA: A site Devoted to Stellar Clusters in the Galaxy and the Magellanic Clouds. http://www.univie.ac.at/webda/webda.html
83. Wiley Online Library. http://onlinelibrary.wiley.com
84. Wolf, C.: Description du groupe des Pleiades et mesure micrometrique des positions relatives des principales étoiles qui le composent., Annales de l'observatoire de Paris, Tome XIV, deuxieme partle, Al–A81 (1877)

Search for Short Transient Gamma-Ray Events in SPI Experiment Onboard INTEGRAL: The Algorithm and Results

Pavel Minaev(✉) and Alexei Pozanenko

Space Research Institute of the Russian Academy of Sciences (IKI), 84/32 Profsoyuznaya Str, 117997 Moscow, Russia
minaevp@mail.ru, apozanen@iki.rssi.ru

Abstract. We consider the possibilities for a searching and analyzing various short transient gamma-ray events in the archival data of the SPI experiment onboard the INTEGRAL observatory. The problems of the raw observational data processing, including the search algorithm and the method of automated classification of detected events based on a set of various criteria are discussed. The results of the analysis of the SPI/INTEGRAL archived data obtained for the period 2003–2010 are presented.

Keywords: Search · Classification · Catalog · Transient · Gamma-rays Gamma-ray burst · GRB · Terrestrial gamma-ray flash · TGF Soft gamma repeater · SGR · Anomalous X-ray pulsar · AXP SPI · IBIS/ISGRI · SPI-ACS · INTEGRAL

1 Introduction

One of the actual problems in modern astrophysics is connected with investigation of various transient events in gamma-ray energy range and especially the investigation of the two most mysterious ones: cosmological gamma-ray bursts (GRB) and terrestrial gamma-ray flashes (TGF). GRBs are the most powerful explosions in the Universe and observed as sporadic flashes of gamma-ray emission with duration of 0.1–100 s in energy range above 10 keV [1]. TGFs are much shorter (<1 ms) and are supposed to be generated in Earth' atmosphere and associated with thunderstorm activity [2, 3].

Most space based gamma-ray telescopes (including SPI/INTEGRAL) operate in photon-by-photon mode – they record time, energy, position at detector plane and other quantities of individual photons [1, 4]. Photon-by-photon mode brings many advantages in data analysis. Among them, one could emphasize the developing of advanced methods of automated searching and interpreting of events in real time data flow. Automated methods are very actual for searching of short transients like TGF with duration less than 1 ms and for analysis of huge data arrays collected during several years of observations (>10 Tb of data for SPI experiment) because of enormous number of false instrumental triggers, connected with fluctuations and instrumentation effects (>15000 events per year for SPI).

L. Kalinichenko et al. (Eds.): DAMDID/RCDL 2017, CCIS 822, pp. 128–138, 2018.
https://doi.org/10.1007/978-3-319-96553-6_10

The original algorithms for searching and classifying detected events in the archived data of the SPI/INTEGRAL experiment, as well as interpretation the obtained results are considered. The article continues studies [5] and a detailed analysis of instrumental effects is carried out. In particular we discuss the effect of saturation of SPI detectors due to these events. Number density (registration rate) of different types of detected events was used for physical classification of the events. The number density figure for instrumentation events was added. All presented in [5] figures were updated.

2 SPI/INTEGRAL Experiment

The INTEGRAL observatory was launched on October 17, 2002 to a high-elliptical orbit (the perigee of the initial orbit is 9 thousand km, the apogee is 153 thousand km) with a period of 72 h [6]. The Observatory has two main gamma-ray telescopes (IBIS/ISGRI, SPI) and several other telescopes (JEM-X, OMC, SPI-ACS). All aperture telescopes (SPI, IBIS/ISGRI, JEM-X, and OMC) are aligned, but the shape and size of the fields of view are different.

The SPI gamma-ray spectrometer consists of 19 hexagonal detectors made of ultrapure germanium, with a total geometric area of 508 cm^2 [4]. A coded mask made of tungsten is used for imaging. The spectral resolution of the SPI/INTEGRAL spectrometer reaches 2.2 keV @ 1.33 MeV - one of the best at the time of the observatory's launch (2002). The energy range is 20 keV–8 MeV. The full (zero coding) field of view of the telescope is 32°.

To increase the sensitivity of the SPI telescope, by eliminating the background associated with the interaction of the equipment with cosmic rays, an anti-coincidence shield SPI-ACS consisting of 91 bismuth germanate crystals (BGO) with an effective area of 0.3 m^2 is used [7].

3 Searching Algorithm

We developed the triggering algorithm for searching events in the archived raw data of the SPI/INTEGRAL experiment, received for the period from July 12, 2003 to January 23, 2010.

Gamma-spectrometer SPI/INTEGRAL registers individual gamma-ray photons by recording their moment of registration, energy and the position on the detectors plane. The search for events was carried out in the energy range (20–650, 2000–8000) keV. The range (650, 2000) keV was eliminated, since a significant portion of the counts in this range are related to the noise of the electronics.

The events were selected at time scales of 0.001, 0.01, 0.1, 1 and 10 s with thresholds of 20, 6, 5, 5 and 4 σ, respectively. The threshold of 20 σ for the 0.001 s interval was chosen to minimize the number of fluctuations, so that only high intensity events are triggered at the initial selection. Only such intense short events of 0.001 s duration could be detected in the data of other experiments (with a lower temporal resolution).

When searching for events, three types of counts were used. SGL is an "ordinary" count, registered in one detector. PSD count is recorded in one detector whose pulse shape confirms its photon nature. DBL counts are recorded simultaneously in two different detectors due to Compton scattering of the original photon inside one of the detectors.

In total, this algorithm selected more than 100000 events at different time scales (from 1 ms to 10 s). For each detected event, a light curve and an energy diagram were constructed, the distribution of counts on a detectors plane was also analyzed. All these steps were used to classify events.

4 Classification of the Detected Events

To classify the events found in the SPI experiment, the data of the anti-coincidence shield SPI-ACS and the telescope IBIS/ISGRI, placed on the INTEGRAL observatory, were also used.

Three classes of events were formed: fluctuations, candidates for "real" gamma-ray events (e.g., gamma-ray bursts), and three types of instrumentation effects, possibly associated with the interaction of the detectors with charged particles. Events of the "fluctuation" type, as a rule, were found at the threshold of significance, and were not confirmed by the data of other space telescopes (primarily the IBIS/ISGRI and SPI-ACS experiments) and therefore excluded from the analysis.

Let us consider the criteria used for the classification.

4.1 Duration

For cosmological gamma-ray bursts (GRB), the duration usually lies within the limits (0.1, 100) s, for terrestrial gamma-ray flashes (TGF) - within (0.1, 1) ms, for SGR and AXP flares - in the range (0.01, 3) s. The instrumentation events, depending on the type, have a duration in the range (0.001, 2) s.

4.2 Hardness Ratio

The hardness ratio of counts between (100, 1000) keV and (20, 100) keV energy ranges was used to characterize the hardness of the energy spectrum. For SGR and AXP flashes, this parameter is significantly less than one. Spectra of cosmological gamma-ray bursts are more diverse - the value of the hardness ratio can vary over a wide range and is, on average, about one. For the instrumentation events, depending on the type, the value of this parameter is either much less than one, or significantly greater than one.

4.3 Distribution of Counts on Detectors Plane

To quantify the distribution of counts on a detectors plane, the ratio of the maximum count rate in one detector to the average count rate was used. For "real" gamma-ray events (GRB, TGF, SGR and AXP flashes) occurring in the field of view of the SPI telescope (including events at the edge of the field of view), the value of this parameter lies in the interval (1, 3). For instrumentation events of the first type, the value of this criterion is much higher than 3.

4.4 Spectral Evolution Behavior

It was an additional criterion for the selection of instrumentation events. For most "real" investigated in the paper gamma-ray events, the evolution of emission from hard to soft is typical. The evolution of the spectrum of events associated with the first type of the instrumentation effects is opposite and very pronounced (see Fig. 1A). Events related to the second type of the instrumentation effects are observed as spectral lines with energies of 53, 66, and 198 keV, corresponding to nuclear reactions of thermal neutron capture by Ge nuclei [8], when thermal neutrons are presumably produced as the result of cascade reactions (see Fig. 2A). To select events of this type, additional parameter was also introduced - the ratio of the counts in the interval (0, 50) ms relative to the trigger to the counts in the interval (−50, 0) ms in a narrow energy range (195, 201) keV. This criterion was applied only for the selection of events at time scales of 10 and 100 ms, since events of this type have a duration of, on average, about 50 ms.

4.5 Registration Rate Features

Most of the SGR and AXP flashes were observed during the activity of the corresponding source, which lasts typically about a month. An extremely uneven registration rate at the time scale of days is characteristic for events related to the first two types of instrumentation effects. The registration rate of cosmological gamma-ray bursts is uniform.

4.6 Detection of the Event in Data of Other Experiments

More than 90% of the "real" gamma-ray events were also found in the IBIS/ISGRI data, and about a third of the events were detected in the SPI-ACS experiment. The detection and localization of an event in IBIS/ISGRI data is a reliable sign of a "real" gamma-ray event. As for instrumentation events, 30% of the second type and 70% of the third type were detected in SPI-ACS data. For candidates in real gamma-ray events, searches for confirmation have been carried out in known catalogs of gamma-ray transients [9–11].

5 The Results

Let us examine in detail the properties of various types of detected events.

5.1 Fluctuations

This event type is characterized by a quasi-uniform distribution of counts on a detectors plane and background-like energy spectrum. The time profile also does not reveal any

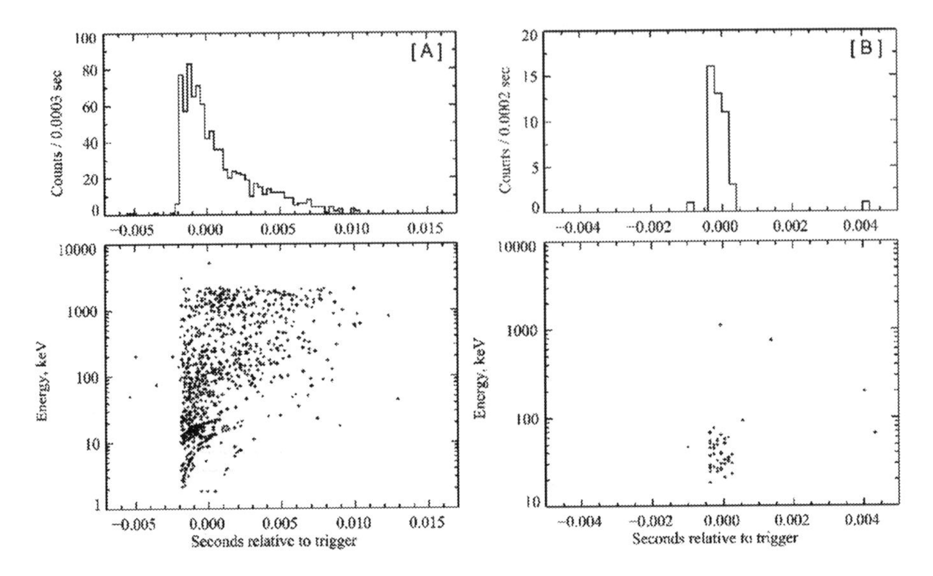

Fig. 1. Light curve in the energy range (20, 8000) keV (top) and energy diagram (bottom) of the two types of instrumentation effects: type I on the left [A] and type II on the right [B]. Each point at energy diagram represents individual count (SGL, PSD, or DBL).

features. This category also includes triggers that can not be attributed to another class of events. All events of this type were excluded from further analysis.

5.2 Instrumentation Effects, Type I

Triggers of this type are easy to identify, since they have several observational features at once. An example of such an event is shown in Fig. 1A. The average duration is about 10 ms. The shape of the light curve is usually asymmetrical and is characterized by rapid growth and slow decay. The main feature of the type is the extremely uneven distribution of counts on a detectors plane. The energy spectrum, as a rule, is hard, with a sharp break of 2 MeV, the hardness ratio increases with time. These events typically are very bright, containing hundreds of counts (see Fig. 1A) but none of them were found in data of SPI-ACS and IBIS/ISGRI experiments.

The registration rate of these events is highly irregular, there are several short episodes (1–2 days long) of active registration (up to 100 events per day), in the intervals between these periods events were not registered (Fig. 2A). It is important that registration rate correlates with SPI annealing periods[1], during which Ge detectors are heated up to about 100 °C for a few days so that their crystal structure can re-arrange and correct itself for most of the accumulated radiation damages. The nature of this type of events and of their connection with annealing periods is unknown, but there is no doubt in their instrumentation origin.

[1] http://www.isdc.unige.ch/integral/download/osa/doc/10.2/osa_um_spi/node70.html.

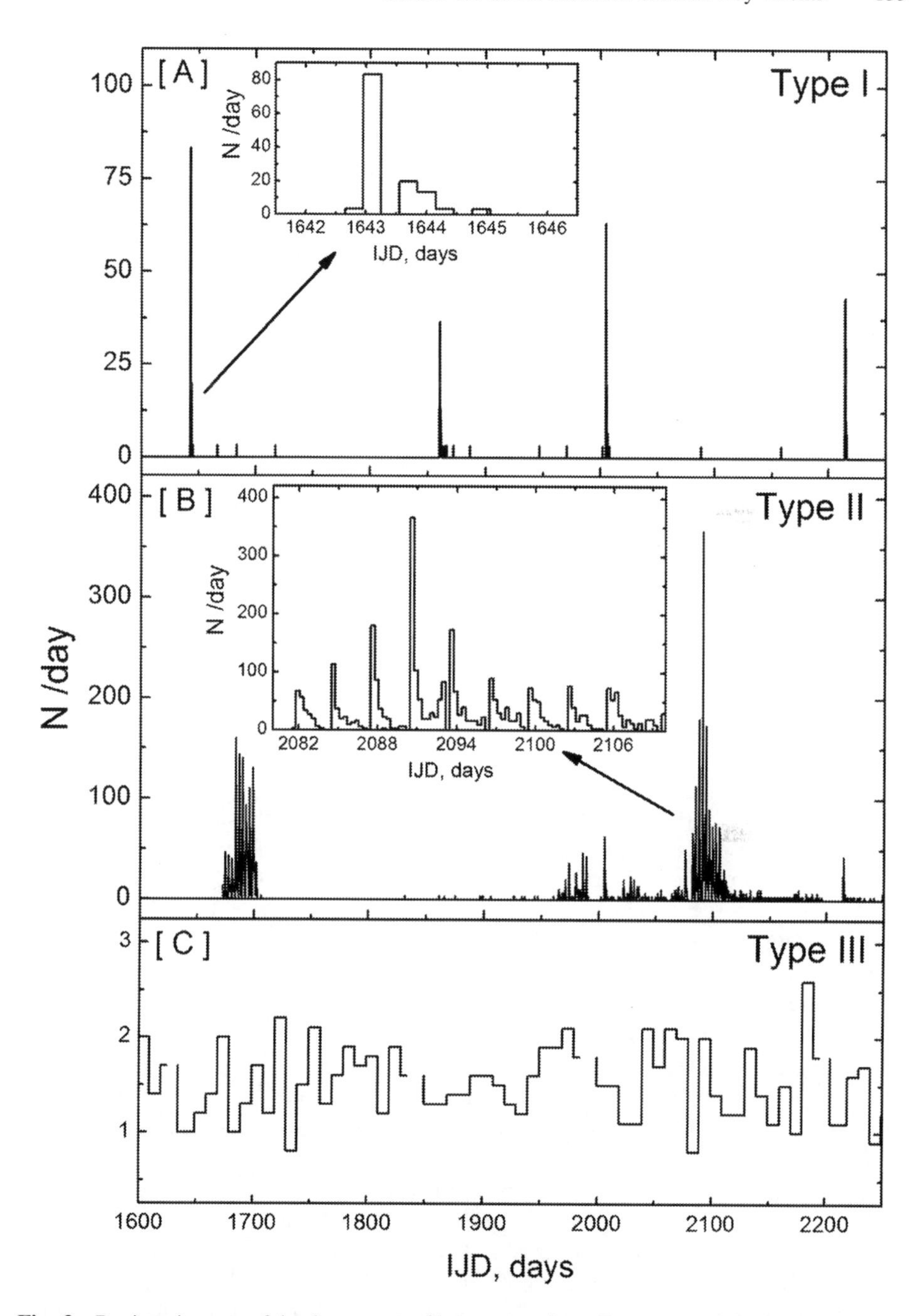

Fig. 2. Registration rate of the three types of instrumentation effects: type I at the top [A], type II in the middle [B] and type III at the bottom [C]. For the types I and II of events periods of active registration are presented at insets. Time in units of INTEGRAL Julian Date (IJD) is on x-axis. Number of registered events per day is on y-axis.

5.3 Instrumentation Effects, Type II

An example of an event of this type is shown at Fig. 1B. Duration in most cases does not exceed 1–2 ms, but events consisting of individual millisecond pulses are possible with total duration up to 10 ms. The shape of the pulse time profiles is quasi symmetrical in most cases. The energy spectrum is soft, with a cut-off at 100 keV. The distribution of counts at detectors plane is uniform. About a third of the events were found in the SPI-ACS data as short 50 ms long pulses (time resolution of SPI-ACS).

The registration rate of these events is irregular and does not correlate with one of the previous type of instrumental effect, several periods of active registration lasting about a month each are observed (Fig. 2B). At time intervals between these episodes, events were rare. The rate of registration during active periods is periodic and varies with a period of 3 days - the period of revolution of the INTEGRAL observatory around the Earth (see inset at Fig. 2B). It is possible that these events are related to the interaction of the detectors with electron beams of the Earth's outer radiation belt, and they are recorded at the moments when the orbit of the observatory crosses the tail of the Earth's magnetosphere.

5.4 Instrumentation Effects, Type III

This type of events is associated with a short-term significant increase in the count rate in spectral lines of 198 keV, 53 keV and 66 keV, caused by the following nuclear thermal neutron capture reactions:

$$^{70}\mathrm{Ge} + \mathrm{n} > {}^{71\mathrm{m}}\mathrm{Ge}\ (\mathrm{half} - \mathrm{life}\ 20.4\,\mathrm{ms}) > {}^{71}\mathrm{Ge} + 2\gamma(175 + 23 = 198\,\mathrm{keV})\ [8].$$
$$^{72}\mathrm{Ge} + \mathrm{n} > {}^{73\mathrm{m}}\mathrm{Ge}(\mathrm{half} - \mathrm{life}\ 0.5\,\mathrm{s}) > {}^{72}\mathrm{Ge} + 2\gamma(53 + 13 = 66\,\mathrm{keV})\ [8].$$

The duration of transient emission in the 198 keV line is tens of ms, correlating with half-life of the corresponding nuclear reaction (Fig. 3A). Transient radiation in the 53 keV and 66 keV spectral lines is not observed in all cases. The duration of transient radiation in these lines can reach 1–2 s, also correlating with half-life of the corresponding nuclear reaction.

The strong correlation between half-life and duration of observed intense emission indicates almost instantaneous generation and capture of hundreds of thermal neutrons by Ge detectors, which could be connected with a powerful cascade of secondary particles, generated during interaction of the equipment with energetic cosmic ray particle.

Events of this class in 25% of cases are accompanied by "saturation" of SPI detectors - the absence of a signal in one or several neighboring detectors within a few seconds. The nature of saturation is not clear. It could be connected with detectors overflow, also caused by a cascade of charged particles.

The rate of registration is quasi-constant of about 1.5 events/day (Fig. 2C), indicating their connection with the most energetic galactic cosmic rays.

More than 70% of the events are also observed in the SPI-ACS data. It was shown in [12] that the BGO crystals, of which SPI-ACS consists, emit secondary neutrons as a

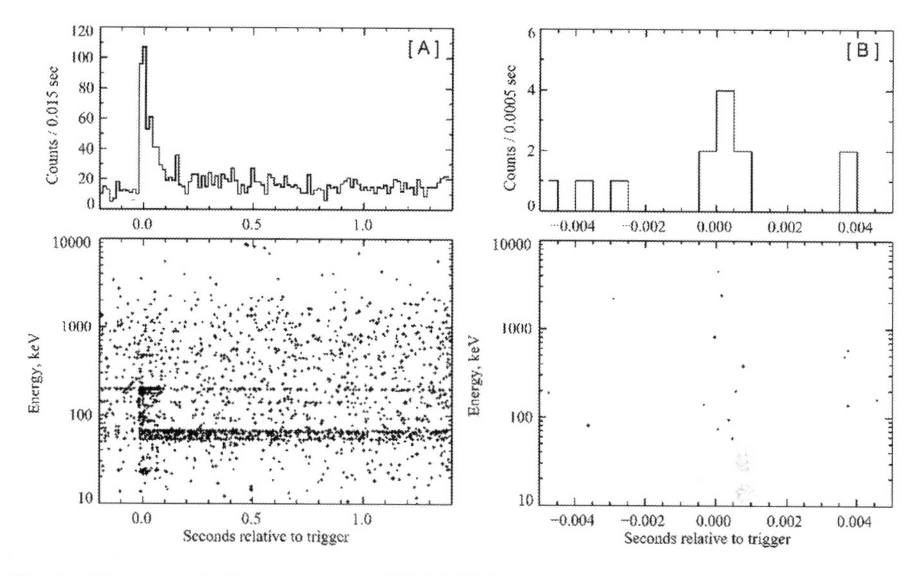

Fig. 3. Light curve in the energy range (20, 8000) keV (top) and energy diagram (bottom) of the third type of instrumentation effects on the left [A] and TGF candidate on the right [B]. Each point at energy diagram represents individual count (SGL, PSD or DBL).

result of interaction with cosmic rays (nuclear cleavage reactions), which in turn are thermalized and trapped by the Ge nuclei of the SPI detectors.

To conclude, most probably, third type of instrumentation effects is related with the most energetic galactic cosmic rays, which generate a powerful cascade of secondary particles in the SPI-ACS, including thermal neutrons, registered by SPI detectors.

5.5 TGF Candidates

To study the diffuse x-ray background by the INTEGRAL observatory, a number of observations were made, when the Earth entered the aperture of SPI and IBIS telescopes onboard the observatory. We used these data with a total duration of 496 ks to search for terrestrial gamma-ray flashes (TGF).

In total, 28 candidates were selected based on the known properties of TGF: duration ≤ 1 ms, hard energy spectrum, presence of counts with energy >1 MeV, the shape of the light curve is quasi symmetrical, not significant spectral evolution, the quasi-uniform distribution of registered counts on a detectors plane. One of the candidates is shown in Fig. 3B.

A detailed study of TGF candidates, found in the SPI/INTEGRAL data, see in [13].

5.6 GRB Catalog

48 cosmological gamma-ray bursts confirmed by other space-based experiments were found (see Tables 4–6 in [1]). Figure 4A shows the light curve and the energy diagram for the GRB 050525.

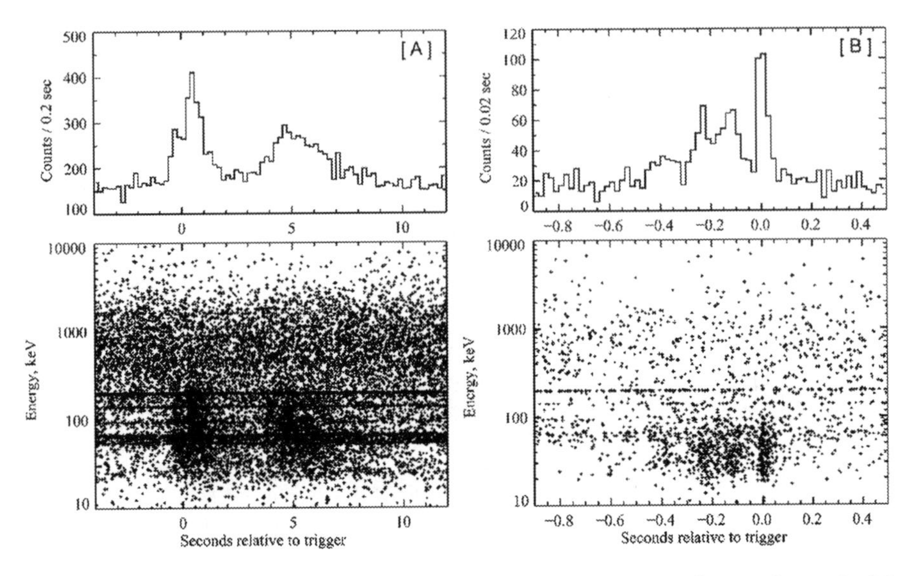

Fig. 4. Light curve in the energy range (20, 8000) keV (top) and energy diagram (bottom) of the GRB 050525 on the left [A] and SGR 1806-20 flash on the right [B]. Each point at energy diagram represents individual count (SGL, PSD or DBL).

In addition to the confirmed gamma-ray bursts, 160 candidates were selected (see Table 7 in [1]), including 151 short ones with a duration of less than 2 s. Candidates were selected in accordance with the observed properties of gamma-ray bursts: 1. the duration of individual pulses is more than 5 ms; 2. a hard energy spectrum of events with a duration of less than 2 s (this criterion also allows the screening of events associated with the SGR flares); 3. The distribution of the detectors is close to uniform.

Details of the study of gamma-ray bursts detected by the SPI/INTEGRAL experiment, see in [1, 14].

5.7 SGR and AXP Catalog of Flashes

223 flashes of the soft gamma repeater SGR 1806-20 (Fig. 4B) and 23 flares of the anomalous X-ray pulsar AXP 1E_1547.0-5408 found in SPI data were also detected and localized in the IBIS/ISGRI experiment (see Tables 1–3 in [1]). Some events were also found in the SPI-ACS data.

In addition, a list of 90 candidates for SGR and AXP flashes was formed, selected in accordance with the observed properties of confirmed bursts of SGR-type sources: 1. The duration of individual event pulses is within (0.01–3) s range; 2. The fraction of counts with energies above 200 keV is negligible due to soft energy spectrum; 3. The distribution of the counts on a detectors plane is close to uniform.

6 Conclusions

A comprehensive study of the archival raw observational data of the SPI gamma spectrometer accumulated over 7 years of the INTEGRAL observations was carried out.

An algorithm for searching of transient events at various time scales from 1 ms to 10 s was proposed, which detected more than 100 000 events.

A technique for classifying detected events based on a series of criteria has been developed, which can be applied automatically to real time observational data flow.

Three classes of events were distinguished: 1. fluctuations; 2. "real" gamma events (gamma-ray bursts of cosmological (GRB) and terrestrial (TGF) origin, flashes of SGR and AXP sources); 3. Three types of instrumental phenomena, two of them are possibly associated with the interaction of the detector with charged particles (interactions with electron beams and galactic cosmic rays of high energies).

The catalogs of gamma-ray bursts and flares of SGR 1806-20 and AXP 1E_1547.0-5408 sources have been compiled.

Acknowledgements. The work was supported by the Russian Foundation for Basic Research (grant no. 17-02-01388).

References

1. Minaev, P.Yu., Pozanenko, A.S., Molkov, S.V., Grebenev, S.A.: Catalog of short gamma-ray transients detected in the SPI/INTEGRAL experiment. Astron. Lett. **40**, 235 (2014)
2. Gurevich, A.V., Milikh, G.M., Roussel-Dupre, R.: Runaway electron mechanism of air breakdown and preconditioning during a thunderstorm. Phys. Lett. A **165**, 463–468 (1992)
3. Fishman, G.J., Bhat, P.N., Mallozzi, R., et al.: Discovery of intense gamma-ray flashes of atmospheric origin. Science **254**, 5163 (1993)
4. Vedrenne, G., Roques, J.-P., Schönfelder, V., et al.: SPI: the spectrometer aboard INTEGRAL. Astron. and Astrophys. **411**, L63–L70 (2003)
5. Minaev, P., Pozanenko, A.: Short gamma-ray transients in SPI/INTEGRAL: search, classification and interpretation. In: Kalinichenko, L.A., Manolopoulos, Y., Skvortsov, N.A., Sukhomlin, V.A. (eds.) Selected Papers of the XVIII International Conference on Data Analytics and Management in Data Intensive Domains (DAMDID/RCDL 2017) and CEUR Workshop Proceedings, Moscow, vol. 2022, pp. 37–42, (2017)
6. Winkler, C., Courvoisier, T.J.-L., Di Cocco, G., et al.: The INTEGRAL mission. Astron. Astrophys. **411**, L1–L6 (2003)
7. von Kienlin, A., Beckmann, V., Rau, A., et al.: INTEGRAL spectrometer SPI's GRB detection capabilities. GRBs detected inside SPI's FoV and with the anticoincidence system ACS. Astron. Astrophys. **411**, L299–L305 (2003)
8. Weidenspointner, G., Harris, M.J., Sturner, S., et al.: MGGPOD: a Monte Carlo suite for modeling instrumental line and continuum backgrounds in gamma-ray astronomy. Astrophys. J. Suppl. **156**, 69 (2005)
9. Chelovekov, I.V., Grebenev, S.A.: Hard X-ray bursts recorded by the IBIS telescope of the INTEGRAL observatory in 2003-2009. Astron. Lett. **37**, 597 (2011). arXiv:astro-ph.HE/1108.2421
10. IBAS IBIS/ISGRI triggers Homepage. http://ibas.iasf-milano.inaf.it/

11. Hurley, K.: Masterlist Homepage. http://www.ssl.berkeley.edu/ipn3/chronological.txt
12. Jean, P., von Ballmoos, P., Vedrenne, G., Naya, J.E.: Performance of advanced Ge-spectrometer for nuclear astrophysics. In: Ramsey, B.D., Parnell, T.A. (eds.) Gamma-Ray and Cosmic-Ray Detectors, Techniques, and Missions 1996, Society of Photo-Optical Instrumentation Engineers (SPIE) Conference Series, vol. 2806, p. 457 (1996)
13. Minaev, P., Pozanenko, A., Grebenev, S., Molkov, S.: The search for terrestrial gamma-ray flashes (TGF) in data of SPI/INTEGRAL (2018, to appear)
14. Minaev, P.Yu., Grebenev, S.A., Pozanenko, A.S., et al.: GRB 070912 - a gamma-ray burst recorded from the direction to the galactic center. Astron. Lett. **38**, 613 (2012)

Ontology Population in Data Intensive Domains

Development of Ontologies of Scientific Subject Domains Using Ontology Design Patterns

Yury Zagorulko(✉), Olesya Borovikova, and Galina Zagorulko

A.P. Ershov Institute of Informatics Systems, Siberian Branch of the Russian Academy of Sciences, Acad. Lavrentjev avenue 6, 630090 Novosibirsk, Russia
{zagor,olesya,gal}@iis.nsk.su

Abstract. As developing ontologies of subject areas is a rather complex and time-consuming process, various methods and approaches have been proposed to simplify and facilitate it. Over the past few years, the approach based on the use of ontology design patterns has been intensively developing. The paper discusses the application of ontology design patterns in the development of ontologies of scientific subject areas. Such patterns are designed to describe the solutions of typical problems arising in ontology development. They are created in order to facilitate the process of building ontologies and to help the developers avoid some highly repetitive errors occurring in ontology modeling. The paper presents the ontology design patterns resulting from solving the problems that the authors have encountered in the development of ontologies for such scientific subject areas as archeology, computer linguistics, system studies in power engineering, active seismology, etc.

Keywords: Scientific subject domain · Ontology · Ontology design patterns
Methodology for ontology development
Thematic intelligent scientific internet resource

1 Introduction

Today, ontologies are the most popular and effective means of conceptualizing and formalizing scientific subject areas [1]. They are widely used to present and record some common knowledge shared by all experts (or a group of experts) about such areas. The formalization of the semantics of a subject area in the form of ontology not only contributes to its compact and consistent description, but also forms a conceptual basis for the representation of the whole body of knowledge about it. For example, in the system of information support of research activity (SISRA) [2], the semantics of the data and information resources used in it can be described in terms of ontology; the expert rules, precedents and other components of the knowledge base of the expert system or decision support system can also be described in terms of ontology [3].

The SISRA must provide the user with a representation of all the necessary information about his/her area of expertise, its components (divisions/subdivisions of science, objects, methods and techniques of research, etc.), as well as subjects (participants) of research activity (personalities, groups, communities and other organizations involved in the research process). In the SISRA, ontology defines a formal

L. Kalinichenko et al. (Eds.): DAMDID/RCDL 2017, CCIS 822, pp. 141–156, 2018.
https://doi.org/10.1007/978-3-319-96553-6_11

description of the knowledge area on the basis of which such information is systematized and relevant information resources and documents are integrated into a single information space. A user interface that provides meaningful access to knowledge and data integrated into the information space of the system is also constructed on the basis of the ontology. In particular, in such an interface the user can use the ontology as a guide to navigate through this space, as well as to formulate search queries that have concepts and relations of ontology as the main elements.

At present, ontologies are widely used for the conceptual modeling of subject areas with intensive use of data [4]. The development and application of infrastructures to support the research based on the conceptual specifications of these areas makes it possible to avoid the dependence of programs on the structure of data sources, to ensure the interoperability of various data processing methods working together, and to increase the reliability of the results obtained through the use of formal consistent specifications.

The development of the ontologies of scientific subject areas is a rather complex and time-consuming process. To simplify and facilitate it, various methods and approaches to the development of ontologies have been proposed. Over the past ten years, the approach based on the application of ontology design patterns (ODPs) has been developing intensively [5, 6].

The ODPs are the documented descriptions of the proven solutions of ontological modeling problems. By now, several catalogs of patterns have been created and are developing [7, 8]. It should be noted that since such catalogs are, as a rule, oriented either to some subject area or group of developers, they are not complete or versatile.

The paper discusses the ontology design patterns resulting from solving the problems that the authors have encountered in the process of developing ontologies for various scientific subject areas [9–12]. We describe the problems and patterns in the context of the ontology development methodology for thematic intelligent scientific Internet resources (ISIR) [13], intended for information and analytical support of research activity in the specified areas of knowledge.

This paper is an extended version of the paper presented at the DAMDID/RCDL’2017 conference [14]. The review of ontology design patterns has been substantially expanded (Sect. 2) and a description of the presentation patterns used in the methodology of building ontology for ISIR [13] has been added.

2 A Review of Ontology Design Patterns

The progenitors of ontology design patterns are design patterns, widely used in software development. In this area of activity, a design pattern is a description of a well-tested, generalized scheme for solving a frequently recurring development problem that arises in a certain context. The patterns have become part of the daily practice of object-oriented design. With their help, specific design tasks are solved, which has resulted in object-oriented design becoming more flexible, elegant, and reusable [15].

By analogy with design patterns, ontology design patterns are used to describe the solutions of typical problems that arise in ontologies development. Patterns are created in order to facilitate the process of building ontologies and help developers avoid some of the highly repetitive errors of ontological modeling [16]. In this capacity, the ODPs were first introduced, independently from each other, by Aldo Gangemi [5] and Eva Blomqvist with Kurt Sandkuhl [17].

The main catalog of ontology design patterns is presented on the portal of the Association for Ontology Design & Patterns (ODPA) [8], created within the NeOn project [18]. Within the framework of this project, the typology of the patterns presented in Fig. 1 was proposed [19].

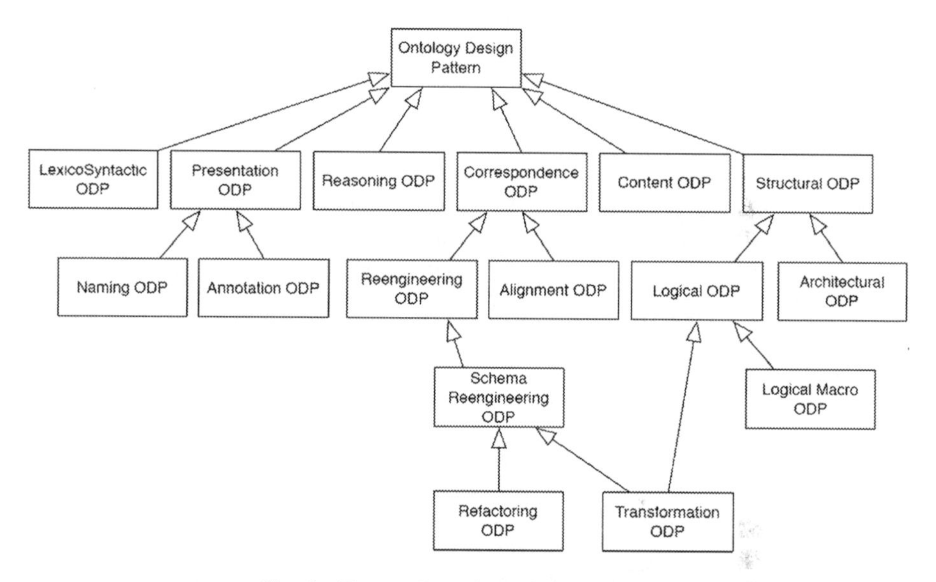

Fig. 1. Types of ontology design patterns.

Depending on the problems for which ontology design patterns are designed, we distinguish between structural patterns, correspondence patterns, content patterns, reasoning patterns, presentation patterns and lexico-syntactic patterns [20].

It should be noted that currently there is no unique standard for the description of patterns [21]. As a rule, however, they are described in the format proposed on the portal of the ODPA association [8]. This pattern description schema includes a graphical representation, text description, a set of scenarios and examples of use, links to other patterns in which it is used, as well as general information about the pattern name, its author and application area. According to the eXtreme Design methodology [22], each content pattern is also supplied with a set of competency (qualification) questions that determine its content.

2.1 Structural Patterns

Structural patterns either fix the ways to solve problems caused by the limitations of the expressive capabilities of ontology description languages or specify the general structure and overall shape of an ontology. Patterns of the first type are called logical patterns, and patterns of the second type are called architectural patterns. They contain proposals for ontology organization in general, including, for example, structures such as taxonomy and modular architecture.

A most acute problem of using the OWL language for constructing ontologies is that OWL can only provide a representation of simple entities, while ontologies need to represent complex concepts and relations. To solve this problem, structural logical patterns are used. These include, for example, patterns solving the problems of representing n-ary relations, relations of partonomy, lists, trees, etc.

Note that structural patterns are subject-independent; they can serve as a basis for constructing the fragments of ontology that are part of the content patterns.

2.2 Reasoning Patterns

The reasoning patterns are built on the basis of structural logical patterns and are designed to produce specific results using a logical inference machine. Such patterns make it possible to provide not only the inference of implicit knowledge in the ontology (patterns of classification, categorization, inheritance, etc.), but also information about the state of ontology, execution of queries to ontology, and its evaluation and normalization (elimination of class anonymity and their instances, an explicit representation (reification) of the class hierarchy, normalization of names, etc.) [23].

As noted in [24], such patterns can be applied to searching, viewing, filtering, integrating, and personalizing ontology elements encountered in Semantic Web applications.

2.3 Content Patterns

The content patterns define the ways of representing the typical fragments of ontology, based on which ontologies of a whole class of subject domains can be built.

In the ODPA association catalog, different kinds of basic patterns involved in many subject areas, such as *Person*, *EventCore*, *Action*, *Participation*, *TimeInterval*, *Location*, *SpatioTemporalExtent*, *Situation*, and *Trajectory*, are represented as content patterns. They can be specialized for specific subject areas.

Based on these patterns, we can describe more complex (composite) content patterns, supplementing them with missing components and building additional relations between the main classes representing the pattern.

The content patterns also include patterns defining structures for the representation of different types of relationships between ontology elements, such as *Classification*, *Collection*, *Set*, *Bag* (a set with repeating elements), *List*, *PartOf*, *Componency* (a special kind of "part of" relation), etc.

To build and use patterns, the ODPA community develops the methodology of extreme ontology design (eXtreme Design methodology, XD) [22], developed within the framework of the NeOn project, whose main principles are iterativity, involvement of the customer in ontology building, and ontology development management based on the requirements. According to this methodology, each content pattern should be provided with a set of competency questions defining its content, as well as a set of contextual statements and reasoning requirements necessary for the implementation of competency questions. Based on these questions, an ontology can be tested by building SPARQL queries to the ontology.

Structural and content patterns are the most popular types of patterns which can be used to describe the fragments of a domain ontology. Unlike ontology repositories, catalogs of patterns do not provide collections of ready-made ontologies, but sets of proposed solutions (i.e., well documented ontology fragments) for ontology design, which can be repeatedly used by developers to create their own ontologies in various subject areas.

2.4 Presentation Patterns

The presentation patterns define recommendations for the naming, annotating and graphical representation of ontology elements so as to increase the ontology readability and usability. The paper [22] notes the importance of a meaningful naming of ontology elements for the purpose of its better understanding by the user, although it does not improve the technical capabilities of the ontology to answer questions and perform a logical inference.

Naming patterns include conventions on rules for naming the namespaces declared for ontologies, files, and ontological elements. For example, it is encouraged to use the basic URIs of the organization that publishes an ontology for constructing the namespace, capital letters and the singular in the name of a class, and the suffix of the name of the parent class in the names of its subclasses.

The annotation patterns include annotation rules for ontology elements, such as providing classes and their properties with comments (*rdfs: comment*) and labels (*rdfs: label*) in several languages.

2.5 Correspondence Patterns

Correspondence patterns are required to perform the reengineering (transformation) or alignment (mapping) of ontologies. The first group of patterns is used when we need to build a new ontology, and the initial model is not necessarily ontological. The second group of patterns is used to determine correspondence between the concepts and individuals of two ontologies [25] so as to ensure interoperability without modifying existing models.

2.6 Lexico-Syntactic Patterns

Lexico-syntactic patterns are used to facilitate the construction (completion) of ontologies based on texts in a natural language. They set the mapping of language structures into ontological structures.

The idea of this type of patterns is not new; it is based on the concept of lexico-syntactic patterns of language constructs for the automatic extraction of linguistic units from the text [26]. Elements of lexico-syntactic patterns can be groups of words and word combinations that correspond to ontological constructions defined in both the ontology description language and structural (logical) patterns.

3 Patterns of the Ontology Constructing Methodology for Thematic ISIRs

In this section, we will consider the ontology design patterns used in the methodology of building ontologies for thematic intellectual scientific Internet resources (ISIR) [13]. This methodology, developed with the participation of the authors, uses the Semantic Web technology tools [27]. In particular, ontologies within this methodology are developed in the OWL language [28] using the Protégé editor [29]. These tools help solve many ontological engineering problems, including ontology validation and reuse, but their use, in turn, creates new problems.

3.1 Structural Logical Patterns

The use of structural logical patterns in the methodology of building ontologies for thematic ISIRs became necessary because the OWL lacked expressive means for representing complex entities and constructions needed for building the ontology of thematic ISIRs, in particular, the areas of admissible values, and n-ary and attributed relations (binary relations with attributes).

First, let us consider the pattern of representing the range of admissible property values. This pattern was introduced because of the absence in the OWL of special tools for specifying the ranges of values called domains in the relational data model and characterized by a name and a set of atomic values. Domains are convenient to use in the descriptions of possible values of properties of a class, when the entire set of such values is known in advance. Using domains will not only allow us to control the input of information; it can also make this operation more convenient by providing users with the opportunity to select property values from a given list of values.

The solution to this problem is to define the domain by using an enumerated class, a descendant of a specially introduced *Domain* class. Each specific domain does not have descendants and consists of a finite set of different individuals (objects or instances of the class) that determine the possible values of a particular property (*ObjectProperty*) of the objects of the class in question (see Fig. 2).

Examples of such domains are "Geographic type", "Position", "Type of organization", "Type of publication", which include, respectively, the types of settlements, positions, organizations and publications. (A description of the "Type of organization" domain is shown at the bottom of Fig. 2).

Note that in the figures of patterns shown in the paper, classes are designated in the form of ellipses, and their individuals and attributes are represented in the form of rectangles. An *ObjectProperty* type connection is shown by a solid straight line, and a *DataProperty* type connection is shown by a dash line. At the same time, obligatory classes, attributes and individuals are represented by figures surrounded by a thick line.

Another common problem in the development of an ontology is the need to present the attributed relations between two objects. For these purposes, as a rule, ordinary binary relations provided with attributes that specialize the connection between the arguments of the relation [30] are used. Since the OWL language does not provide the

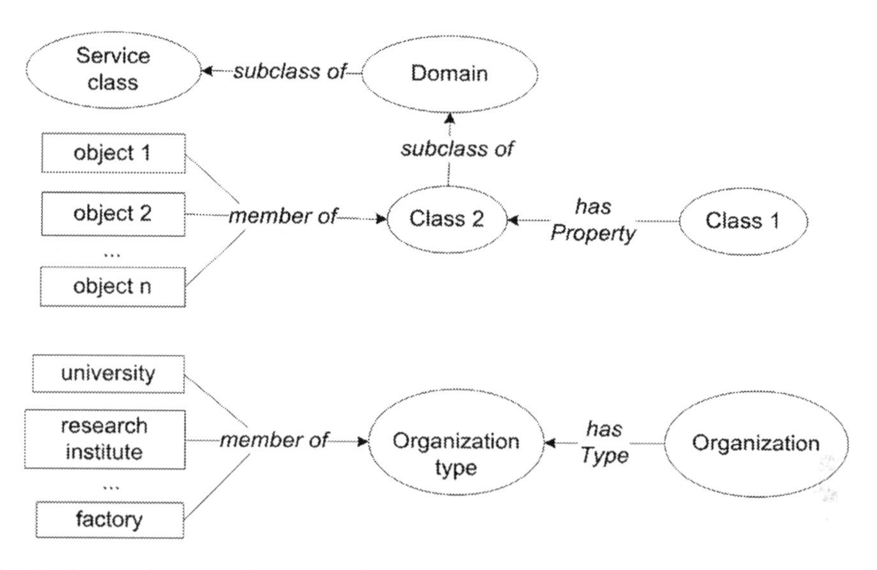

Fig. 2. Structural pattern of representation of the range of admissible values and an example of its use.

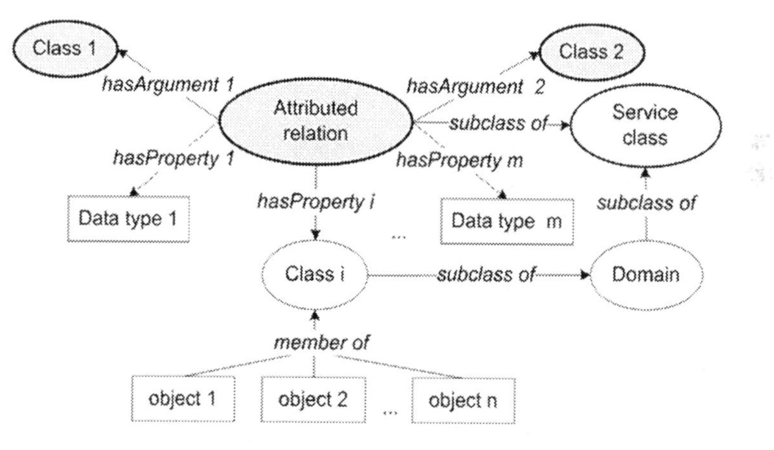

Fig. 3. Structural pattern of the binary attributed relation.

possibility to specify attributes for a relation, a structural pattern that provides for the introduction of the service class *Attributed relation* has been proposed (see Fig. 3).

To represent a specific relation type with attributes, a new class, which is its successor, is defined. An instance of this class is associated with each argument and attribute of the attributed relationship. In this case, it is required to set constraints on the obligatoriness and uniqueness of the arguments, while the restrictions on the number of attributes (properties) are not specified.

Note that this pattern, in contrast to the Qualified Relation pattern introduced in [30] allows specifying explicitly the order of the arguments of the attributed relation, preserving information about its orientation, which is important for providing the user with complete information about the nature of the relation between objects.

Figure 4 shows an example of using this pattern to specify a relation describing a person's participation (*Person* class) in an activity (*Activity* class). The pattern allows us to specify the start and end dates of the person's participation in an activity, as well as his/her role in it.

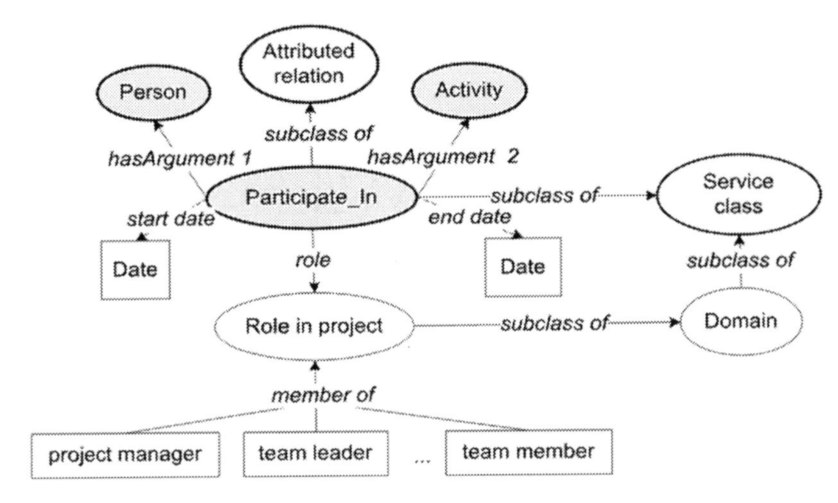

Fig. 4. The pattern of the binary attributed relation "participates".

Similarly, we can build a pattern for an n-ary relation. In this case, we have to specify the order of the arguments of the relation in addition to the properties of the obligatoriness and uniqueness of its arguments (see Fig. 5).

Let us remark that in contrast to the structural pattern of the n-ary relation presented in [31] and the content pattern *Situation* presented in [32], which is proposed as a basis for the description of n-ary relations, in the pattern proposed in this work we can specify its properties (*ObjectProperty*) and attributes (*DataProperty*), in addition to the arguments of the relation and their order. This greatly enhances the expressive possibilities of this pattern.

Structural patterns are subject-independent; due to this, they can serve as a basis for specifying ontology elements for content patterns.

3.2 Content Patterns

As mentioned above, content patterns define the ways of representing typical ontology fragments that can underlie the construction of the ontologies of modeled subject domains. In fact, the content patterns proposed in the paper are fragments of the base ontologies provided by the above-mentioned methodology of building ontology for thematic ISIRs. After the specialization of the concepts contained in these fragments and expansion of new concepts, these fragments become constituent parts of the ontologies of specific subject areas.

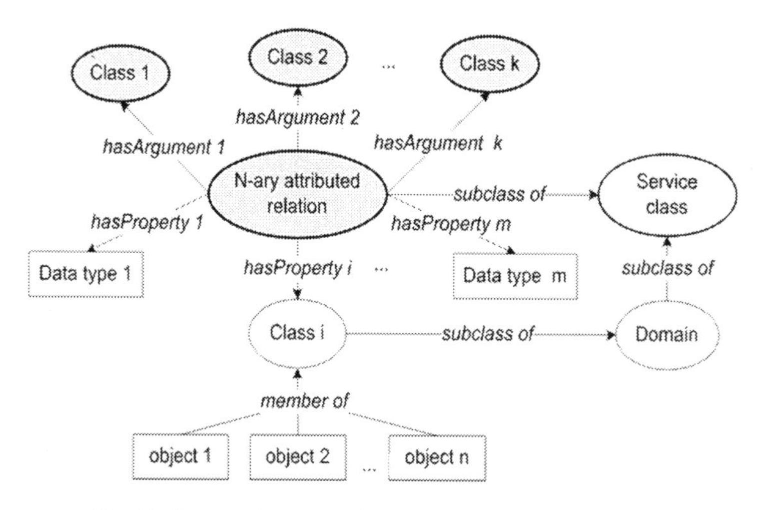

Fig. 5. Structural pattern of an n-ary relation with attributes.

The ontology of the ISIR domain is built on the basis of the following base ontologies: the ontologies of scientific knowledge and research activity, base ontology of tasks and methods, and base ontology of scientific information resources [13].

The ontology of scientific knowledge contains classes that define structures for describing concepts falling into any scientific field of knowledge. Such concepts are *Division of Science*, *Object of Research*, *Subject of Research*, *Research Method*, *Scientific Result*, etc. Using these classes, it is possible to identify and describe the divisions and subdivisions important for the modeled area of knowledge, to specify the typification of methods and objects of research, and to describe the results of scientific activity.

The ontology of research activity is based on the ontology proposed in [33], which is designed to describe research projects. This ontology includes classes of concepts related to the organization of scientific and research activity, such as *Person* (*Researcher*), *Organization*, *Event*, *Activity*, *Project*, *Publication*, etc. This ontology also includes relations that allow us to connect notions of this ontology not only among themselves, but also with the concepts of the ontology of scientific knowledge.

The base ontology of scientific information resources includes the *Information Resource* class as the main class, since this concept plays an important role in any scientific area. The set of attributes and relations of this class is based on the Dublin core standard [34]. Its attributes are *resource name*, *resource language*, *resource topic*, *resource type*, *date of resource creation*, etc. To represent information about resource sources and its creators, as well as related events, organizations, persons, publications and other entities, the ontology contains special relations linking the *Information Resource* class with the classes of other base ontologies.

The base ontology of tasks and methods includes classes such as *Task*, *Solution Method*, and *Web Service*. The concepts and relations of this ontology are meant to describe the tasks which are to be solved using the ISIR, methods for solving these tasks, and web services implementing these methods.

Building a consistent description of scientific subject areas depends on the ability to uniformly represent the concepts used in them. For this purpose, patterns representing the basic concepts and relations of base ontologies were developed. Let us show how patterns for describing the ontology of scientific knowledge look.

The pattern presented in Fig. 6 is intended to describe the research methods used in scientific activity.

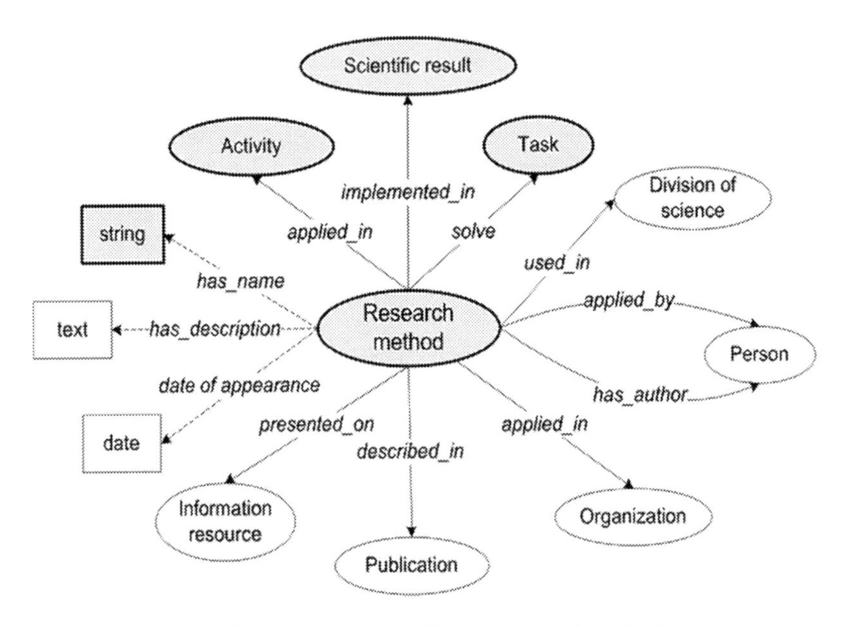

Fig. 6. Pattern for describing the research method.

Elements of the description of the pattern of the research method are represented by such obligatory classes of ontology as *Activity*, *Scientific Result*, *Task* and others, and such relations as *used_in*, *implemented_in*, *solve*, etc.

Let us give a set of competency questions representing the content of this pattern:

- What objects is the method applied to?
- In what activity is the method used?
- What tasks are solved using the method?
- In what divisions of science is the method used?
- In what scientific results is the method implemented?
- Who is the author of the method?
- Who applies the method?
- Who develops the method?
- In what organizations is the method applied?
- In what publications is the method described?
- On what resources is the method presented?

Figure 7 shows an example of using the pattern considered above to describe the method of subdefinite calculations [35] proposed by A.S. Narinyani in 1982 and implemented in the UniCalc solver [36].

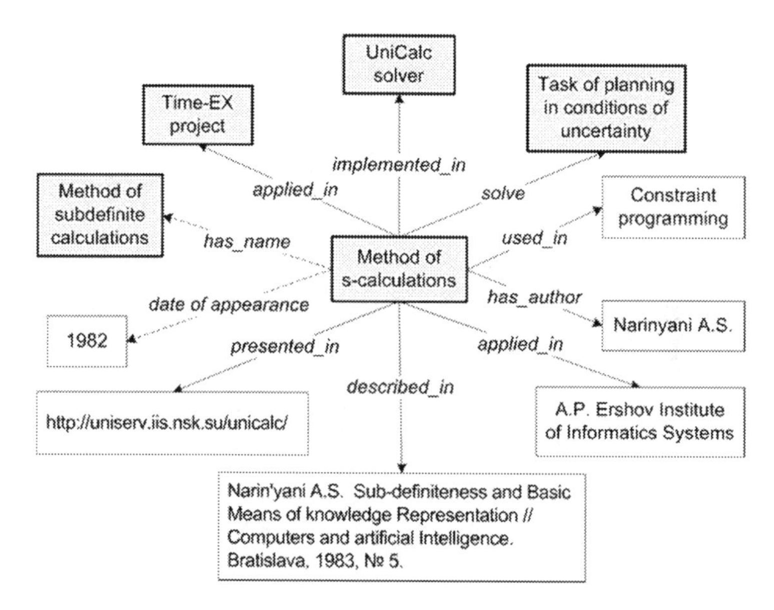

Fig. 7. Example of using the pattern for describing the research method.

Figure 8 shows a pattern for describing the object of research, which includes the following classes as obligatory: *Subject of Research*, *Activity*, and *Division of Science*. The instances of these classes must be connected with the object of research by relations such as *has_Aspect*, *investigated_In* and *studied_In*, respectively. The object of research can be structural (i.e. include other objects of research).

The pattern of the subject of research must necessarily include a reference to the object of research, an aspect of which it is. The subject of research, as well as the object of research, can be structural (include other subjects of research).

The scientific results occupy an important place in the description of scientific activity. The pattern for describing the scientific result is shown in Fig. 9. This pattern sets the requirement that when describing the scientific result, a reference to the activity under which it was obtained should be made.

Note that in the patterns described above, not only "central" concepts of patterns, but also concepts from adjacent patterns are used. For example, in the pattern for describing the scientific result, in addition to the concept of *Scientific Result* such concepts as *Activity*, *Subject of Research*, and *Division of Science* are also used. This allows us to give a related description of the modeled area.

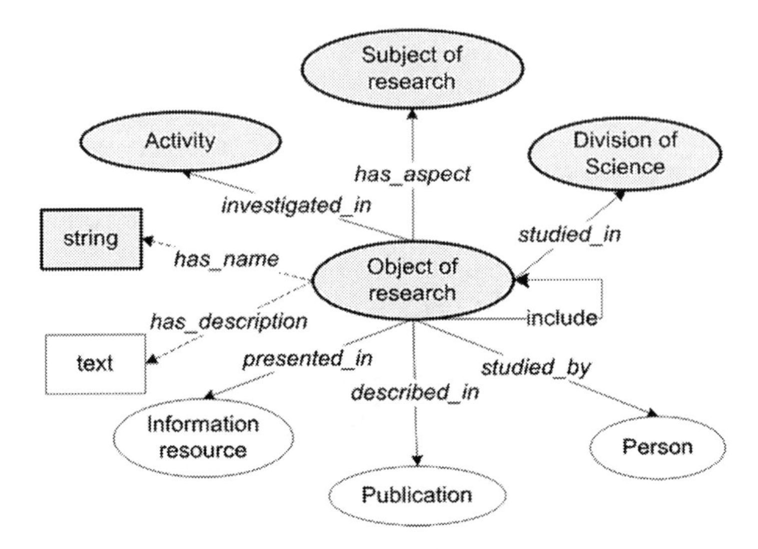

Fig. 8. Pattern for describing the object of research.

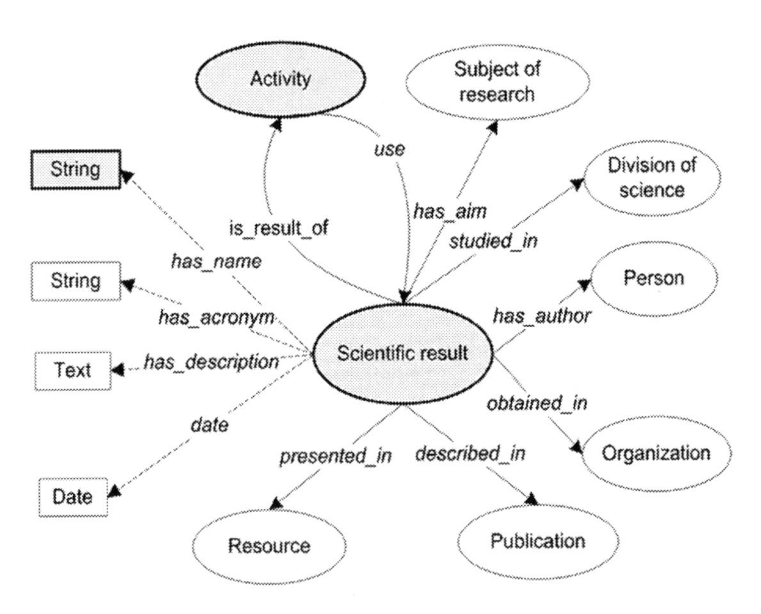

Fig. 9. Pattern for describing the scientific result.

3.3 Presentation Patterns

As mentioned above, presentation patterns define recommendations for naming, annotating, and graphical representation of ontology elements.

The technology of the development of thematic ISIR allows users to customize the display of information objects (instances of ontology class) included in the ISIR content on the monitor screen. To solve this problem, presentation patterns have been proposed. They allow us to fix the order in which the properties of information objects are displayed, specify the way of the dynamic naming of information objects using meaningful references to them, and store these settings directly in the ontology.

The desired order of showing the attributes of the information object is set for each property of its class using the annotation property (*owl: AnnotationProperty*) introduced especially for this purpose and called *order*. Its value is an integer.

When visualizing information from the ISIR content, we can represent information objects by full or short references. Full references are used when a list of objects of a given class is displayed; short ones are used to refer to one object within the description of another object. For creating full references, the *link* annotation property is introduced, for creating short links the *shortlink* annotation property is used. The values of these properties are also integers specifying the order in which the attributes (*DataProperty*) or relations (*ObjectProperty*) of the object enter into the full or short reference.

4 Conclusion

The paper discusses the questions of application of ontology design patterns for the development of ontologies of scientific subject areas. The classification of patterns proposed by the ODPA Association is presented, and the most important types of patterns from this classification are described. In addition, a detailed consideration of the patterns used by authors of the paper in the development of ontologies for a set of scientific subject areas is given.

Ontology design patterns serve to provide a uniform and consistent representation of all the entities of the ontology being developed. The use of ontology design patterns by experts and knowledge engineers helps to save resources and avoid errors in the development of ontologies.

The work was carried out with the partial financial support of the Russian Foundation for Basic Research (grant No. 16-07-00569).

References

1. Sharman, R., Kishore, R., Ramesh, R. (eds.): Ontologies: A Handbook of Principles, Concepts and Applications in Information Systems. Springer, New York (2007). https://doi.org/10.1007/978-0-387-37022-4
2. Zagorulko, Yu., Zagorulko, G.: A role of ontology in information systems for support of scientific and production activity. In: Fujita, H., Marik, V. (eds.) New Trends in Software Methodologies, Tools, and Techniques. Proceedings of the Eighth SoMeT_09, pp. 413–427. IOS Press, Amsterdam (2009)
3. Zagorulko, Yu., Zagorulko, G.: Ontology-based approach to development of the decision support system for oil-and-gas production enterprise. In: Fujita, H. (eds.) New Trends in Software Methodologies, Tools and Techniques. Proceedings of the 9th SoMeT_10, pp. 457–466. IOS Press, Amsterdam (2010)

4. Skvortsov, N., Kalinichenko, L., Kovalev, D.: Conceptual modeling of subject domains in data intensive research. In: Kalinichenko, L., Manolopoulos, Y., Kuznetsov, S. (eds.) Selected Papers of the XVIII International Conference on Data Analytics and Management in Data Intensive Domains (DAMDID/RCDL 2016). CEUR Workshop Proceedings, vol. 1752, pp. 7–15. CEUR-WS.org (2016). http://ceur-ws.org/Vol-1752/paper03.pdf. Accessed 1 Jan 2018
5. Gangemi, A.: Ontology design patterns for semantic web content. In: Gil, Y., Motta, E., Benjamins, V.R., Musen, M.A. (eds.) ISWC 2005. LNCS, vol. 3729, pp. 262–276. Springer, Heidelberg (2005). https://doi.org/10.1007/11574620_21
6. Hitzler, P., Gangemi, A., Janowicz, K., Krisnadhi, A., Presutti, V., (eds.): Ontology Engineering with Ontology Design Patterns: Foundations and Applications. Studies on the Semantic Web, vol. 25, IOS Press/AKA, Amsterdam (2016)
7. Ontology Design Patterns (ODPs) Public Catalog (2018). http://odps.sourceforge.net. Accessed 1 Jan 2018
8. Association for Ontology Design & Patterns (2018). http://ontologydesignpatterns.org. Accessed 1 Jan 2018
9. Borovikova, O., Globa, L., Novogrudska, R., Ternovoy, M., Zagorulko, G., Zagorulko, Yu.: Methodology for knowledge portals development: background, foundations, experience of application, problems and prospects. Jt. NCC&IIS Bull. Ser. Comput. Sci. **34**, 73–92 (2012). Accessed 1 Jan 2018
10. Braginskaya, L., Kovalevsky, V., Grigoryuk, A., Zagorulko, G.: Ontological approach to information support of investigations in active seismology. In: The 2nd Russian-Pacific Conference on Computer Technology and Applications (RPC), Vladivostok, Russky Island, Russia, 25–29 September, 2017, pp. 27–29. IEEE Xplore digital library (2017). http://ieeexplore.ieee.org/document/8168060. Accessed 1 Oct 2018
11. Zagorulko, Yu., Borovikova, O., Zagorulko, G.: Knowledge portal on computational linguistics: content-based multilingual access to linguistic information resources. In: Fujita, H., Sasaki, J. (eds.) Selected topics in Applied Computer Science. Proceedings of the 10th WSEAS International Conference on Applied Computer Science (ACS'10). Iwate Prefectural University, Japan, 4–6 October 2010, pp. 255–262. WSEAS Press (2010)
12. Zagorulko, Yu., Borovikova, O.: Technology of ontology building for knowledge portals on humanities. In: Wolff, K.E., Palchunov, D.E., Zagoruiko, N.G., Andelfinger, Urs (eds.) KONT/KPP -2007. LNCS (LNAI), vol. 6581, pp. 203–216. Springer, Heidelberg (2011). https://doi.org/10.1007/978-3-642-22140-8_13
13. Zagorulko, Yu., Zagorulko, G.: Ontology-based technology for development of intelligent scientific internet resources. In: Fujita, H., Guizzi, G. (eds.) SoMeT 2015. CCIS, vol. 532, pp. 227–241. Springer, Cham (2015). https://doi.org/10.1007/978-3-319-22689-7_17
14. Zagorulko, Yu., Borovikova, O., Zagorulko, G.: Application of ontology design patterns in the development of the ontologies of scientific subject domains. In: Kalinichenko, L., Manolopoulos, Y., Skvortsov, N., Sukhomlin, V. (eds.) Selected Papers of the XIX International Conference on Data Analytics and Management in Data Intensive Domains (DAMDID/RCDL 2017). CEUR Workshop Proceedings, vol. 2022, pp. 258–265. CEUR-WS.org (2017). http://ceur-ws.org/Vol-2022/paper42.pdf. Accessed 1 Jan 2018
15. Johnson, R., Vlissides, J., Helm, R.: Design Patterns: Elements of Reusable Object-Oriented Software by Erich Gamma. Addison-Wesley Professional, Boston (1994)
16. Hammar, K.: Towards an ontology design pattern quality model. Linköping Studies in Science and Technology. Thesis, vol. 1606. Linköping University Electronic Press, Linköping (2013)

17. Blomqvist, E., Sandkuhl, K.: Patterns in ontology engineering: classification of ontology patterns. In: Chen, C-S., Filipe, J., Seruca, I., Cordeiro, J. (eds.) Proceedings of the Seventh International Conference on Enterprise Information Systems, ICEIS 2005, Miami, USA, 25–28 May 2005, vol. 3, pp. 413–416. Springer, Netherlands (2005)
18. NeOn Project (2010). http://www.neon-project.org. Accessed 1 Jan 2018
19. Presutti, V., Gangemi, A., David, S., Aguado de Cea, G., Suárez-Figueroa, M.C., Montiel-Ponsoda, E., Poveda, M.: D2.5.1: a library of ontology design patterns: reusable solutions for collaborative design of networked ontologies. Technical report, NeOn Project (2007)
20. Gangemi, A., Presutti, V.: Ontology design patterns. In: Staab, S., Studer, R. (eds.) Handbook on Ontologies, International Handbooks on Information Systems, pp. 221–243. Springer, Heidelberg (2009). https://doi.org/10.1007/978-3-540-92673-3_10
21. Karima, N., Hammar, K., Hitzler, P.: How to document ontology design patterns. In: Hammar, K., Hitzler, P., Krisnadhi, A., Ławrynowicz, A., Nuzzolese, A., Solanki, M. (eds.) Advances in Ontology Design and Patterns, Studies on the Semantic Web, vol. 32, pp. 15–28. IOS Press, Amsterdam/AKA Verlag, Berlin (2017)
22. Blomqvist, E., Hammar, K., Presutti, V.: Engineering ontologies with patterns: the eXtreme design methodology. In: Hitzler, P., Gangemi, A., Janowicz, K., Krisnadhi, A., Presutti, V. (eds.) Ontology Engineering with Ontology Design Patterns, Studies on the Semantic Web, vol. 25, pp. 23–50. IOS Press, Amsterdam (2016)
23. Vrandečić, D., Sure, Y.: How to design better ontology metrics. In: Franconi, E., Kifer, M., May, W. (eds.) ESWC 2007. LNCS, vol. 4519, pp. 311–325. Springer, Heidelberg (2007). https://doi.org/10.1007/978-3-540-72667-8_23
24. Van Harmelen, F., Ten Teije, A., Wache, H.: Knowledge engineering rediscovered: towards reasoning patterns for the semantic web. In: The Fifth International Conference on Knowledge Capture, pp. 81–88. ACM, New York (2009)
25. Scharffe, F., Zamazal, O., Fensel, D.: Ontology alignment design patterns. Knowl. Inf. Syst. **40**(1), 1–28 (2014)
26. Maynard, D., Funk, A., Peters, W.: Using lexico-syntactic ontology design patterns for ontology creation and population. In: Proceedings of WOP2009 Collocated with ISWC2009, vol. 516, pp. 39–52. CEUR-WS.org (2009). http://ceur-ws.org/Vol-516/pap08.pdf. Accessed 1 Jan 2018
27. Hitzler, P., Krötzsch, V., Rudolph, S.: Foundations of Semantic Web Technologies. Chapman & Hall/CRC Press, Boca Raton (2009)
28. Antoniou, G., Harmelen, F.: Web ontology language: OWL. In: Staab, S., Studer, R. (eds.) Handbook on Ontologies, pp. 67–92. Springer, Heidelberg (2004). https://doi.org/10.1007/978-3-540-24750-0_4
29. Protege (2010). https://protege.stanford.edu. Accessed 1 Jan 2018
30. Dodds, L., Davis, I.: Linked Data Patterns (2012). http://patterns.dataincubator.org/book. Accessed 1 Jan 2018
31. Hoekstra, R.: Proceedings of the 2009 Conference on Ontology Representation: Design Patterns and Ontologies that Make Sense. Frontiers of Artificial Intelligence and Applications, vol. 197, IOS Press, Amsterdam (2009)
32. Situation Pattern (2010). http://ontologydesignpatterns.org/wiki/Submissions:Situation. Accessed 1 Jan 2018
33. Benjamins, V.R., Fensel, D.: Community is knowledge! in (KA)2. In: 11th Proceedings of Banff Knowledge Acquisition for Knowledge-Based Systems Workshop KAW'98, Banff, Canada, pp. KM.2-1–KM.2-18. SRDG Publications, Department of Computer Science, University of Calgary, Calgary, April 1998

34. DCMI Metadata Terms (2010). http://dublincore.org/documents/dcmi-terms. Accessed 1 Jan 2018
35. Narin'yani, A., Semenov, A., Babichev, A., Kashevarova, T., Leshchenko, A.: A new approach to solving algebraic systems by means of sub-definite models. In: Henry, J., Yvon, J.-P. (eds.) Proceedings of the 16th IFIP Conference on System Modelling and Optimization. Compiegne, France. July, 1993, LNCS, vol. 197, pp. 355–364. Springer, Berlin (1994)
36. Semenov, A.: Solving optimization problems with help of the uniCalc solver. In: Kearfott, R., Kreinovich, A. (eds.) Applications of Interval Computations, pp. 211–225. Kluwer Academic Publishers, Dordrecht (1996)

PROPheT – Ontology Population and Semantic Enrichment from Linked Data Sources

Marina Riga, Panagiotis Mitzias, Efstratios Kontopoulos(✉), and Ioannis Kompatsiaris

Information Technologies Institute, Thessaloniki, Greece
{mriga, pmitzias, skontopo, ikom}@iti.gr

Abstract. Ontologies are a rapidly emerging paradigm for knowledge representation, with a growing number of applications in various domains. However, populating ontologies with massive volumes of data is an extremely challenging task. The field of ontology population offers a wide array of approaches for populating ontologies in an automated or semi-automated way. Nevertheless, most of the related tools typically analyse natural language text, while sources of more structured information like Linked Open Data would arguably be more appropriate. The paper presents PROPheT, a novel software tool for ontology population and enrichment. PROPheT can populate a local ontology model with instances retrieved from diverse Linked Data sources served by SPARQL endpoints. To the best of our knowledge, no existing tool can offer PROPheT's diverse extent of functionality.

Keywords: Ontologies · Ontology population · Semantic enrichment Linked Data · DBpedia

1 Introduction

Ontologies constitute a knowledge representation paradigm for modelling domains, concepts and interrelations in a structured, uniform and effective way, enabling the sharing of information between different systems [1]. The rapidly emerging popularity of ontologies has led to their deployment in various domains, like bioinformatics [2], e-commerce [3] and digital libraries [4]. Nevertheless, in order for ontologies to be more efficiently used at an enterprise level, massive volumes of data are required for populating the underlying models.

If performed manually, this task is extremely time-consuming and potentially error-prone. *Ontology population* attempts to alleviate this problem, by introducing methods and tools for automatically augmenting an ontology with instances of concepts and properties that represent real data/objects [5]. The schema of the ontology itself is not altered but only the realisation of its set of concepts and the asserted relations on the newly introduced instances. This process is part of *ontology learning*, which refers to the automatic (or semi-automatic) construction, enrichment and adaptation of ontologies [6].

L. Kalinichenko et al. (Eds.): DAMDID/RCDL 2017, CCIS 822, pp. 157–168, 2018.
https://doi.org/10.1007/978-3-319-96553-6_12

The vast majority of ontology population tools and methodologies are aimed at textual input, typically extracting knowledge from natural language text [7]. Nevertheless, the unstructured nature of free text drastically increases the efforts for utilising its content in already structured frameworks. Instead, other more structured sources of information could be used alternatively; such an example is *Linked Open Data* (*LOD*, or often referred to simply as *Linked Data*) [8], which builds upon established Web technologies and is a standard for publishing interlinked structured data that are capable also of responding to semantic queries. Linked Data are formalised using controlled vocabulary terms based on ontologies and can be publicly accessible via a SPARQL endpoint [9]. A popular Linked Dataset is *DBpedia*[1], the Linked Data version of Wikipedia.

This paper argues that the rapidly increasing array of published Linked Datasets [10] can serve as the input for scalable ontology population and presents *PROPheT* (*PERICLES*[2] *Ontology Population Tool*), a novel software tool for user-driven ontology population from Linked Data sources. The tool is domain-agnostic and can efficiently handle vast volumes of input data. Upon user request, PROPheT locates realisations of concepts in Linked Data sources and appropriately inserts them into a local schema, preserving its initial structure and semantic representations. To the best of our knowledge, no existing tool can offer PROPheT's extent of functionality.

The work presented here constitutes an extension to previous work of ours [11], including the following new materials over the previous paper:

- An extended related work section, now featuring a number of approaches for ontology population from DBpedia (Sect. 2);
- A more thorough account of PROPheT's technical implementation details: background processes performed through PROPheT's available population mechanisms, typical SPARQL queries submitted to endpoints, etc. (Sect. 3.2);
- Two revised use cases, demonstrating the tool's functionality in diverse scenarios (Sect. 4).

The rest of the paper is structured as follows: Sect. 2 gives an overview of related work approaches. Section 3 presents PROPheT's functionalities and operational workflow, followed by two illustrative use cases that demonstrate the tool's versatility and scalability in Sect. 4. Section 5 reports on evaluating PROPheT, and the paper is concluded with final remarks and directions for future work.

2 Related Work

Ontology population has already been deployed in various domains, like e.g. e-tourism [12], web services [13] and clinical data [14], amongst others. Another recent work deploys ontology population in a Big Data setting [15], indicating a potentially emerging interest in the area. Overall, state-of-the-art ontology population approaches

[1] http://wiki.dbpedia.org/.

[2] The tool has been developed in the context of the PERICLES FP7 project: http://www.pericles-project.eu/.

are mainly addressed to extracting and retrieving possible instances from natural language text, like e.g. product catalogues or university homepages, corpora, other Web sources, etc., and typically involve machine learning, text mining and natural language processing techniques. Other representative approaches besides the ones discussed above are presented in [5, 7].

Another, albeit less popular, direction of ontology population research is aimed at retrieving instances from other types of input, like e.g. CAD files [16], or more structured content, like e.g. spreadsheets [17, 18], and XML files [19].

Regarding ontology population from DBpedia, a recent attempt is presented in [20], where the authors manually map a local ontology to DBpedia classes and run a series of SPARQL queries that retrieve the respective instances; no details are given regarding the specifics of the population process. A similar approach for semantic annotation of news items is presented in [21], while the authors in [22] present a methodology for ontology enrichment based on input from DBpedia and (the now obsolete) Schema.org[3].

PROPheT's similarity to these approaches lies in the use of a LOD source as input for ontology population and enrichment. In this sense, PROPheT could easily be used as the underlying ontology population tool in [20–22]. Nevertheless, no other ontology population tool can currently instantiate new concepts from a LOD source so flexibly, regardless the domain of interest or the content of the source. PROPheT can handle any kind of LOD as an external knowledge source for extracting concepts of interest and for populating them to corresponding resources into the domain ontology.

3 The PROPheT Ontology Population Tool

PROPheT[4] is a novel software tool for ontology population and semantic enrichment that can retrieve instantiations of concepts from Linked Data sources. In this sense, the tool is fully domain-independent and capable to operate with any OWL ontology and any RDF LOD dataset served via a SPARQL endpoint. The retrieved instances are filtered by the user and are then inserted, together with their accompanied/selected properties and values, into a target ontology. As described in the following subsections, PROPheT provides various modes of instance retrieval, and allows establishing user-defined mappings of the respective properties. Through its step-by-step wizard-based interaction mode, the tool is extremely easy to use even by unfamiliarised users.

3.1 Technical Infrastructure

PROPheT's front-end (see Fig. 1) is implemented in Python along with PyQt[5], while specialised Python APIs (RDFLib[6], SPARQLWrapper[7]) are deployed for handling

[3] http://schema.org/.

[4] PROPheT is available at: http://mklab.iti.gr/project/prophet-ontology-populator.

[5] https://riverbankcomputing.com/software/pyqt.

[6] https://github.com/RDFLib/rdflib.

[7] https://github.com/RDFLib/sparqlwrapper.

local and remote ontologies. An SQLite database was also set up in the back-end for storing dynamic data (e.g. settings, user preferences) that are created during the tool's operation.

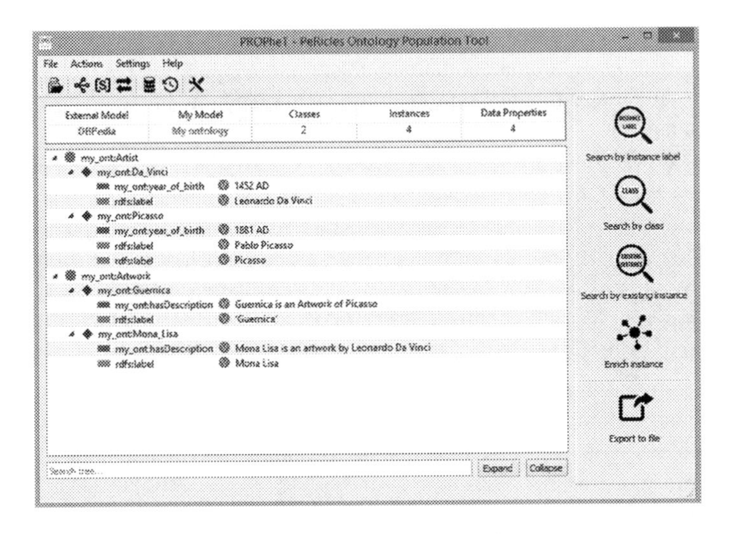

Fig. 1. PROPheT's main window.

3.2 Ontology Population

PROPheT offers the capability of class-based and instance-based ontology population. *Class-based population* retrieves instances from an external source, based on a given class name[8], and inserts them into a local ontology. PROPheT submits appropriate SPARQL queries to the remote endpoint in order to first retrieve a result set of instances belonging to the specified class (declared with a unique *classURI* value in Table 1), and then to derive additional info for each instance, such as its label(s) (`rdfs:label`), data properties and related values defined in the remote ontology. The total number of fetched instances can be bound according to a maximum number of results (*query_limit*) specified by the user. The user may then select the instances to populate under an existing class in the local ontology.

The second method, *instance-based population*, has two different modes:

1. Retrieval instances based on their `rdfs:label` property value, where the match of the retrieved instances is based on specific parameters. More specifically, the user may input the corresponding label field to search for instances defined in the external source, together with additional search options, such as: (i) the exact or partial match (i.e. contains term) of typed text with the label of retrieved instance(s), (ii) the exact match of language code[9] selected by the user with that specified in

[8] E.g. `dbo:Artist` for the DBpedia class representing artists.

[9] Language code can be a two-letter abbreviation as standardized by ISO 639-1. Selection is feasible in PROPheT via the provided list box.

label(s) of retrieved instance(s), and (iii) the ability for the search execution to be performed as case sensitivity or insensitive. Detailed examples of corresponding SPARQL queries are presented in Table 2. Retrieved results can be of any class (`rdf:type`), thus, the user may select any of the derived instances to be populated under a specific class in the local ontology.

Table 1. An example SPARQL query submitted to the endpoint during class-based population.

```
PREFIX rdf:<http://www.w3.org/1999/02/22-rdf-syntax-ns#>
SELECT ?instances WHERE {
    ?instances rdf:type ?classURI
}
ORDER BY ?instances
LIMIT ?query_limit
```

2. Retrieval based on instances similar to an existing instance. More specifically, PROPheT detects the classes in the remote ontology that include an instance with a similar `rdfs:label` property value (exact match) with the input instance. The user may then select specific classes and choose which instances to import into the local ontology, following a similar approach as class-based population but performed for multiple classes results simultaneously.

In all the above cases, after the set of preferred instances has been selected by the user to be populated into the ontology, PROPheT launches the ontology mapping process described in Subsect. 3.4.

3.3 Instance Enrichment

PROPheT also offers the option of semantically enriching instances already existing in the local ontology with properties and values from similar instances in remote ontologies, i.e. instances with similar labels. The similar instances may belong to one or more different classes in the remote ontology, thus, the tool presents the user with the `rdf:type` of each instance. Based on the content and semantics of the derived instances, the user may then decide which property-value pairs he/she will insert from the remote into the local ontology.

3.4 Ontology Mapping

In order for PROPheT to proceed with populating the ontology with the selected instances, the available properties of the retrieved instances have to be selected and mapped to properties defined in the local model. PROPheT displays a list of all datatype properties for the selected instances, so that the user can define suitable mappings to datatype properties already existing in the local ontology; for example, mapping the retrieved property `dbo:birthDate` to the local property

Table 2. Example SPARQL queries submitted to the endpoint during instance-based population, when performing searching by instance label.

Case A. Exact match and language code specified.

```
PREFIX rdf:<http://www.w3.org/1999/02/22-rdf-syntax-ns#>
PREFIX rdfs:<http://www.w3.org/2000/01/rdf-schema#>
SELECT DISTINCT ?instance WHERE {
    ?instance rdf:type ?class .
    ?instance rdfs:label 'example_label'@language_code . }
```

Case B. Contains word(s), case insensitive and language code not specified.

```
PREFIX rdf:<http://www.w3.org/1999/02/22-rdf-syntax-ns#>
PREFIX rdfs:<http://www.w3.org/2000/01/rdf-schema#>
SELECT DISTINCT ?instance WHERE {
    ?instance rdf:type ?class .
    ?instance rdfs:label ?label .
    FILTER regex (?label, 'example_label', 'i') }
```

Case C. Contains word(s), case sensitive and language code specified.

```
PREFIX rdf:<http://www.w3.org/1999/02/22-rdf-syntax-ns#>
PREFIX rdfs:<http://www.w3.org/2000/01/rdf-schema#>
SELECT DISTINCT ?instance WHERE {
    ?instance rdf:type ?class .
    ?instance rdfs:label ?label .
    FILTER regex (?label, 'example_label') .
    FILTER langMatches( lang(?label), 'language_code' ). }
```

`ex:dateOfBirth`. PROPheT stores the mappings in a linked SQLite database and offers suggestions when the same mappings occur again in future occasions.

When ontology mapping is finalised, the instances and their related properties and values can directly be populated as new triples in the local ontology. A relevant SPARQL query is submitted to the endpoint for fetching the property values of the selected properties and thus for populating each corresponding "property-value" pair to the local ontology.

3.5 Semantic Enrichment

The local model may also be semantically enriched by establishing links between properties in the local and the remote ontologies via `owl:equivalentProperty` declarations added into the local model. Similar links between instances are represented via `owl:sameAs` and `rdfs:seeAlso` declarations added to the local ontology, preserving this way the origin and relation between populated instances and those from LOD sources.

4 Use Cases

This section presents two use case scenarios: the former demonstrates PROPheT's versatility in performing ontology population and semantic enrichment from diverse sources, while the latter illustrates the tool's scalability in data-intensive domains.

4.1 Use Case 1: Ontology Population from Different LOD Sources

Suppose that Alice, an avid movie enthusiast, has developed an ontology of actors and films and wishes to initially populate it with an instance of the movie "*The Godfather*" retrieved from LinkedMDB[10]. She loads her model in PROPheT and registers LinkedMDB as the current source. Since the name of the movie is specified a priori, she searches for existing instances through the "*Search by Instance Label*" method. One result is retrieved[11] and Alice adds this instance to her local ontology.

She then wishes to retrieve additional information on the specific movie from another LOD source, DBpedia. Through PROPheT's "*Enrich existing instance*" function, Alice retrieves a set of instances that may belong to different classes, but they all share the same *rdfs:label* with the newly populated instance. At this point, she may select any pairs of datatype properties/values she wants to add to her local instance of "*The Godfather*" movie. After manually mapping the relevant pairs of properties, the data is inserted into the corresponding fields in Alice's ontology.

In case the user wishes to further populate her model with similar resources, she can employ PROPheT's methods "*Search by Class*" or "*Search by Existing Instance*" for any LOD endpoint. For instance, if the former method is selected, Alice should type e.g. *dbo:Film* for DBpedia, or *movie:film* for LinkedMDB. A set of instances will be retrieved, and Alice may then proceed with the selection and mapping process as described previously. If, on the other hand, "*Search by Existing Instance*" is selected, PROPheT will search for alternative classes that contain instances with the same label. Alice can now select one or more classes from which instances will be retrieved and proceed with the selection of instances to be populated (see Fig. 2).

4.2 Use Case 2: Ontology Population in a Data-Intensive Domain

Bob, an employee at a government institution monitoring pollution in rural environments, wishes to create a directory of cities and towns worldwide, including related information, such as population, postal codes, etc., along with the respective pollution levels. Bob deploys a local ontology schema incorporating the necessary classes (e.g. `Town`, `City`, etc.) and properties (e.g. `hasPopulation`, `hasPostalCode`, etc.) and loads it into PROPheT. This ontology now needs to be populated with instances of cities and towns.

---

[10] http://www.linkedmdb.org/.

[11] The retrieved entry in LinkedMDB is http://data.linkedmdb.org/resource/film/43338.

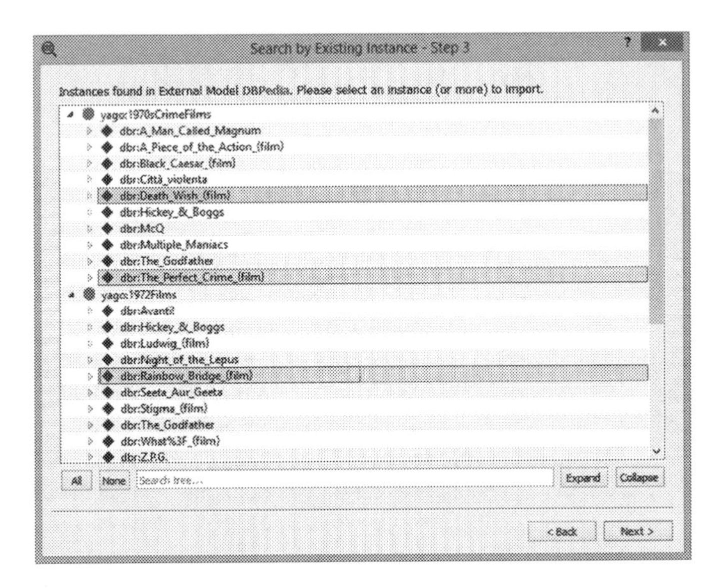

Fig. 2. Selection of movie instances from different classes.

Bob then registers the sources that serve the desired data. Two suitable candidates are *ENVO*[12] and *LinkedGeoData*[13]. Specifically, ENVO's class *City* (ENVO_00000856) and LinkedGeoData's classes *City* and *Town* contain relevant instances. Using PROPheT's class-based instance extraction wizard, Bob populates his ontology with 10 K instances from ENVO's *City* and 10 K instances from LinkedGeoData's *City*, along with an additional 10 K instances from LinkedGeoData's class *Town*. Data property values were also mapped and added. Table 3 displays the population times (in seconds) for the 30 K instances.

Table 3. Instance retrieval and population times.

Ontology	No of instances	Population time (sec)
LinkedGeoData	10,000	120
ENVO	10,000	204
LinkedGeoData	10,000	158

Finally, through PROPheT's "*Enrich Instance*" function, Bob can semantically enrich the major cities' instances (e.g. London, Paris, Amsterdam) with data from different endpoints regarding air pollution levels.

[12] http://www.obofoundry.org/ontology/envo.html.

[13] http://linkedgeodata.org.

5 PROPheT Evaluation

We conducted a user evaluation of the tool, which resulted in very encouraging conclusions by the participants, who distinguished the following aspects of the tool as the most positive ones: attractiveness (93.5%), user-friendliness (93.5%), ease of usage (100%), innovativeness (87.5%), and efficiency (93.5%); the numbers in parentheses correspond to the respective percentages indicating acceptance on behalf of the users. More information on the user evaluation is presented in [23].

Furthermore, we also conducted a qualitative evaluation of PROPheT, based on the criteria for ontology population tools proposed in [7]. The key findings of the evaluation are presented in Table 4, while more information is given in [11].

Table 4. PROPheT's qualitative evaluation.

Criterion	PROPheT's evaluation
Elements extracted	Objects and relations
Initial requirements	Availability of a local OWL ontology – no domain-dependant resources or specialised software is required
Learning approach	Step-by-step ontology population and enrichment; SPARQL querying of Linked Data endpoints
Degree of automation	Retrieval is automated; selection is user-driven, but highly user-friendly
Consistency maintenance	Integrated specialised APIs ensure consistency
Redundancy elimination	The same instance, i.e. those carrying the same URI, cannot be populated multiple times
Domain portability	Totally domain-agnostic
Corpora modality	Limited to LOD sources with a SPARQL endpoint

Finally, considering the fact that the availability and scalability of SPARQL endpoints serving Linked Data is not always guaranteed [9], and in order to demonstrate PROPheT's scalability, we experimented with timing the retrieval and population of instances from several well-known SPARQL endpoints into a local custom ontology model. Our findings are presented in more detail in [11].

6 Conclusions and Future Work

The paper argued that, with the rapidly emerging advent of the use of ontologies in various domains, the process of ontology population becomes increasingly relevant. Most proposed solutions are typically aimed at analysing natural language text, often overlooking other sources of more structured information, like e.g. Linked Data. In this context, we presented PROPheT, a domain independent software tool for ontology population and enrichment from Linked Data sources. Through wizard-based user-driven processes, the tool facilitates the automatic retrieval of instances and their insertion into a local OWL ontology, without the need for technical details of the applied queries in the Linked Data endpoints or of the SPARQL query language's syntax. An advanced mapping process enables the dynamic definition of matching classes and properties between source and target models.

The tool's rich functionality and versatility outweighs any other ontology population tool found in literature, making PROPheT a truly innovative system for populating and enriching ontologies in various domains where populating ontologies from diverse sources poses a formidable challenge. Indicative paradigms include cultural heritage [24][14], telecommunications and news [25], health and biomedicine [26, 27]. This was our main motivation for turning PROPheT into a truly domain-agnostic tool, capable of performing ontology population and enrichment from Linked Data sources in virtually any domain, data-intensive or not.

Nevertheless, there are still a few areas of improvement for the tool. In its current implementation, PROPheT is only limited to handling datatype and not object properties; the latter are significantly more complex to tackle. In this context, we are planning adopting the approach presented in [28]. Additionally, the ability of simultaneous querying in multiple selected endpoints or the handling of direct/indirect imports of ontologies would enrich the size and content correspondingly of the retrieved results, in one single query. A further improvement could be considering additional semantic enrichment associations, like e.g. `skos:narrower` and `skos:broader` from SKOS [29]. Finally, the process of suggesting similar instances or classes to the user during the population and enrichment steps could be suggested by the tool itself, according to appropriate similarity metrics.

Acknowledgements. This research received funding by the European Commission Seventh Framework Programme under Grant Agreement Number FP7-601138 PERICLES.

References

1. Uschold, M., Gruninger, M.: Ontologies: principles, methods and applications. Knowl. Eng. Rev. **11**(02), 93–136 (1996). https://doi.org/10.1017/S0269888900007797
2. Gene Ontology Consortium: The gene ontology (GO) database and informatics resource. Nucleic Acids Res. **32**(suppl 1), D258–D261 (2004). https://doi.org/10.1093/nar/gkh036
3. Hepp, M.: GoodRelations: an ontology for describing products and services offers on the web. In: Gangemi, A., Euzenat, J. (eds.) EKAW 2008. LNCS (LNAI), vol. 5268, pp. 329–346. Springer, Heidelberg (2008). https://doi.org/10.1007/978-3-540-87696-0_29
4. Buckingham Shum, S., Motta, E., Domingue, J.: ScholOnto: an ontology-based digital library server for research documents and discourse. Int. J. Digit. Libr. **3**(3), 237–248 (2000). https://doi.org/10.1007/s007990000034
5. Buitelaar, P., Cimiano, P.: Ontology learning and population: bridging the gap between text and knowledge, vol. 167. Ios Press, Amsterdam (2008)

[14] PROPheT is already being successfully deployed in the DigiArt EU-funded project on cultural heritage (http://digiart-project.eu/).

6. Maedche, A., Staab, S.: Ontology learning from the semantic web. IEEE Intell. Syst. **16**(2), 72–79 (2001). https://doi.org/10.1007/978-1-4615-0925-7
7. Petasis, G., Karkaletsis, V., Paliouras, G., Krithara, A., Zavitsanos, E.: Ontology population and enrichment: state of the art. In: Paliouras, G., Spyropoulos, Constantine D., Tsatsaronis, G. (eds.) Knowledge-Driven Multimedia Information Extraction and Ontology Evolution. LNCS (LNAI), vol. 6050, pp. 134–166. Springer, Heidelberg (2011). https://doi.org/10.1007/978-3-642-20795-2_6
8. Heath, T., Bizer, C.: Linked data: evolving the web into a global data space. Synth. Lect. Semant. Web Theory Technol. **1**(1), 1–136 (2011). https://doi.org/10.2200/s00334ed1v01y201102wbe001
9. Buil-Aranda, C., Hogan, A., Umbrich, J., Vandenbussche, P.-Y.: SPARQL Web-querying infrastructure: ready for action? In: Alani, H., et al. (eds.) ISWC 2013. LNCS, vol. 8219, pp. 277–293. Springer, Heidelberg (2013). https://doi.org/10.1007/978-3-642-41338-4_18
10. Abele, A., McCrae, J.P., Buitelaar, P., Jentzsch, A., Cyganiak, R.: Linking Open Data cloud diagram (2017). http://lod-cloud.net/. Accessed 01 Jan 2018
11. Kontopoulos, E., Mitzias, P., Riga, M., Kompatsiaris, I.: A domain-agnostic tool for scalable ontology population and enrichment from diverse linked data sources. In: Kalinichenko, L. A., Manolopoulos, Y., Skvortsov, N.A., Sukhomlin, V.A. (eds.) Selected Papers of the XIX International Conference on Data Analytics and Management in Data Intensive Domains (DAMDID/RCDL 2017), CEUR Workshop Proceedings 2022, pp. 184–190 (2017)
12. Ruiz-Martınez, J.M., Minarro-Giménez, J.A., Castellanos-Nieves, D., Garcıa-Sánchez, F., Valencia-Garcia, R.: Ontology population: an application for the e-tourism domain. Int. J. Innov. Comput. Inf. Control (IJICIC) **7**(11), 6115–6134 (2011)
13. Reyes-Ortiz, J.A., Bravo, M., Pablo, H.: Web services ontology population through text classification. In: Federated Conference on Computer Science and Information Systems (FedCSIS), pp. 491–495. IEEE (2016). https://doi.org/10.15439/2016f332
14. Mendes, D., Rodrigues, I.P., Baeta, C.F.: Development and population of an elaborate formal ontology for clinical practice knowledge representation. In: Proceedings of the International Conference on Knowledge Engineering and Ontology Development (2013). https://doi.org/10.5220/0004548602860292
15. Knoell, D., Atzmueller, M., Rieder, C., Scherer, K.P.: BISHOP-big data driven self-learning support for high-performance ontology population. In: Proceedings of the LWDA 2016 (FGWM Special Track), Potsdam, Germany, pp. 157–164. University of Potsdam (2016)
16. Häfner, P., Häfner, V., Wicaksono, H., Ovtcharova, J.: Semi-automated ontology population from building construction drawings. In: Proceedings of the International Conference on Knowledge Engineering and Ontology Development, pp. 379–386 (2013). https://doi.org/10.5220/0004626303790386
17. Han, L., Finin, T., Parr, C., Sachs, J., Joshi, A.: RDF123: from spreadsheets to RDF. In: Sheth, A., et al. (eds.) ISWC 2008. LNCS, vol. 5318, pp. 451–466. Springer, Heidelberg (2008). https://doi.org/10.1007/978-3-540-88564-1_29
18. Jupp, S., Horridge, M., Iannone, L., Klein, J., Owen, S., Schanstra, J., et al.: Populous: a tool for building OWL ontologies from templates. BMC Bioinf. **13**(Suppl 1), S5 (2012). https://doi.org/10.1186/1471-2105-13-s1-s5
19. Araujo, C., Henriques, P.R., Martini, R.G.: Automatizing ontology population to drive the navigation on virtual learning spaces. In: 12th Iberian Conference on Information Systems and Technologies (CISTI), pp. 1–6 (2017). https://doi.org/10.23919/cisti.2017.7975754
20. Gavankar, C., Kulkarni, A., Fang Li, Y., Ramakrishnan, G.: Enriching an academic knowledge base using linked open data. In: 24th International Conference on Computational Linguistics, p. 51 (2012)

21. Agarwal, S., Singhal, A.: Autonomous ontology population from DBpedia based on context sensitive entity recognition. In: Fourth International Joint Conference on Advances in Engineering and Technology, AET, pp. 580–589 (2013). https://doi.org/10.1145/2797115.2797127
22. Tiddi, I., Mustapha, N.B., Vanrompay, Y., Aufaure, M.-A.: Ontology Learning from Open Linked Data and Web Snippets. In: Herrero, P., Panetto, H., Meersman, R., Dillon, T. (eds.) OTM 2012. LNCS, vol. 7567, pp. 434–443. Springer, Heidelberg (2012). https://doi.org/10.1007/978-3-642-33618-8_59
23. Mitzias, P., et al.: User-driven ontology population from linked data sources. In: Ngonga Ngomo, A.-C., Křemen, P. (eds.) KESW 2016. CCIS, vol. 649, pp. 31–41. Springer, Cham (2016). https://doi.org/10.1007/978-3-319-45880-9_3
24. Wacker, M.: Linked Data for Cultural Heritage, edited by Ed Jones, Michele Seikel. ALA Editions, Chicago (2016)
25. Belam, M.: What is the value of linked data to the news industry? In: The Guardian (2010). https://www.theguardian.com/help/insideguardian/2010/jan/25/news-linked-data-summit. Accessed 01 Jan 2018
26. Callahan, A., Cruz-Toledo, J., Dumontier, M.: Ontology-based querying with Bio2RDF's linked open data. J. Biomed. Semant. **4**(1), S1 (2013). https://doi.org/10.1186/2041-1480-4-s1-s1
27. Jupp, S., Malone, J., Bolleman, J., Brandizi, M., Davies, M., Garcia, L., et al.: The EBI RDF platform: linked open data for the life sciences. Bioinformatics **30**(9), 1338–1339 (2014). https://doi.org/10.1093/bioinformatics/btt765
28. Booshehri, M., Luksch, P.: An ontology enrichment approach by using DBpedia. In: Proceedings of the 5th International Conference on Web Intelligence, Mining and Semantics WIMS 2015 (2015). https://doi.org/10.1145/2797115.2797127
29. Isaac, A., Summers, E.: SKOS Simple Knowledge Organization System. Primer, World Wide Web Consortium W3C (2009)

Ontological Description of Applied Tasks and Related Meteorological and Climate Data Collections

Andrey Bart[1], Vladislava Churuksaeva[1], Alexander Fazliev[2](✉), Evgeniy Gordov[2,3], Igor Okladnikov[2,3], Alexey Privezentsev[2], and Alexander Titov[2,3]

[1] National Research Tomsk State University, Tomsk 634050, Russia
bart@math.tsu.ru, chu.vv@mail.ru
[2] V.E. Zuev Institute of Atmospheric Optics SB RAS, Tomsk 634055, Russia
{faz, remake}@iao.ru, {gordov, oig, titov}@scert.ru
[3] Institute of Monitoring of Climatic and Ecological Systems SB RAS, Tomsk 634055, Russia

Abstract. The use of the OWL-ontology of climate information resources on the web-GIS of the Institute of Monitoring of Climatic and Ecological Systems, Siberian Branch, Russian Academy of Sciences (IMCES SB RAS) for building an A-box of knowledge base used in an intelligent decision support system (IDSS) is considered in this work. A mathematical model is described, which is used for solution of the task of water freezing and ice melting on the Ob' river. An example is given of the reduction problem solution with ontological description of the related input and output data of the task.

Keywords: Ontology description of object domains
Systematization of domain data · Climate and meteorological data

1 Introduction

Investigations of ongoing and projected climate change lead to appearance of huge amounts of observation and modeling data sets describing the Earth climatic system and its components [1], which are now available to researchers. Results of climatic numerical simulation, weather observations and forecasts, or reanalysis of meteorological fields are collections of meteorological parameters that characterize the state of the climatic system. Large meteorological centers use original meteorological models for calculation of climate and meteorological parameters, which can differ both in the level of detail and set of calculated values of physical parameters.

To deal with petabytes of this valuable information targeted community needs in special IT infrastructure enabling a whole spectrum of stakeholders: from scientists and researchers to decision and policy makers at the highest level, to collaborate, share, analyse and visualise data over the internet. In recently formed branch of science called data science dealing with data intensive domains of science such infrastructures usually

L. Kalinichenko et al. (Eds.): DAMDID/RCDL 2017, CCIS 822, pp. 169–182, 2018.
https://doi.org/10.1007/978-3-319-96553-6_13

referred as Virtual Research Environments (VREs) [2]. Sometimes instead of VRE in US the term Science Gateways is used while in UK e-infrastructure is more popular.

A topical VRE prototype focused on support of basic and applied climate change research aimed at ensuring the secure and sustainable availability of natural resources and understanding natural hazards has been developed [3] at the Institute of Monitoring of Climatic and Ecological Systems SB RAS. It provides inter-disciplinary working and sharing of large amounts of data across diverse geographic locations and science disciplines to work towards solution of targeted problems. The VRE prototype is providing data processing environment dealing with collections of meteorological data, which are described by sets of metadata that characterize physical parameters included into collections.

When using climate data from different collections of numerous data manufacturers, the problem of ambiguous identification of physical parameters from these collections arises. They are represented by data arrays in common formats, e.g., grib [4], netCDF [5], HDF5 [6], etc. The meaning of each physical parameter in the collections agrees with the corresponding physical parameter advised by World Meteorological Organization (WMO). They are described in the taxonomy of the WMO ontology Codes Registry [7], as well as in the taxonomy of the ontology of the GRIB Discipline Collection [8] intended for the use in the Climate Information Platform for Copernicus (CLIPC).

An approach to apply data collections of IMCES SB RAS to construct a knowledge base of the intellectual decision support system (IDSS) is presented in this work. The fact part (A-box) of the knowledge base is formed with solutions of applied tasks, presented in the form of OWL-ontology. A task of the Ob' river ice formation, evolution and melting is the example of such applied task.

Semantic description of this task and data collections that are used in the task as input data is designed for:

1. Choosing data arrays from data collections of the IMCES web-GIS [9];
2. Building knowledge base of IDSS, which is aimed making decisions on transportation across the iced or open water;
3. Describing the input information that is used to solve a computational problem and it's solution in an explicit and consistent way.
4. Controlling the quality of input information and the solution.

The OWL-ontology created characterizes data collections of the IMCES web-GIS and input and output data, connected to the problem being solved. Meaningful part of the solution of the applied problem, which contains the information about the river that is essential to make decisions about transportation (especially river transport) is exported to the data base of IDSS.

2 Virtual Data Processing Environment

Approaches used in the creation of the prototype of a topical virtual data processing environment for the analysis, estimation, and forecast of the impacts of global climate changes on the natural environment and climate of a region were mainly developed during the design of the "Climate" web-GIS [9, 10]. This subject web-GIS has been designed with the use of up-to-date information and communication technologies based on the conceptions of spatial data infrastructure (SDI) [11, 12]. It grounds a software infrastructure for the complex use of geophysical data and information support of integrated multidisciplinary scientific researches in the modern quantitative meteorology. We have selected it as a topical component of VRE for Earth sciences. A web geoportal [13, 14] is a single access point to topical spatial data, processing procedures and results [15, 16]. The portal allows a user to search for geoinformation resources in metadata catalogues, to form samples of spatial data according to their characteristics (access functionality), and to manage tools and applications for data processing and mapping.

The web-GIS Web Client [17] is the main tool of the user's desktop. It ensures the fulfillment of OGC (Open Geospatial Consortium) requirements for web services: spatial data visualization (Web Map Service—WMS), data representation in vector (Web Feature Service—WFS) and bitmap formats (Web Coverage Service—WCS), and their geospatial processing. It provides the access to collections of climate data and tools for their analysis and visualization of the results via typical graphical web browser. The Web Client satisfies the general requirements of INSPIRE standards and allows selection of data set, processing type, geographic region for the analysis of processes, and representation of the processing results of spatial data sets in the form of WMS/WFS map layers in bitmap (PNG, JPG, GeoTIFF), vector (KML, GML, Shape), and binary formats (NetCDF).

Today, the VRE prototype combines data collections (reanalyses and climate simulation results and weather station measurements) within the unified geoportal, supports the statistical analysis of archive and required data, and provides access to the WRF and «Planet Simulator» models. In particular, a user can run a VRE-integrated model, preprocess the results, process them numerically and analyze, and gain the results in graphical representation. The prototype provides prompt tools for integral study of climate and ecological systems on the global and regional scales for specialists that participate in a multidisciplinary research process. With these tools, a user that does not know programming is able of processing and graphically representing multidimensional observation and simulation data in the unified interface with the use of the web browser.

3 VRE Prototype Capabilities

To study the results, the user is provided for a possibility of selecting a geographical region, scaling, getting values from all layers at a point, additionally processing earlier results (e.g., comparison between data from different layers). In addition to the direct analysis of geophysical data, a user can carry out joint researches with other user, share the results, and use proper data collections in the processing. In general, this hardware-software complex provides for distributed access, processing and visualization of large collections of geospatial data with the use of cloud technologies.

3.1 Data Processing

1. Statistical characteristics of meteorological parameters: sample mean, variance, excess, median, minimum and maximum, and asymmetry.
2. Derived climate parameters: vegetation period duration, sum of effective temperature, Selyaninov hydrothermal coefficient.
3. Periodic variations: standard deviation, norms, aberrations, amplitudes of diurnal and annual variations.
4. Non-periodic variations: duration and repeatability of atmospheric phenomena with meteorological parameters below or above the limits specified at different time points.

The data processing environment "Climate" developed at IMCES SB RAS limits possibilities of users by local software applications. A current task is to extend the environment by external user applications. For this, the corresponding problems should be specified in general. Below we describe one of possible classes of problems connected with decision-making.

4 General Definition of the Problem

The virtual information platform under development "Climate+" includes collections of meteorological and climate data. It is intended for the data representation with the use of GIS technologies. Its further development is oriented to providing researchers possibilities of using selected data sets or their parts as input data. Most collections include data related to some (not all) spatiotemporal objects of the Earth; different collections often include different sets of physical parameters. To search for required spatiotemporal objects and their meteorological and clime characteristics, it was necessary to create a corresponding expert system on the basis of a knowledge base on spatial objects of the data collections and their parameters.

Figure 1 shows a simplified block-diagram which is a basis of the "Climate+" platform modification. There are three groups of subsystems: meteorological and climate data collections; subsystem for work with knowledge bases (expert system for selecting input data for applied tasks and decision-making support system), and applied tasks with their input and output data.

In this work, we discuss questions of creation of a knowledge base for the IDSS. The main problem which has been solved is substantiation of the reduction problem solution [18] or, in other words, construction of typical individuals of an OWL-ontology of the applied task that characterize properties of spatiotemporal object (Ob' river). The development of the conceptual part of this ontology (T- and R-box) is connected in our solution with IDSS problems.

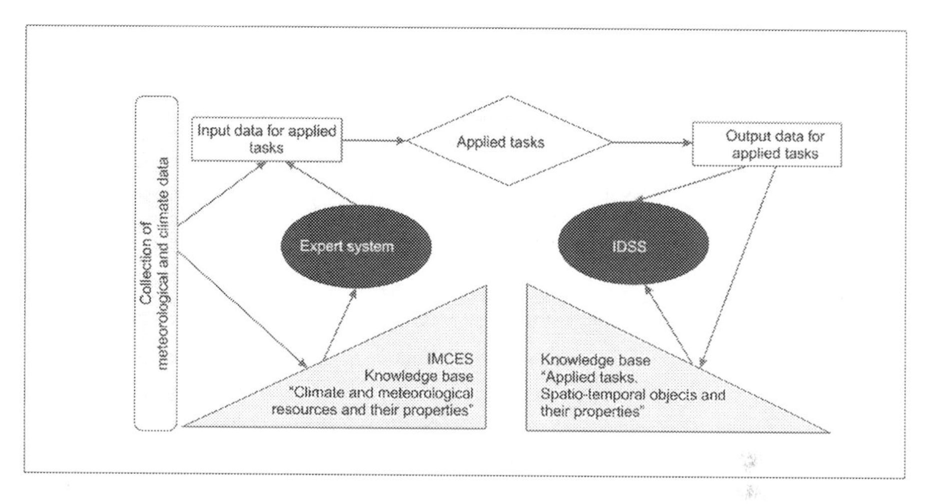

Fig. 1. Simplified block-diagram of "Climate+" platform modification. The data representation services are omitted.

5 Mathematical Model of the Applied Task

There are both one-dimensional and two-dimensional mathematical models for the thermal growth of the ice cover. As in the case considered the river flow is modeled on the long bend and there is no data about river bathymetry, and time scale is of order a year the one-dimensional model is considered. The process modeled consists of three consecutive stages: river freeze-up when the temperature drops the freezing point; thermal growth of the ice cover during winter; and ice cover melting and decay (and further break-up) in spring.

5.1 Thermodynamic Processes Considered in the Model

The coupled thermodynamic model of the atmosphere-ice-water-bed system should be considered to model the river ice processes. The rate of ice thickness growth or decay is governed by heat exchanges at the top and bottom surfaces of the ice cover and heat conduction in it [19].

In general heat exchange between open water or an ice cover and the atmosphere is equal to $\Sigma\varphi = \varphi_s + \varphi_b + \varphi_e + \varphi_c + \varphi_p$, where φ_s is the shortwave radiation; φ_b is the longwave radiation; φ_e is the evapo-condensation flux; φ_c conductive heat flux; φ_p precipitation (As the snow cover is not considered in the model, φ_p is equal to 0).

The components of the heat exchange processes are defined as functions of meteorological data [20] (water and air temperature, wind velocity, cloudiness and other) and also physical characteristics of the river.

Shortwave Radiation φ_s

$\varphi_{ri} = (a - b(lat - 50))(1 - 0.0065C^2)$, where φ_{ri} is the incoming shortwave radiation, *lat* is the latitude in degrees; *C* is the cloud cover in tenths; *a*, *b* are the table values, that represent variation of the solar radiation under a clear sky. $\varphi_s = (1-\alpha)\varphi_{ri}$, where α is the albedo. Albedo could be also calculated for different types of ice with semi-empirical formulas.

Penetrating of shortwave radiation through the ice cover is considered as a heat source term in the equation that describes the heat transfer in the ice cover.

Longwave Radiation φ_b

$\varphi_b = \varphi_{bs} - \varphi_{bn} 1.1358 \cdot 10^{-7}\left[T_{sk} - (1+k_c C^2)(c+d\sqrt{e_a})T_{ak}^4\right]$ is the effective terrestrial (longwave) radiation.

Evapo-Condensation Flux φ_e

$\varphi_e = (1.65K_n + 6.0V_a)(e_s - e_a)$ is the rate of heat loss due to evaporation. Here V_a is the wind velocity at 2 m; e_s is the saturated vapor pressure at temperature T_s; $K_n = 8.0 - 0.35(T_s - T_a)$ accounts for the free convection (T_s is the temperature of the water surface, T_a is the air temperature at 2 m.

Conduction φ_c

$\varphi_c = (K_n + 3.9V_a)(T_s - T_a)$ is the heat loss due to conduction for the case of open water.

Heat Flux From Water to Ice Cover

$q_{wi} = h_{wi}(T_w - T_f)$ is the turbulent heat flux from water to ice, where T_w is the water temperature; T_f is the freezing point; heat conduction intensity h_{wi} is calculated as follows $h_{wi} = C_{wi}(U_w^{0.8}/D^{0.2})$, where U_w is the velocity of the flow, D is the water depth, $C_{wi} = 1.622$.

5.2 Thermal Growth of the Ice Cover

The freeze-up of a river starts with the formation of border ice along the banks and in low velocity zones [21] where supercooling of water surface leads to skim ice formation. With further heat loses from the water surface the area and the thickness of ice cover grows and frazil ice in the areas of more intensive turbulent flow forms.

The procedure presented in this work is based on the mathematical model proposed by Shen [20]. The following assumptions were made to build the model used:

- the river is open at the initial moment (there is neither skim nor frazil ice in the river stream);
- flow characteristics do not vary much within the river cross-section;
- the flow is well-mixed;
- river ice grows only on the water-ice border as there is no water on the ice cover (as Wang et al. [19] reported water on melting ice does not significantly affect the dissipation rate);
- temperature in the ice cover has linear distribution;
- the flow in the river is calm and almost stagnated, so frazil ice is not forming during supercooling period.

Within the assumptions made the following model correctly describes skim ice formation on a calm plains river or in slow-flowing regions of rivers.

The longitudinal distribution of the thermal energy in the river flow could be described by the following unsteady convection-diffusion equation

$$\frac{\partial}{\partial t}(\rho_w C_p A T_w) + \frac{\partial}{\partial x}(Q\rho_w C_p T_w) = \frac{\partial}{\partial x}\left(AE_x\rho_w C_p \frac{\partial T_w}{\partial x}\right) + B\sum \varphi \tag{1}$$

where A is the cross-sectional area of the river; B is the river width; Q is the river discharge; ρ_w is the density of water; C_p is the specific heat of water; E_x is the longitudinal dispersion coefficient. The last term in the right part of the equation expresses total heat influx per unit surface area of the river. It is equal to:

$\sum \varphi = q_{gw} + \varphi_s - \varphi_b - \varphi_e - \varphi_c$ for the river without ice cover;

$\sum \varphi = q_{gv} - q_{wi} + \varphi_{sp}$ for the river with ice cover (in the following cases);

The initial and boundary conditions for the Eq. (1) are:

$T_w(x,0) = g(x)$ is the initial distribution of the water temperature.

$T_w(0,t) = T_{w\,L}(t)$ is known temperature distribution at the upstream boundary.

$\frac{\partial T_w}{\partial x}(X,t) = 0$ is the downstream boundary condition (X – is the length of the section studied).

When water temperature passes the freezing point, ice cover should be taken into account. Within the assumptions made heat conduction in the ice cover is described by one-dimensional heat conduction equation

$$\rho_i c_i \frac{\partial T_i}{\partial t} = \frac{\partial}{\partial z}\left(k_i \frac{\partial T_i}{\partial z}\right) + A(z,t) \tag{2}$$

where T_i is the temperature in the ice cover; ρ_i is the ice cover density; c_i is the specific heat of ice; k_i is the thermal conductivity of ice; A is the rate of internal heating per unit volume due to the adsorption of shortwave radiation.

Boundary condition at the ice-air interface is:

$$k_i \frac{\partial T_i}{\partial z} = \sum \varphi - \rho_i L_i \frac{d\theta}{dt}. \tag{3}$$

Here $\sum \varphi$ is the total heat loss at the ice-air interface; L_i is the specific heat of ice; θ is the ice cover thickness. Boundary condition at the ice-water interface is:

$$k_i \frac{\partial T_i}{\partial z} = q_{wi} - \rho_i L_i \frac{d\theta}{dt}. \tag{4}$$

Here $q_{wi} = h_{wi}(T_w - T_f)$ – heat flux from water to ice cover, $h_{wi} = C_{wi}\frac{U_w^{0.8}}{D^{0.2}}$ – flow velocity, D is the water depth; $C_{wi} = 1.622$; θ is the ice cover thickness. Empirical considerations for h_{wi} are also given in [22].

For quasi-steady linear distribution of the temperature in the ice cover $\frac{\partial T}{\partial z} = \frac{T_f - T_s}{\theta}$, Eq. (4) will be $k_i \frac{T_f - T_s}{\theta} = q_{wi} - \rho_i L_i \frac{\Delta\theta_\omega}{\Delta t}$, which gives the following numerical formula for change in the ice thickness:

$$\Delta\theta_\omega^{(k)} = \frac{\Delta t}{\rho_i L_i} \left\{ k_i \frac{T_f - T_s^{(k)}}{\theta^{k-1}} - h_{wi} \left[T_w^{(k)} - T_f \right] \right\}. \tag{5}$$

Here $\Delta\theta_\omega^{(k)}$ is the change in the ice thickness at the k-th time step.

The model described allows defining the time t_{freeze} when the ice cover appears on the river ($\theta > \varepsilon$, where is the predefined small value) and the thickness of the ice cover by a particular time. In this case, a one-dimensional model is used, i.e. the ice thickness is averaged over the cross section.

During the winter, the ice cover thickness will change with the continuous energy exchanges with the atmosphere and the flow and can be computed by the same procedure (Eqs. (2)–(5)).

5.3 Thermal Decay of the Ice Cover

In spring the ice cover deteriorates due to the absorption of shortwave radiation.

According to the literature, river breakup is drove by both mechanical and thermal processes. If the river flow is relatively steady (as in the case considered), the cover remains stable until the eventual melt out. In contrast, mechanical breakup scenarios are characterized by limited ice cover deterioration or melt prior to breaking and mobilization, which likely leads to severe ice jamming and impeded ice runs. In reality, most river ice breakup scenarios are neither entirely thermal nor completely mechanical. There are several mostly empirical criteria developed for the threshold between mechanical and thermal breakup [23]. But even if the breakup is mostly mechanical thermal erosion of the ice and decay of the ice sheet are important factors which can significantly reduce the strength and mass of the ice cover in the river [24].

Thus solving Eqs. (2)–(5) for the spring period allows defining the moment t_{melt} when ice on the water surface melts out or ice sheet becomes thin and weak to be destroyed by the mechanical energy of the flow.

6 Primitive Ontology of the Applied Task

Key elements of OWL-ontologies are classes, properties, and individuals. The individuals are defined during the solution of a reduction problem, as well as most properties. Some of the ontology classes are used for specification of the ontology properties domains and ranges, and other ones are used for the solution of applied problems.

Two applied ontologies are described below:

- WMO taxonomy of physical quantities (wmo:);
- ontology of information resources of applied task (task of growth and decay of ice cover of Ob' river (GDRIC)) (t2:).

The brief description of these applied OWL-ontologies is given below. Ontology of meteorological and climate information of web portal (http://climate.scert.ru/) is represented in our paper [25]. The classes and properties of the ontologies are in the Tables 1, 2 and 3. Table 2 presents the main classes. The namespace prefix is placed before the class name; the class name abbreviation used in the text below is given after the class name in parentheses. Properties of the applied ontologies are presented in Tables 2 and 3. The first three columns of the tables contain the domain, the object (datatype) property and the range, respectively; the fourth column - the property name abbreviation, used in the text and in Fig. 2 below.

The basic classes of the ontology of the applied task (t2:) are (t2:T, t2:InD, t2: OuD), including individuals that characterize the task, classes that characterize physical parameters (t2:MP, t2:TP), and a numerical model of a spatiotemporal object (t2:STG) used in the task. Object properties and data types properties of this ontology are specified in Tables 3 and 4. Two individuals are connected with the task that describe its input and output data. The input data include those connected with a spatiotemporal grid (t2:Spatio-temporal_River) and values of (climate, meteorological, physical, etc.) parameters on this grid. For simplicity, the spatiotemporal grid of the Ob river is selected as a part of the spatiotemporal grid used in calculation in INM CM4 model. Physical parameters are divided into two sets, where parameter values are taken from the data collection (φ_s и φ_b) and from other sources ($\varphi_c, \varphi_e, A, B, Q, C_W, E_x, \rho_W$) (see Sect. 5).

Table 1. Classes of applied ontologies

Class	Class	SubClassOf
wmo:Products	wmo:Land_surface_products	wmo:Products
	wmo:Soil_category	wmo:Land_surface_products
	wmo: Meteorological_products	wmo:Products
	wmo:Temperature_category	wmo: Meteorological_products
t2:Task (T)	t2:InputData (InD)	t2:Data
t2:Data (D)	t2:OutputData (OuD)	t2:Data
t2:MeteorologicalParameter (MP)	t2:ObjectParameter (TP)	
t2:SpatioTemporalObject	t2:SpatioTemporalGrid (STD)	
t2:Data_array		

Table 2. Object properties of the ontology of climate information resources

Domain	ObjectProperty	Range	
iao:Collection	iao:has_organization (do)	iao:Organization	o01
iao:Collection	iao:has_data_set	iao:Data_set	o02
iao:Data_set	iao:has_scenario	iao:Scenario	o03
iao:Data_set	iao:has_spatial_resolution	iao:Spatial_resolution	o04
iao:Data_set	iao:has_time_step	iao:Time_step	o05
iao:Data_set	iao:has_data_array	iao:Data_array	o06
iao:Physical_data	iao:has_physical_quantity (dpq)	iao:Physical_quantity	o07
iao:Physical_data	iao:has_unit	iao:Unit	o08
iao:Data_array	iao:has_spatiotemporal_grid	iao:Spatiotemporal_grid	o09
iao:Spatiotemporal_grid	iao:has_longitudes_list	iao:Longitudes_list	o10
iao:Spatiotemporal_grid	iao:has_latitudes_list	iao:Latitudes_list	o11
iao:Spatiotemporal_grid	iao:has_height_levels_list	iao:Height_levels_list	o12
iao:Spatiotemporal_grid	iao:hat_times_list	iao:Times_list	o13
t2:Task	t2:hasInputData	t2:InputData	o14
t2:Task	t2:hasOutputData	t2:OutputData	o15
t2:SpatioTemporalObject	t2:hasObjectParameter	t2:ObjectParameter	o17
t2:Data_array	t2:hasRiverCharacteristic	t2:RiverCharacteristics	o18
t2:InputData	t2:has_data_array	t2:Data_array	o19
t2:Data_array	t2:has_physical_quantity (dpq_t)	t2:Physical_quantity	o20
t2:Data_array	t2:has_spatiotemporal_grid	t2:SpatioTemporalGrid	o21

Table 3. Datatype property of the ontology of climate information resources

Domain	DatatypeProperty	Range	
iao:Physical_data	iao:has_number_of_values (dn)	int	d01
iao:Physical_data	iao:has_minimum_value (dmiv)	float	d02
iao:Physical_data	iao:has_maximum_value (dmav)	float	d03
iao:Physical_data	iao:has_value (dv)	float	d04
iao:Physical_data	iao:has_step_value	string	d05
iao:Times_list	iao:has_initial_time (dit)	string	d06
iao:Times_list	iao:has_final_time (dft)	string	d07
t2:Task	t2:hasMathematicalStatement	anyURI	d08
t2:Task	t2:hasPhysicalStatement	anyURI	d09
t2:OutputData	t2:hasCoordinate	float	d10
t2:OutputData	t2:hasTime	string	d11
t2:OutputData	t2:hasTemperatureOfWater	float	d12
t2:OutputData	t2:hasTemperatureOfIce	float	d13
t2:OutputData	t2:hasHeightOfIceCover	float	d14
t2:OutputData	t2:hasSpeedOfGrowthOfIceCover	float	d15

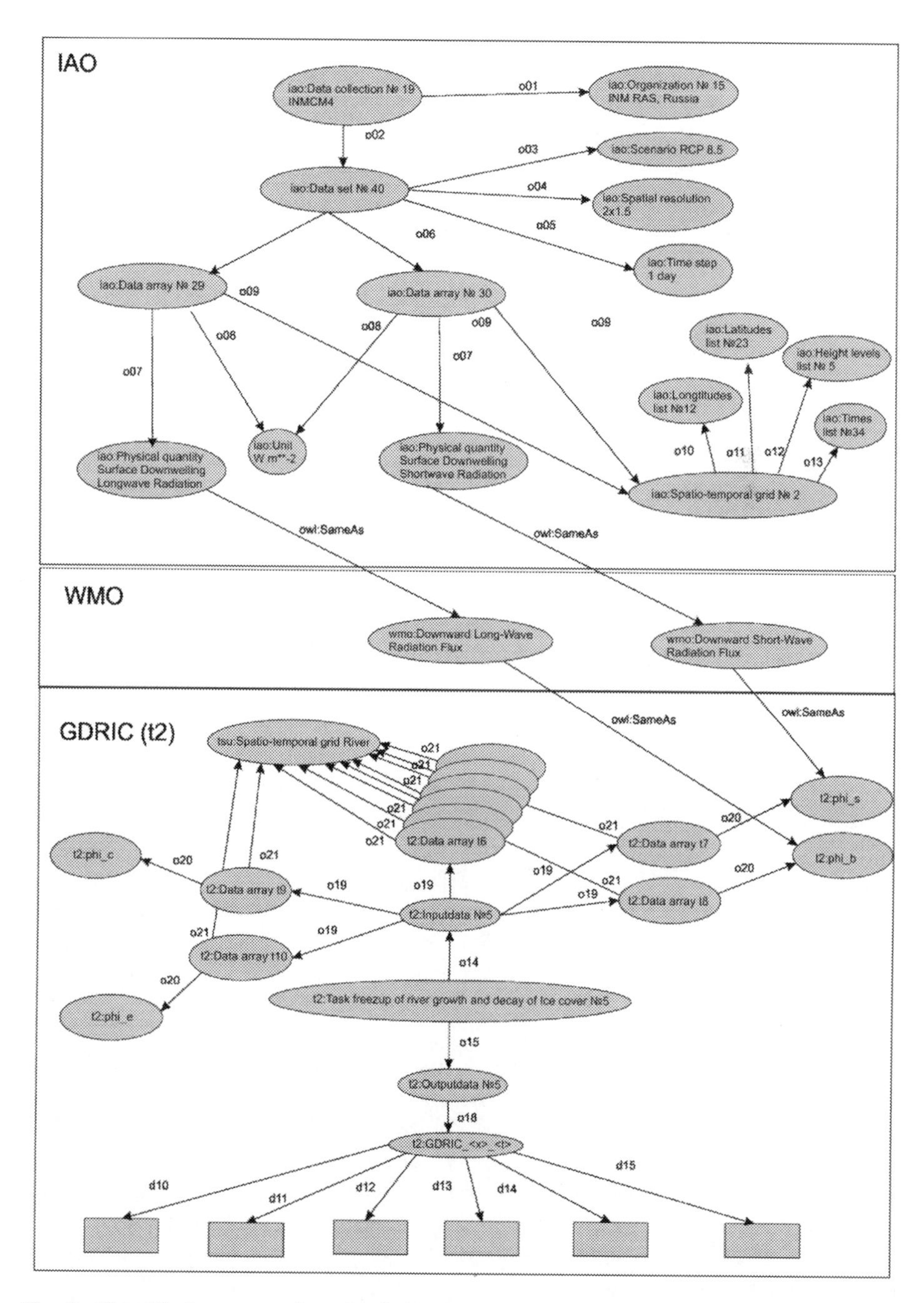

Fig. 2. Simplified representation of individual describing INMCM4 data collections and task of decay and growth of ice cover.

The output data of the applied task (t2:) are defined by several sets of individuals. For the decision making, each individual characterizes a part of the river at a critical time point (e.g., t_{melt} or t_{freeze}) and includes values of physical parameters (t2: (Coordinate, Time, TemperatureOfWater, TemperatureOfIce, HeightOfIceCover, SpeedOfGrowthOfIceCover).

Figure 2 exemplifies a simplified individuals of the OWL-ontology of climate information resources, used in the description of a INMCM4 data collection, within the formal description of RDF resources [26].

Individuals of the OWL-ontology are shown in ovals; literal values are given in rectangles; the arrows show properties with unique identifiers in small rectangles, taken from Table 3. Three arrows mean probable property cardinality higher than unity. Three overlapped ovals mean some of individuals of the OWL-ontology larger than unity.

7 Conclusions

The prototype of topical virtual data processing environment has been developed to provide for researchers, specialists, and people that make decisions an access to different geographically distributed and georeferenced resources and climate data processing services via a typical web browser. It includes a geoportal, systems for distributed storage, processing, and providing of spatial data and results of their processing. In particular, it allows the simultaneous analysis of several subject sets of climate data with the use of up-to-date statistical methods and, thus, revealing the impacts of climate changes on ecological processes and human activity. After finishing the work on the prototype, different interactive web tools are to be developed for the profound analysis of climatic variables and their derivatives provided by the subject geoportal.

The developed software is used for processing spatial datasets, including observation and reanalysis data, for the spatiotemporal analysis of recent and probable climate changes, with the special focus on extreme climate phenomena in northern latitudes.

Collections of climate and meteorological data of the IMCES web-GIS have been used for the solution of applied computational task. A primitive ontology description of input and output data for the task of river freezing and melting has been constructed in the work with the aim of building a knowledge base for IDSS. This IDSS knowledge base is represented in the OWL language [27]. During the ontology construction, the main attention is given to the reduction problem solution. The primitiveness of the description is caused by a low detail level of the processes of ice evolution in the river, ignorance of precipitation, and a simplified spatiotemporal grid in the mathematical model of the Ob' river.

The further study is connected with a T-box IDSS knowledge base for making decisions on the terms of organization of river and road traffic in autumn, winter, and spring.

Acknowledgments. The authors thank the Russian Science Foundation for the support of this work under the grant 16-19-10257.

References

1. Stocker, T.F., Qin, D., Plattner, G.-K., Alexander, L.V., Allen, S.K., Bindoff, N.L., Bréon, F.-M., Church, J.A., Cubasch, U., Emori, S., Forster, P., Friedlingstein, P., Gillett, N., Gregory, J.M., Hartmann, D.L., Jansen, E., Kirtman, B., Knutti, R., Krishna Kumar, K., Lemke, P., Marotzke, J., Masson-Delmotte, V., Meehl, G.A., Mokhov, I.I., Piao, S., Ramaswamy, V., Randall, D., Rhein, M., Rojas, M., Sabine, C., Shindell, D., Talley, L.D., Vaughan, D.G. and Xie, S.-P.: 2013: Technical Summary. In: Stocker, T.F., Qin, D., Plattner, G.-K., Tignor, M., Allen, S.K., Boschung, J., Nauels, A., Xia, Y., Bex, V., Midgley, P.M. (eds.) Climate Change 2013: The Physical Science Basis. Contribution of Working Group I to the Fifth Assessment Report of the Intergovernmental Panel on Climate Change, pp. 33–115. Cambridge University Press, Cambridge (2013)
2. Virtual Research Environment: A Guide. http://www.allianzinitiative.de/en/core_activities/virtual_research_environments/definition
3. Gordov, E.P., Okladnikov, I.G., Titov, A.G.: Application of web mapping technologies for development of information-computational systems for georeferenced data analysis. Vestnik NGU, Ser.: Inf. Technol. **9**(4), 94–102 (2011)
4. Guide to the WMO Table Driven Code Form Used for the Representation and Exchange of Regularly Spaced Data In Binary Form: FM 92 GRIB Edition 2 (World Meteorological Organization Extranet. 2003). http://www.wmo.int/pages/prog/www/WMOCodes/Guides/GRIB/GRIB2_062006.pdf
5. Network Common Data Form (NetCDF). https://www.unidata.ucar.edu/software/netcdf/
6. HDF Group - HDF5: https://support.hdfgroup.org/HDF5/
7. WMO Codes Registry: (2013). http://codes.wmo.int/grib2
8. The GRIB Discipline Collection: http://vocab-test.ceda.ac.uk/collection/grib/Discipline
9. Gordov, E.P., Okladnikov, I.G., Titov, A.G.: Information and computing Web-system for interactive analysis of georeferenced climatic data sets. Vestnik NGU, Ser.: Inf. Technol. **14** (1), 13–22 (2016)
10. Gordov, E.P., Lykosov, V.N., Krupchatnikov, V.N., Okladnikov, I.G., Titov, A.G., Shulgina, T.M.: Computational-information technologies for monitoring and modeling of climate change and its consequences, Nauka, Novosibirsk (2013). (in Russian)
11. van der Frans, J.M.: Wel: Spatial data infrastructure for meteorological and climatic data. Meteorol. Appl. **12**(1), 7–8 (2005)
12. Koshkarev, A.V., Ryakhovskii, A.V., Serebryakov, V.A.: Infrastructure of distributed environment of spatial data storage. Search and Processing. Open Educ. **5**, 61–73 (2010)
13. Becirspahic, L., Almir Karabegovic, A.: Web portals for visualizing and searching spatial data. In: 2015 38th International Convention on Information and Communication Technology, Electronics and Microelectronics (MIPRO), pp. 321–327 (2015)
14. Koshkarev, A.V.: Geoportal as a tool to control geospatial data and services. Geospatial Data **2**, 6–14 (2008)
15. Gordov, E., Shiklomanov, A., Okladnikov, I., Prusevich, A., Titov, A.: Development of Distributed Research Center for analysis of regional climatic and environmental changes. In: IOP Conference Series: Earth and Environmental Science, vol. 48, p. 012033 (2016)

16. Okladnikov, I.G., Gordov, E.P., Titov, A.G.: Development of climate data storage and processing model. In: IOP Conference Series: Earth and Environmental Science, vol. 48, p. 012030 (2016)
17. Gordov, E.P., Okladnikov, I.G., Titov, A.G., Fazliev, A.Z.: Some aspects of development of virtual research environment for analysis of climate change consequences. In: Kalinichenko, L., Manolopoulos, Y., Kuznetsov, S. (eds.) CEUR Workshop Proceedings, vol. 1752, pp. 195–201 (2016)
18. Zinov'ev, A.A.: Foundations of the logical theory of scientific knowledge (Complex Logic). D. Reidel Publishing Company, Dordrecht-Holland (1973). Revised and Enlarged English Edition
19. River ice processes. In: Wang, L.K., Yang, C.T., Wang, M.H.S. (eds.) Advances in Water Resource Management, pp. 483–530. Springer (2016)
20. Shen, H.T., Chiang, L.A.: Simulation of growth and decay of river ice cover. J. Hydraul. Eng. **110**(7), 958–971 (1984)
21. Svensson, U., Billfalk, L., Hammar, L.: A mathematical model of border-ice formation in rivers. Cold Reg. Sci. Technol. **16**, 179–189 (1989)
22. Hamblin, P.F., Carmack, E.C.: On the rate of heat transfer between a lake and an ice sheet. Cold Reg. Sci. Technol. **18**(2), 173–182 (1990)
23. Beltaos, S.: Threshold between mechanical and thermal breakup of river ice cover. Cold Reg. Sci. Technol. **37**, 1–13 (2003)
24. Jasek, M.: Ice jam release surges, ice runs, and breaking fronts field measurements. Can. J. Civ. Eng. **30**, 113–127 (2003)
25. Bart, A., Churuksaeva, V., Fazliev, A., Privezentsev, A., Gordov, E., Okladnikov, I., Titov, A.: Ontological description of meteorological and climate data collections. In: Kalinichenko, L., Manolopoulos, Y., Skvortsov, N., Sukhomlin, V. (eds.) Selected Papers of the XIX International Conference on Data Analytics and Management in Data Intensive Domains (DAMDID/RCDL 2017), vol. 2022, pp. 266–272. CEUR (2017)
26. Resource Description Framework (RDF): Concepts and Abstract Syntax, W3C Recommendation 10 February 2004. In: Klyne, G., Carroll, J.J. (eds) http://www.w3.org/TR/2004/REC-rdf-concepts-20040210/
27. Motik, B., Patel-Schneider, P.F., Grau, B.C. (eds.): OWL 2 Web Ontology Language Direct Semantics, W3C Recommendation 27 October 2009. http://www.w3.org/TR/2009/REC-owl2-direct-semantics-20091027/

Heterogeneous Data Integration Issues

Integration of Data on Substance Properties Using Big Data Technologies and Domain-Specific Ontologies

Adilbek Erkimbaev(✉), Vladimir Zitserman, Georgii Kobzev, and Andrey Kosinov

Joint Institute for High Temperatures, Russian Academy of Sciences, Moscow, Russia
adilbek@ihed.ras.ru, vz1941@mail.ru, gkbz@mail.ru, kosinov@gmail.com

Abstract. A new technology for storage and categorization of heterogeneous data on the properties of matter is proposed. Availability of a multitude of heterogeneous data from a variety of sources justifies the use of one of the popular toolkit for Big Data processing, Apache Spark. Its role in the proposed technology is to manage with extensive data warehouse in text files of the JSON format. The first stage of the technology involves the conversion of primary resources (relational databases, digital archives, Web-portals, etc.) to a standardized form of the JSON document. Advantages of JSON-format - the ability to store data and metadata within a text document, accessible perceptions of a person and a computer and support for the hierarchical structures needed to represent complex and irregular data structure. The presence of such data structures is associated with the possible expansion of the subject area: new types of materials, expansion of the nomenclature of properties, and so on. For the semantic integration of resources converted to the JSON format a repository of subject-oriented ontologies is used. The search for data in the JSON document store is implemented through a combination of SPARQL and SQL queries. The first one (addressed to the ontology repository) provide the user with the ability to view and search for adequate and related concepts. The second, accessing the JSON document sets, retrieves the required data from the document body using the capabilities of Apache Spark SQL. The efficiency of the developed technology is tested on the problems of thermophysical data integration with a characteristic for them complexity of the logical structure.

Keywords: Thermophysical properties · Semi-structured data
JSON format · Ontology

1 Introduction

The constantly increasing volume and complexity of the data structure on the substances and materials properties imposes stringent requirements for the information environment that integrates diverse resources belonging to different organizations and states. In contrast to the earth science or medicine, here the source of data is the growing

L. Kalinichenko et al. (Eds.): DAMDID/RCDL 2017, CCIS 822, pp. 185–197, 2018.
https://doi.org/10.1007/978-3-319-96553-6_14

publication flow. In so doing the volume of data is determined not so much by the number of objects studied, as by the unlimited variety of conditions for synthesis, measurement, morphological and microstructural features, and so on. It can be said that of the three defining dimensions of Big Data (the so-called "3 V-Volume, Velocity, Variety" [1]), it is the latter plays a decisive role, that is, an infinite variety of data types.

In the previous work [2], we proposed a set of solutions borrowed from Big Data technology, allowing to overcome with minimum expenses two main difficulties in the way of integration of resources. The first one is the variety of accepted schemes, terminologies, types and formats of data and so on, and the second is the need for permanent adaptation of the created structure to the emerging variations in the nomenclature of terms (objects, concepts etc.) not provided at the design stage. The need for variation in the data structure can be associated with the expansion of the range of substances (e.g. by including nanomaterials), the range of properties (e.g. by including state diagrams), or by changing the data type, say with the transition from constants to functions.

Earlier in the report [2] only main ideas of the proposed infrastructure were formulated. On their basis here specific technical solutions are proposed as well as their use for the storage and processing of thermophysical data. Their specificity, namely the variety of objects, models, measurements methods, sample conditions justifies the use of the Big Data approach, which can cover an unlimited variety of data types. This work, supplementing the previous [2], contains a description of the infrastructure, oriented specifically to work with thermophysical data.

The solutions proposed in this paper to handle the thermophysical data are based on the joint use of known and widely used technologies:

- data interchange standard in the form of text-based structured documents, each of which is treated as an atomic storage unit;
- ontology-based data management;
- general framework Apache Spark for large-scale data processing.

Section 5 of the paper deals with the integration of thermophysical data, expanding this work in comparison with the previous one [2]. In order to assess the possibilities of the proposed technology, we considered in detail the integration of two thermophysical databases with significant differences in the content and structure of the data.

2 Data Integration - General Scenario

The general scheme of data preparation (Fig. 1) assumes as an initial material a large body of external resources, thematically related, but arbitrary in terms of volume, structure and location. Among them are sources of structured data which include factual SQL databases (DB), document-oriented DB, originally structured files in ThermoML [3] or MatML [4] standards, numerical tables in CSV or XLS formats and so on. The second group (possibly dominant in terms of volume) is formed by unstructured data: text, images, raw experimental or modeling data etc.

The first stage of data preparation is the unloading of records from external sources with their subsequent conversion to the standard form of JSON documents [2]. In so doing, the conversion of structured documents can be entrusted to software whereas the unstructured part is subject to "manual" processing with the extraction of relevant information from the texts. Finally, the control element in this scheme is the repository of subject specific (domain) and auxiliary ontologies.

The distinctive characteristic of the proposed approach is that the starting data sources remain "isolated" and unchanged. Resource owners periodically download data to JSON files by templates linked with ontological models.

In so doing they determine themselves the composition, amount and relevance for the "external" world of the data being download. This type of interaction is passive, in contrast to active, when client can use the JDBC or ODBC interface to access databases.

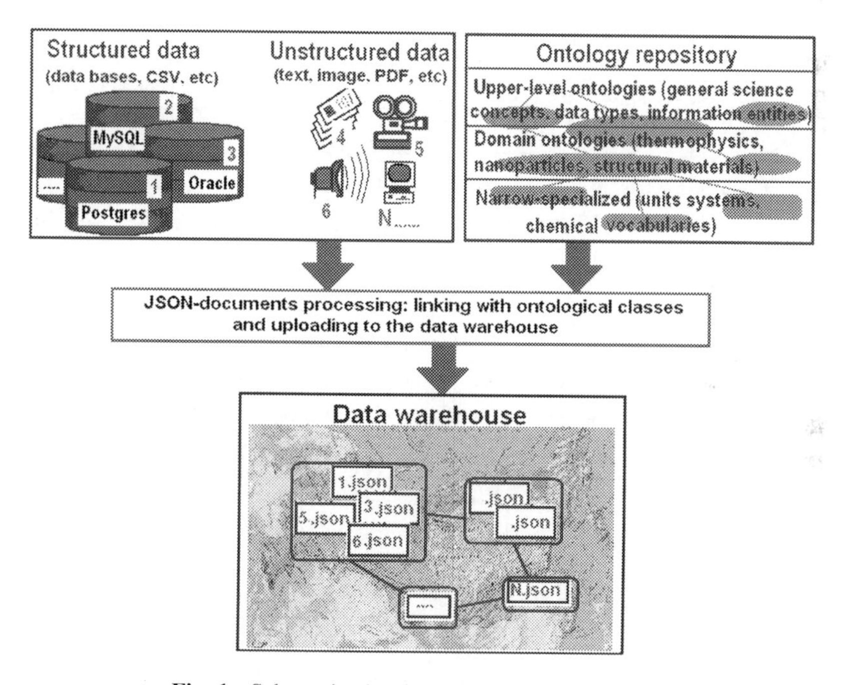

Fig. 1. Schematic sketch of initial data processing.

3 JSON-Documents and Ontologies

3.1 JSON Format

The basic unit of storage is a structured text document recorded in JSON format, one of the most convenient for data and metadata interchange propose [5]. The advantage of JSON-document - text-based language-independent format, is easy to understand and quickly mastered, a convenient form of storing and exchanging arbitrary structured information. Previously, structured text based on the XML format was proposed as a means of storing and exchange thermophysical data in the ThermoML project [3] and data on the properties of structural materials in the MatML project [4].

Here, a text document is proposed as the main storage unit, written in JSON format, which is less overloaded with details, simplifying the presentation of the data structure, reducing their size and processing time. In particular, the JSON format is shorter, faster read and written, can be parsed by a standard JavaScript function, rather than a special parser, as in the case of the XML format (https://www.w3schools.com/js/js_json_xml.asp).

Among other advantages of the format, one can note a simple syntax, compatibility with almost all programming languages, as well as the ability to record hierarchical, that is, unlimited nested structures such as "key-value". By way of ***value*** may be accepted object (unordered set of key-value pairs), ***array*** (ordered set of values), string, number (integer, float), ***literal*** (true, false, null). It is also important that the JSON format is a working object for some platforms, in particular for Apache Spark, allowing for the exchange, storage and queries for distributed data.

The rich possibilities of JSON-format as a means of materials properties data interchange attracted the attention of developers of the Citrination platform [6]. They proposed JSON-based hierarchical scheme PIF (physical information file), detailing the object, its properties, manufacturing technology, data sources etc. The top level in this hierarchy is occupied by the object ***System***, whose fields include three data groups, explaining what an object is (name, ID), how it was created/synthesized and what its characteristics are. According to the developers' concept [6], the generality of the created scheme should be sufficient for storing objects of arbitrary complexity, "from parts in a car down to a single monomer in a polymer matrix". Flexibility of the PIF-scheme is achieved due to additional fields and objects, as well as the introduction of the concept of ***category***. This concept is nothing but a version of the scheme, oriented to a certain kind of objects, say substances with a given chemical identity.

3.2 Ontology-Based Data Management

The second stage of data preparation is the linking of extracted metadata with concepts from ontologies and dictionaries assembled into a single repository. The management of the repository is entrusted to an ***ontology-based data manager***, which allows for the search and edit terms (class) of ontologies, as well as their binding to JSON documents, Fig. 2.

This means that when the particular source schema is converted to a JSON format, terms from ontologies, rather than source attributes, are used as its keys. It is also possible to use additional keys for a detailed description of the data source itself, for example, indicating the type of DBMS, name and format of text or graphical file, authorship and other official data, "sewn up" in the atomic "unit" of storage.

The role of ontologies is to introduce semantics (a common interpretation of meaning) into documents, as well as the ability to adjust the data structure of the JSON-documents by editing the ontology. Linking documents with ontologies allows to perform semantic (meaningful) data search (more precisely, metadata) using SPARQL queries, which makes it possible to reveal the information of the upper and lower levels (super and sub-classes) and side-links (related terms), without knowing the schema of the source data. Thus, the user can view and retrieve information without being familiar

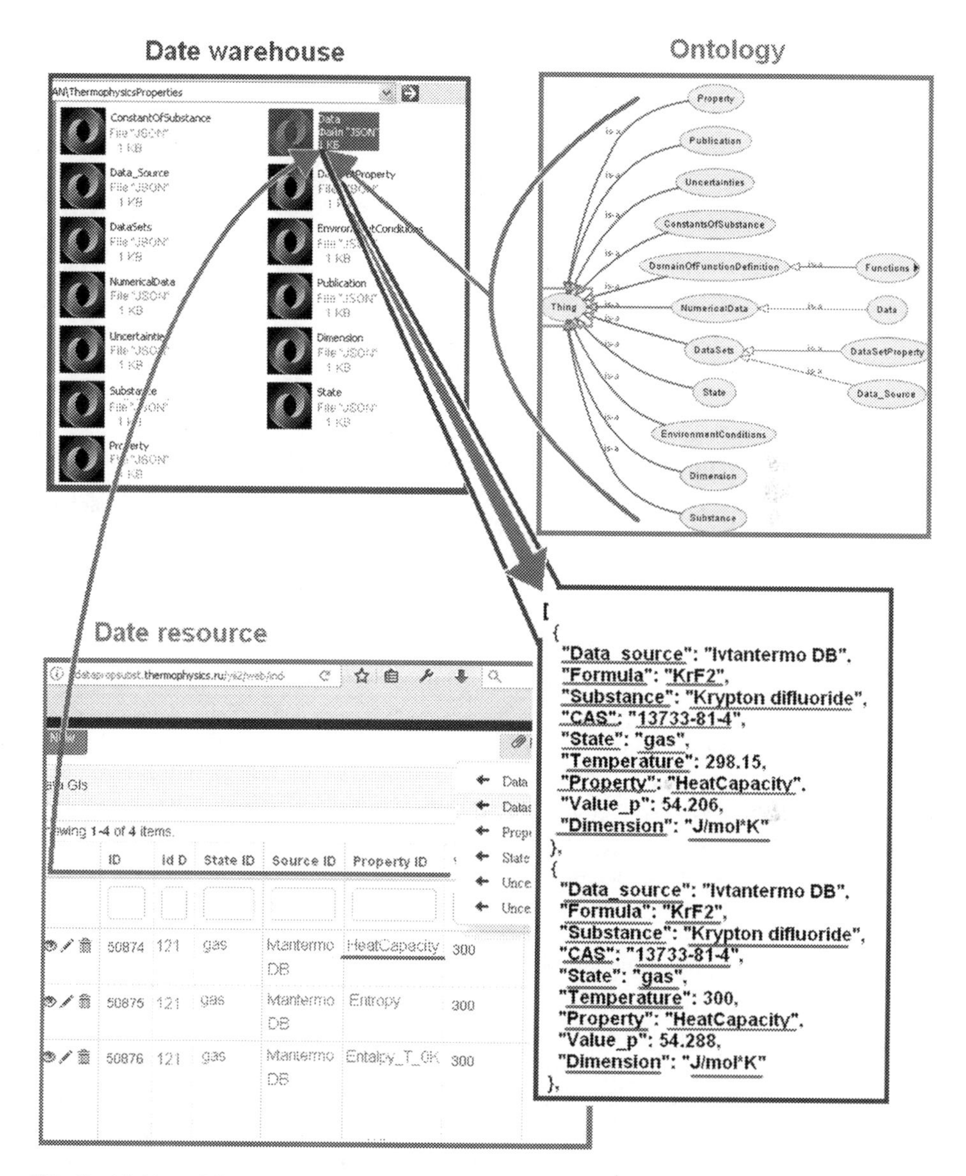

Fig. 2. Linking JSON documents to ontology classes using the example of the ontology for the domain of thermophysical properties.

with the conceptual schema of a particular DB or the metadata extracted from unstructured sources.

The repository should include three types of ontologies and controlled vocabularies: upper-level, domain and narrow-specialized. The first type is scientific top-level ontology, which introduces the basic terminology used in different fields, for example such concepts as ***substance***, ***molecule***, ***property***, ***state***, as well as informational entities that reflect the representation of data: ***data set***, ***data item***, ***document***, ***table***, ***image***, etc. Most of these terms and links between them can be borrowed from ontologies presented on the server Ontobee [7], for example SIO (Semanticscience Integrated ontology) or CHEMINF (Chemical Information ontology). The second type of ontology (domain ontology) should cover the terminology of certain subject areas, for example, thermophysics, structural materials, nanostructures, etc. For each of the domains, as a rule, some ontologies previously created and presented in publications or Web are already available on the basis of which it is possible to build and further maintain its own subject-specific ontology [8, 9]. Finally, the third type (narrow-specialized) should include ontologies or vocabularies for systems of units (for example, UO on the above portal Ontobee) or chemical dictionaries, for example ChemSpider [10] and the like. Figure 2 illustrates the binding of terms from a JSON document to ontological terms.

The proposed technology already at the stage of data preparation provides:

- consistency with accepted standards regardless of the structure and format of the original data;
- semantic integration of created JSON-documents;
- inclusion of previously not provided objects and concepts by expanding classes or introducing new ontologies and dictionaries.

The scheme of the generated data is determined by the initial data scheme with subsequent correction in the process of linking with the terms and structure of the corresponding ontological model. It should be noted that JSON-documents are objects with which one can operate using external API. In so doing, there is always the possibility of accessing keys in JSON documents not currently linked with a particular ontology term.

At the same time, it seems justified to identify or bind not only keys, but also values with ontological terms. For example, the key/value pair "Property": "Heat Capacity" is presented in Fig. 2. This will allow in the future to facilitate the formation of SQL query, relying on the information received from the ontologies repository.

The experience of using ontology in the data interchange through text documents has already been implemented in a special format CIF (Crystallographic Information File) [11], intended for crystallographic information.

In other cases of using a JSON document for the storage of scientific data (for example, in the mentioned Citrination system [6]), the categorization and introduction of new concepts is carried out by a special commission without linking with the concepts of ontological models.

4 Big Data Technique of Storage and Access to Data

Given the increasing volume and distributed nature of the data on the properties, some of the Big Data technologies would be appropriate for infrastructure design. Their advantage is due not so much to high performance in parallel computing, but rather to a pronounced orientation to work with data (storage, processing, analysis and so on) in a distributed environment (when data sources are located on remote severs). Among the available open-source means, the Apache Spark high-performance computing platform [12] is offered here. Along with other technological features, it is distinguished by the presence of built-in libraries for complex analytics including running SQL-queries. By means of SQL-queries one can access the contents of structured JSON documents. It is the ability of SQL-queries to data plays a key role in the task of their integration. The efficiency of Spark in the storage and processing of data is also associated with its ability to maintain interaction with a variety of store types: from HDFS (Hadoop Distributed Files System) to traditional database on local servers. We should also note the built-in library GraphX – an application for processing graphs, which provides our project with our own tools for working with ontologies.

All kinds data transformations executed by Apache Spark (loading, unloading, creation, modification, calculations, etc.) are reduced to few RDD (Resilient Distributed Dataset) transformations, where RDD is a fundamental data structure of Spark [12].

The main task (access and search among structured and semistructured data) is implemented by the Spark SQL module, which is one of the built-in Apache Spark libraries. Spark SQL defines a special type of RDD called SchemaRDD (in recent versions, the term DataFrame is used). The SchemaRDD class represents a set of objects row, each of which is a normal record. The SchemaRDD type is called a schema (a list of data fields) of records. SchemaRDD supports a number of operations that are not available in other sets, in particular, execution of SQL-queries. SchemaRDD sets can be created from external sources, (JSON - documents, Hive tables), query results and from ordinary RDD sets.

Three main features of Spark SQL:

- It can download data from different sources;
- It requests data using SQL within Spark programs and from external tools related to Spark SQL through standard mechanisms for connecting to databases via JDBC/ODBC;
- It provides an interface between SQL and ordinary code (when used within Spark SQL programs), including the ability to connect sets of RDDs and SQL tables.

It is possible to configure Spark SQL to work with structured data Apache Hive. The Apache Hive store is specifically designed for querying and analyzing large data sets, both inside HDFS and in other storage systems. JDBC/ODBC standards are also supported by Spark SQL, which allows executing a direct SQL query to external relational databases, in the case of the above defined active type of interaction.

The main scenario that uses the Apache Spark features is shown in Fig. 3. As a result of uploading data to JSON documents according to the above procedure, we will have data sets with a single classifying key system identical to terms from ontologies.

The organization of data requests in JSON documents will always be based on definitions from this single system. Thus, one can form the SQL query of interest in the interface of the data processing system. In this case, it always remains possible to access the ontology repository to refine or supplement the terms of the query using the SPARQL query (Fig. 3). Then the SQL-request coming from the user interface initiates the work of the Spark SQL module. As a result of the work of Spark SQL module, RDD or DataFrame sets are created, including the selected records, which can be

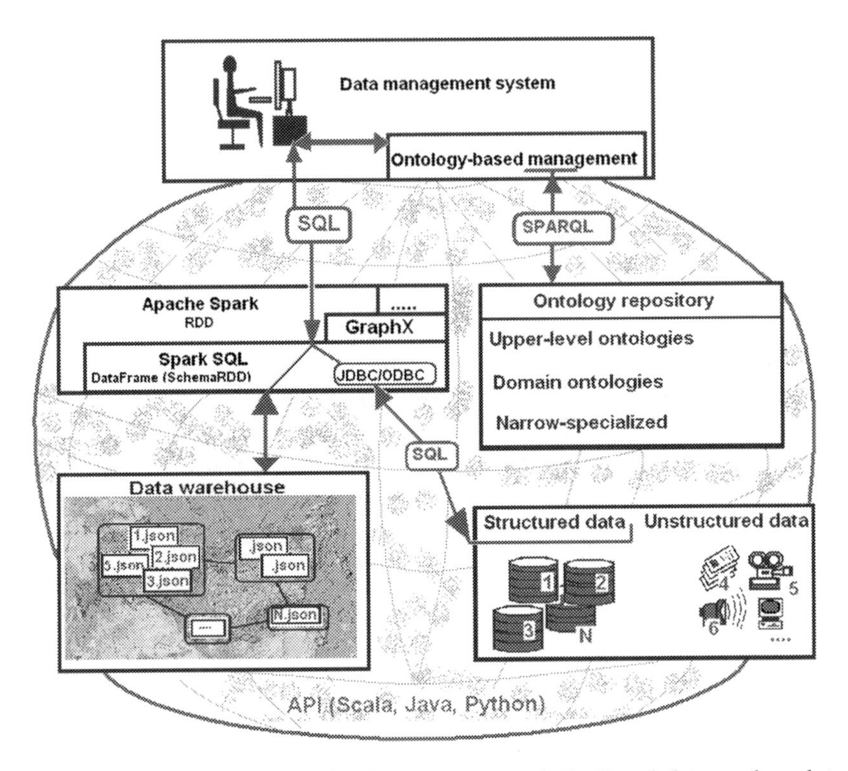

Fig. 3. Web-environment for managing heterogeneous and distributed data on the substance properties (databases, unstructured and semistructured files and so on).

processed by the system's service functions for further use. In fact, the user's work consists of two phases: viewing ontologies terms with the choice of adequate for the formation of SQL-query; access to the repository of processed data with a SQL query. Thus, the main scenario involves unifying heterogeneous data by converting them to JSON documents and processing them using Spark SQL.

Other scenarios are also justified, if we take into account the diversity of the source data. For example, the Spark SQL module allows direct query to relational databases without their conversion to the JSON format. On the other hand, you can provide access to JSON documents by collecting them in a file system using other Big Data tools. The first and main feature of JSON data collection based on ontological models terms - the unambiguous interpretation of the content and type of data. In this case users and external programs can freely work with data, because the ontology term, mapped to a key or value in the body of the JSON file, has available and accepted definitions and properties. For example, links to various types of files (graphics, multimedia, exe-files, etc.) can be described adequately and functionally using keys-terms from ontologies describing data formats. The second feature, as it is not strange, is the possibility of including in the exchange of such data sources that do not allow active access or changes due to various reasons. Then unloading the data to an external JSON file solves this problem, providing independent data storage and their full description via ontological models.

The listed technologies, supported by Apache Spark, provide unlimited productivity and variety of opportunities to handle complex data, which include data on the properties of substances, including traditional materials and nanostructures.

5 Example of Heterogeneous Thermophysical Data Integration

The testing and analysis of the new possibilities were performed by us on thermophysical data, widely represented in various databases, including those supported by the Joint Institute for High Temperatures [13][1]. Thermophysical data characterizing the volumetric, thermochemical, and transport properties are presented in the world databases for the widest range of substances: pure and solutions, organic and inorganic, structural materials and nanostructures.

Specific physical models and associated vocabularies of concepts and data structures are inherent in each of these classes. Another peculiarity of these data is their preferential use for the work of computational applications that ensure the calculation of chemical and phase equilibria, thermal and mass transfer, modeling of various natural and manufacturing processes. As a consequence, the need for methods for the exchange of heterogeneous data, differing in format and structure, arose in thermophysics long before the time when the problem of data integration became focal for the information community [14].

In the previous works [8, 9], authors was used the Semantic Web means to integrate thermophysical data which involves the conversion of relational data to the RDF format and their subsequent publication in the LOD (Linked Open Data) space. A new approach recent proposed by us [2] has a number of advantages, especially noticeable when working with data of a complex and irregular structure, which is typical for the representation of thermophysical properties. First of all, proposed technology as sources for integration includes not only relational databases, but also resources of arbitrary type: structured XML files, tables in CSV or XLS formats, unstructured data in the form of texts, images etc. Secondly, the initial data, regardless of the type of source, is converted to the structured JSON file, accepted as a storage and exchange standard. Earlier (see Sect. 3.1) was mentioned a number of advantages of JSON format, in particular support of hierarchical structures. In so doing the JSON file is easily handled by Spark SQL with the ability to set up and run SQL queries.

Finally, the inclusion of ontologies as a key element in data management opens up unlimited possibilities in semantic integration and adjustment of the data structure to permanent variations of the domain.

Technology capabilities have been tested on the long-standing problem - integration of two qualitatively different databases operating under the auspices of the Joint Institute for High Temperatures, IVTANTHERMO and THERMAL [13, 15], Fig. 4.

[1] Section "Databases at the Joint Institute for High Temperatures, Russian Academy of Sciences" on page 47 of the review [13].

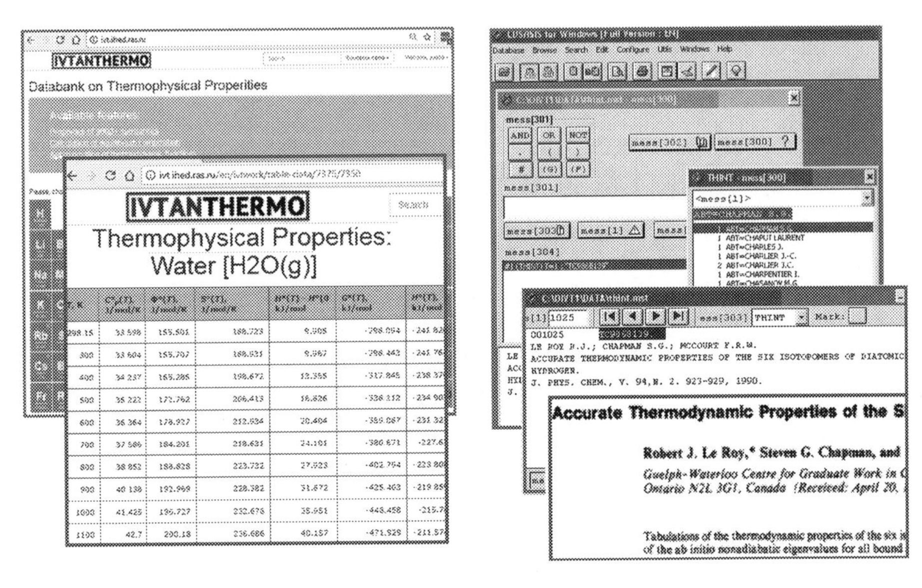

Fig. 4. Heterogeneous thermophysical data from databases IVTANTHERMO and THERMAL.

The first of these is one of the most popular databases in the chemical thermodynamics. It contains tables of basic thermodynamic properties (heat capacity, enthalpy, entropy, Gibbs function) for individual substances, mainly inorganic. In so doing, its application is limited only to standard states: an ideal gas or condensed phases at a pressure of 1 atm. Another of our databases (THERMAL) covers a much broader scope of substances, properties, states, etc. In particular, apart from traditional thermophysical properties (heat capacity and the like), THERMAL includes data on mechanical, optical and other physical characteristics. However, the content of THERMAL is limited in a more serious respect - it is a bibliographic database. This means that by requesting data for definite substance it gives detailed information only about the source, but not the tables of the experimental or calculated data itself. As a compromise, a database for a series of entries offers a hyperlink to the full text of the publication without selecting the required data. Obviously, combining the capabilities of both databases will dramatically increase the potential of each database, opening access simultaneously to bibliographic and factual data.

In accordance with the adopted scenario, the first phase in solving this problem is "unloading" data from each database in JSON-format. A data fragment from IVTANTHERMO is shown in Fig. 5. It also shows the data schema viewing, the data itself and the operation of loading them into the temporal database.

It is significant that after this, the Spark SQL (see Sect. 4) implements the traditional SQL query. This confirms the adequacy of replacing the original relational database with a document in JSON format. A similar procedure in application to the IVTANTHERMO and THERMAL results in a couple of independent JSON files. The initial difference in the scheme of both databases is naturally reflected in the difference in the list of attributes. The ontology of thermophysical properties adopted at the testing

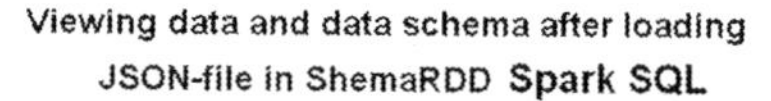

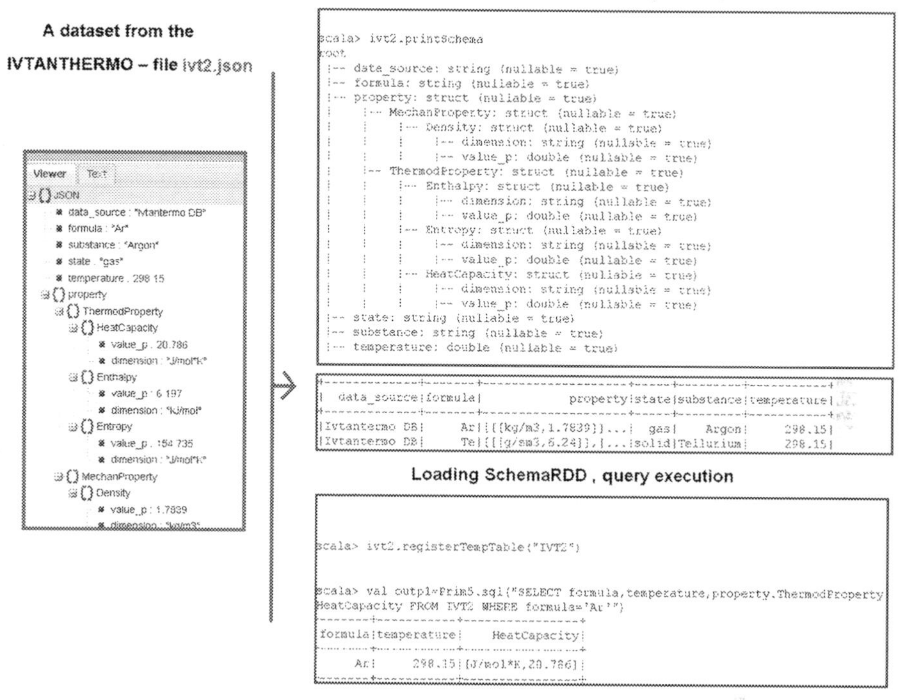

Fig. 5. Processing a JSON-file using Spark SQL; file ivt2.json was obtained by converting the relational database IVTANTHERMO.

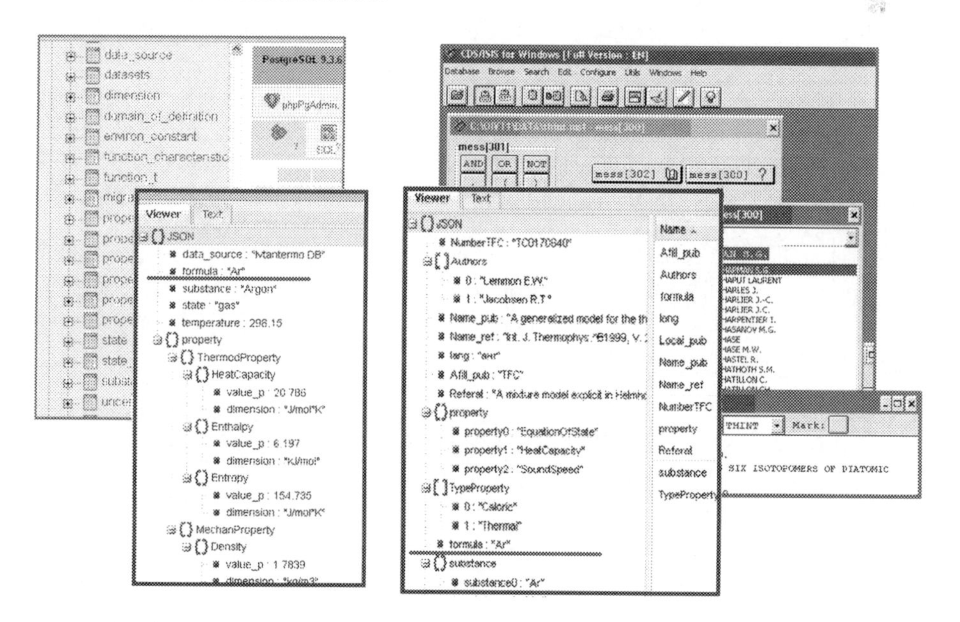

Fig. 6. Unloading data from IVTANTHERMO and THERMAL to JSON-files.

stage [8, 9] allows, in principle, to form their single list in accordance with the terms of ontology. The example in Fig. 6 shows the presence of a coincidental attribute ***"formula"***, which is included in the ontology. The ability to generate typical SQL queries in the Spark environment allows you to search in both files. A corresponding example with the search for records for a particular substance (***formula*** = "Ar") is shown in Fig. 7.

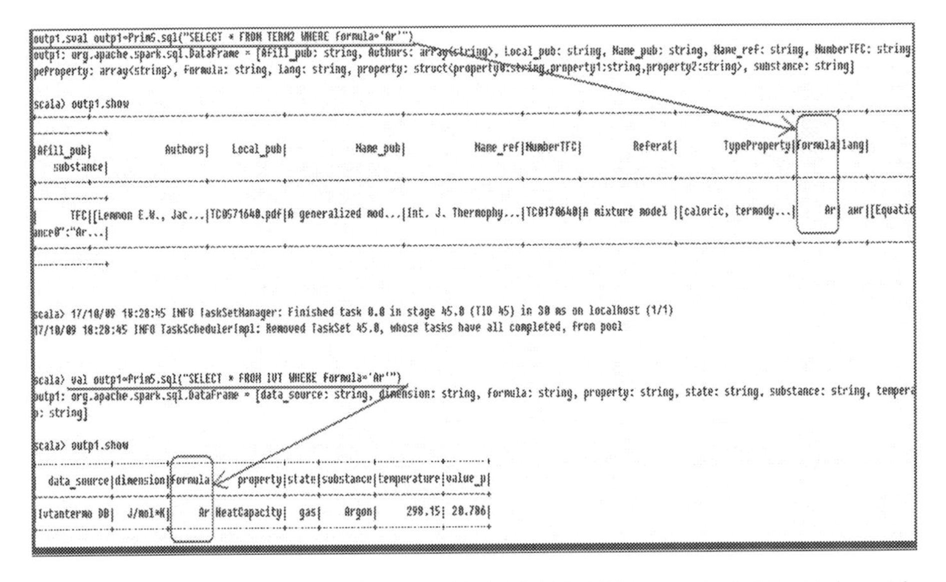

Fig. 7. SQL query to data from JSON-files: IVTANTHERMO.json and THERMAL.json. The query from Spark SQL uses the common attribute ***"formula"***.

As can be seen, the data found during the search corresponds to different sets of attributes. Naturally, this enriches the result, supplementing the standard thermodynamic properties with references to publications with a wider range of physical properties. To integrate both databases it is necessary that all the attributes used in the search are associated with the terms of the ontology, which implies its extension. Moreover, with the expansion of the domain (new substances, properties, states, etc.), new classes of the ontology are introduced for them, which allows for permanent adjustment of the data structure.

Acknowledgments. The work is supported by Russian Scientific Foundation, grant 14-50-00124.

References

1. WhatIs.com (a reference and self-education tool about information technology). http://whatis/techtarget.com/definition/3Vs
2. Erkimbaev, A.O., Zitserman, V.Y., Kobzev, G.A., Kosinov, A.V.: Standardization of Storage and Retrieval of Semi-structured Thermophysical Data in JSON-documents Associated with the Ontology. In: CEUR –WS 2022, urn: nbn:de:0074-2022-6 (2017). http://ceur-ws.org/Vol-2022/paper36.pdf
3. Frenkel, M., Chirico, R.D., Diky, V., et al.: XML-based IUPAC standard for experimental, predicted, and critically evaluated thermodynamic property data storage and capture (ThermoML). Pure Appl. Chem. **78**, 541–612 (2006). https://doi.org/10.1351/pac200678030541
4. Sturrock, C.P., Begley, E.F., Kaufman, J.G.: NISTIR 6785. MatML – Materials Markup Language Workshop Report, U.S. Department of Commerce. National Institute of Standards and Technology (2001)
5. Introducing JSON. http://json.org/index.html
6. Michel, K., Meredig, B.: Beyond bulk single crystals: A data format for all materials structure–property–processing relationships. MRS Bull. **41**, pp. 617–623. https://doi.org/10.1557/mrs.2016.166
7. Ontobee: A linked data server designed for ontologies. http://www.ontobee.org
8. Erkimbaev, A.O., Zhizhchenko, A.B., Zitserman, V.Yu, Kobzev, G.A., Son, E.E., Sotnikov, A.N.: Integration of databases on substance properties: approaches and technologies. Autom. Documentation Math. Linguist. **46**, 170–176 (2012). https://doi.org/10.3103/S000510551204005X
9. Ataeva, O.M., Erkimbaev, A.O., Zitserman, V.Yu. et al.: Ontological Modeling as a Means of Integration Data on Substances Thermophysical Properties. In: 15th All-Russian Science Conference "Electronic Libraries: Advanced Approaches and Technologies, Electronic Collections", s1_3. Yaroslavl (2013). http://rcdl.ru/doc/2013/paper/s1_3.pdf
10. ChemSpider. http://www.chemspider.com
11. Hall, S.R., McMahon, B.: The implementation and evolution of STAR/CIF ontologies: interoperability and preservation of structured data. Data Sci. J. **15**(3), 1–15 (2016). https://doi.org/10.5334/dsj-2016-003
12. Apache Spark. http://spark.apache.org
13. Kiselyova, N.N., Dudarev, V.A., Zemskov, V.S.: Computer information resources of inorganic chemistry and materials science. Rus. Chem. Rev. **79**, 145–166 (2010). https://doi.org/10.1070/RC2010v079n02ABEH004104
14. Frenkel, M.: Global communications and expert systems in thermodynamics: Connecting property measurement and chemical process design. Pure Appl. Chem. **77**, 1349–1367 (2005). https://doi.org/10.1351/pac200577081349
15. Belov, G.V., Iorish, V.S., Yungman, V.S.: IVTANTHERMO for Windows-database on thermodynamic properties and related software. Calphad **23**, 173–180 (1999). https://doi.org/10.1016/s0364-5916(99)00023-1

Rule-Based Specification and Implementation of Multimodel Data Integration

Sergey Stupnikov(✉)

Institute of Informatics Problems, Federal Research Center "Computer Science and Control" of the Russian Academy of Sciences, Moscow, Russia
sstupnikov@ipiran.ru

Abstract. An approach for rule-based specification of data integration using RIF-BLD logic dialect that is a recommendation of W3C is presented. The approach allows to combine entities defined in different sources represented in different data models (relational, XML, graph-based, document-based) in the same rule. Logical semantics of RIF-BLD provides for unambiguous interpretation of data integration rules. The paper proposes an approach for implementation of RIF-BLD rules using IBM High-level integration language (HIL) as well. Thus data integration rules can be compiled into MapReduce programs and executed over Hadoop-based distributed infrastructures.

Keywords: Data integration · Data models · Logic rules · Rule implementation

1 Introduction

Data warehouses are one of the main components of infrastructures for business analytics nowadays. Data are extracted from various collections, transformed into representation conforming a warehouse schema, integrated and loaded into the warehouse. A layer of applications for data analysis is constructed over the warehouse. Applications use mathematical statistics, machine learning etc. and output reports for users. Significant growth of number, volume and heterogeneity of data sources keeps urgent issues of specification and implementation methods for scalable data integration. The methods are to be implemented over distributed infrastructures for data storage, management and computing like Apache Hadoop [1].

Legacy industrial solutions for data integration and warehousing like IBM InfoSphere Information Server [2] offer visual tools for development of data integration workflows. These tools manipulate entities and operations of relational model, generate SQL programs for data transformation and integration as usual. Specific visual tools like InfoSphere FastTrack [2] are applied for matching schemas of source data collections and data warehouse schema (called *target* schema). Schema matching tools allow generating parts of data integration workflows. Separate components are applied for integration and transformation of data represented in different data models into relational model. For instance, XML transformation component is provided as a part of IBM InfoSphere DataStage [3].

L. Kalinichenko et al. (Eds.): DAMDID/RCDL 2017, CCIS 822, pp. 198–212, 2018.
https://doi.org/10.1007/978-3-319-96553-6_15

Alongside with the industrial solutions methods for formal specification of data integration rules are investigated. A classic example in this direction is the *data exchange* approach [9]. The approach allows defining data integration within relational model using logic rules. Semantics of rules are defined applying first order predicate logic.

This particular paper considers an approach for specification of data integration rules using the RIF-BLD logic dialect [6] that is a recommendation of W3C. RIF-BLD is a dialect of the RIF-FLD framework for logic dialects [5] intended for unification of syntax and semantics of logic rule languages. RIF-FLD itself is a part of Rule Interchange Format (RIF). RIF-BLD includes a wide range of specification constructs: *positional terms* and *terms with named arguments* to generalize terms of first order logic, *frame terms* to define assertions on structure of *objects, classification terms, equality terms, external terms* used to reference entities considered as "black boxes" within a specification.

Application of RIF-BLD allows to combine entities defined in various sources represented in different data models in the same rule. Rules considered in the paper can contain target entity predicates in the heads and source entity predicates in the bodies only.

Logical semantics of RIF-BLD [6] provides for unambiguous interpretation of data integration rules and allows their implementation by means of various declarative (like SQL) and imperative (like Java) languages. The paper proposes an approach for implementation of RIF-BLD rules using High-level integration language (HIL) [7, 10] developed by IBM. HIL is supplied as a part of BigInsights 3.0 [11] Hadoop solution as well as a part of InfoSphere Big Match for Hadoop [12]. Distributed execution of HIL programs over Hadoop is achieved by their compilation into Java programs and subsequent execution using MapReduce [14] computational model.

Approaches for specification and implementation of data integration rules are illustrated by examples of integration of several heterogeneous collections of data on Arctic region into a data warehouse intended for support of search and rescue operations. Specific data collections to be integrated were chosen [16] and unified data warehouse schema was developed [20]. The paper considers collections represented in relational model, XML, MongoDB document model. Neo4j graph model data integration is considered in [21]. Data warehouse schema is represented in relational model.

The structure of the paper is as follows. Sections 2, 3 and 4 illustrate approaches for specification of integration rules for data represented in different models (Sect. 2 – XML and relational model, Sect. 3 – document model, Sect. 4 – graph-based model) and their implementation using HIL. Section 5 generalizes principles applied for specification and implementation of data integration rules.

This work is an extension of [21]. Comparing it with the previous work in [21], examples in Sects. 2 and 3 are refined; Sect. 4 is extended with basic patterns of implementation of data integration rules using HIL. To deal with implementation issues of RIF-BLD an *abstract syntax* of RIF-FLD was defined [24] applying Ecore metamodel (that is an implementation of OMG's Essential Meta-Object Facility) used in Eclipse Modeling Framework [22]. Concrete syntax of RIF-FLD [24] binding syntactic sugar and abstract syntax was formalized applying EMFText framework [23].

2 Specification and Implementation of Relational and XML Data Integration Rules with Conflict Reconciliation

This section considers two examples of data integration. Both examples manipulate entities represented in XML and relational data model.

The first example considers a rule mapping XML data into relational data conforming warehouse schema (target schema). The rule includes constructs applied for reconciliation of structure, name and value conflicts.

The second example considers a rule that combines XML source collection and relational source collection.

2.1 Object Track Data Integration

The left column of the Table 1 contains an example of XML data on a track of a rescue helicopter. The data are extracted from the MoRe information system [18] intended for monitoring and classification of vessels. The track (*ISSKOI_Track* element) consists of the track points (*ISSKOI_TrackPoint* element). For every point of a track coordinates (*pos*), time (*Time*) and height (*BarAltitude*) are defined.

Table 1. Data on a track of an object and the respective elements of the target schema

Source data example (XML)	Target schema elements
<ISSKOI_Track> <Id>56473</Id> <TrackName>copter-1</TrackName> <ISSKOI_TrackPoints> <ISSKOI_TrackPoint id="uuid-2b7ca14"> <Position> <Point id="uuid-859bef91"> <pos>33.8957 246.37</pos> </Point> </Position> <Time>2016-12-12T13:33:11</Time> <BarAltitude>533.89</BarAltitude> <HSpeed>108.1</HSpeed> <VSpeed>2</VSpeed> </TrackPoint> </TrackPoints> </Track>	Track(PK trackId, name) TrackPoint(PK pointId, FK path, time, height, latitude, longitude)

The right column of the table contains elements of the target schema that correspond to source data. *ISSKOI_Track* element corresponds to *Track* relation of the target schema; *Id* nested element corresponds to the primary key (PK) *Track.trackId*; *TrackName* nested element corresponds to the *Track.name* attribute. *ISSKOI_-TrackPoint* element corresponds to *TrackPoint* relation of the target schema; attribute *id* corresponds to the primary key *TrackPoint.pointId*; nested elements *Time* and *BarAltitude* correspond to *TrackPoint.time* and *TrackPoint.height* attributes; *pos* nested element corresponds to *TrackPoint.logtitude* and *TrackPoint.latitude* attributes.

Example considered above as well as further examples in the paper consider only a part of source and target schema elements.

Several conflicts have to be resolved within specification of track data integration. These are name conflicts (for instance, *BarAltitude* element and *height* attribute possess the same semantics but different names), structure conflicts, value conflicts (for instance, value of a *pos* element includes both *logtitude* and *latitude* values).

The left column of the Table 2 contains a RIF-BLD rule for integration of data on tracks. According to RIF-BLD notation *Forall* denotes universal quantifier, « :- » sign denotes logical implication, *And* denotes conjunction. Identifiers like *?X* denote variables.

Table 2. A specification and implementation of a rule for integration of data on tracks

RIF-BLD	HIL
Forall ?t ?Id ?TrackName	declare ISSKOI_Track: ?;
?TrackPoints ?TrackPoint	declare Track: ?;
?Position ?Time ?Height	declare TrackPoint: ?;
?Point ?pos	
(And(	declare get_latitude:
Track(trackId->?Id	function string to double;
name->?TrackName)	declare get_longitude:
TrackPoint(pointId->?pid	function string to double;
path->?Id time->?Time	
height->?Height	@jaql{ get_latitude = fn($s)
latitude->	convert(substring($s, 0,
External(get_latitude(?ppos))	strPos($s, ' ')-1), schema double);
longitude->	get_longitude = fn($s)
External(get_longitude(?ppos))	convert(substring($s,
))	strPos($s, ' ')+1, strLen($s)),
:-	schema double);
And(?t#ISSKOI_Track	}
?t[Id->?Id	
TrackName->?TrackName	insert into Track
TrackPoints->?TrackPoint]	select [trackId: t.Id,
?tpt#?TrackPoint	name: t.TrackName]
?tpt[id->?pid Time->?Time	from ISSKOI_Track t;
Position->?ps	
BarAltitude->?Height]	insert into TrackPoint
?ps#Position	select [path: t.Id, time: p.Time,
?ps[Point->?pt]	height: tpt.BarAltitude,
?pt#Point	hSpeed: tpt.HSpeed,
?pt[pos->?ppos]	vSpeed: tpt.VSpeed,
))	course: tpt.Course,
	latitute: get_latitude(
	tpt.Position.Point.pos),
	longitude: get_longitude(
	tpt.Position.Point.pos)]
	from ISSKOI_Track t,
	t.TrackPoints tpt;

The rule operates three kinds of predicates. The head of the rule includes *Track* and *TrackPoint* predicates with named arguments [6]. The predicates correspond to relations of target schema.

The body of the rule incudes *membership predicates* [6] (for instance, predicate *?t#ISSKOI_Track* turns true if and only if the value of *?t* variable is an *ISSKOI_Track* element) and *frame predicates* [6], reflecting XML structure of source data. For instance, predicate *?pt[pos-> ?pos]* turns true if and only if an element, that is the value of *?pt* variable, has a nested element *pos,* and the value of the nested element is *?pos.* Conjunction of body predicates corresponds to a structure of arbitrary *ISSKOI_Track* XML-element.

Source and target attribute values are bound by variables, thus name and structure conflicts are resolved.

Value conflicts can be resolved with the help of functions like *get_latitude.* Such functions are represented in RIF-BLD by *external terms* [6]. Semantics of external terms is not defined in RIF-BLD explicitly, it have to be defined during implementation of the rule. For instance, semantics of the *get_latitude* function is to return the first number from input string that is composed by two numbers.

Considered RIF-BLD rule possess natural logical semantics. Let *t* be an arbitrary tuple of values of variables bound by universal quantifier. Let values from *t* turn all the predicates from rule body into true. Then target relations have to include tuples that correspond to predicates from rule head with the same values of universally bound variables. According to the considered example, target relation have to include tuples *Track (trackId: 56473, name: "copter-1")* and *TrackPoint (pointId: "uuid-2b7ca14", path: 56473, time: "2016-12-12T13:33:11", height: 533.89, latitude: 33.8957, longitude: 246.37).*

The right column of the Table 2 contains an implementation of the rule for integration of data on tracks in HIL.

HIL program *declares* source entity (*ISSKOI_Track* element) and target entities for *Track* and *TrackPoint* relations. « ? » sign in a declaration means that the structure of an entity is not predefined but inferred from the program. Conflict resolution functions *get_latitude, get_longitude* are declared and their signatures are defined. Functions are implemented using the Jaql language [4] in sections marked by *@jaql* directives. Built-in functions *substring, strPos* are used for the implementation.

Insert statements are defined for target relations *Track* and *TrackPoint.* The statements generate tuples to be included into relations. A *from* clause of an *insert* statement defines source entities for the statement. Every membership predicate of a rule body (for instance, *?t#ISSKOI_Track*) corresponds to a definition in a *from* clause (for instance, *ISSKOI_Track t*).

Source relation attributes and their value expressions are defined in *select* clauses of statements. Attribute value expressions correspond to predicates in rule head and frame predicates in rule body. For instance, head predicate *Track (trackId- > ?Id)* and body frame predicate *?t[Id-> ?Id]* correspond to a definition *trackId: t.Id* of the *select* clause of the *insert into Track* statement.

Note that the HIL language manipulates JSON [13] data. Thus transformation of source XML data on tracks into target schema using HIL rule requires preliminary transformation of source data into JSON. Built-in *xmlToJson* function of Jaql [4] can be applied for that. JSON documents generated by HIL rules are further loaded into a relational warehouse over Hadoop (*Hive* [15] is used).

2.2 Ship Data Integration

The left column of the Table 3 contains an example of XML data on ships (*ERRTableShips*) extracted from the Sea Rescue system [19] and an example of relational data on ships extracted from ESIMO Unified State System of Information on the Global Ocean [17]. Relational data are extracted in CSV format and transformed into JSON.

Table 3. Data on ships and the respective elements of the target schema

Source data example (XML, relational model)	Target schema elements (relational model)
<ERRTableShips> <Id>64694571</Id> <Name>Rostov the Great</Name> <Callsing>UBZG5</Callsing> <Dates> <StartDate>2017-01-08 </StartDate> <EndDate>2017-01-09</EndDate> <Dates> </ERRTableShips> [{ "PlatformName": "Rostv the Great", "CountryName": "Russia", "OrganizationName": "Baltic Basin Emergency Rescue", "Designator": "UBZG5", "ImoNumber": "9586796" }]	SARUnit(beginDuty, endDuty, vehicle) Vessel(PK vehicleId, name, call, country, FK owner, imoNumber) LegalEntity(PK entityId, name)

Both collections contain complementary data on the same ships (a ship can be identified by name and call sign).

The right column of the Table 3 contains elements of the target schema that correspond to source data. Data on a ship as a transport facility are concentrated in the *Vessel* relation. Data on a ship as a rescue unit are represented in the *SARUnit* relation. *LegalEntity* relation contains data on ships as legal entities.

The left column of the Table 4 contains a RIF-BLD rule for integration of data on tracks. Head of the rule combines predicates that correspond to target relations. Head predicates are framed by existential quantification on *?owner* variable. Value of this variable is the primary key *entityId* of the *LegalEntity* relation and a foreign key of the *Vessel* relation and is not defined in source data. Strictly speaking, existential quantifier in the head of a rule is not allowed in RIF-BLD, but it is allowed in general framework RIF-FLD.

The body of the rule combines *Ship* predicate from ESIMO collection, and membership and frame predicates that correspond to structure of XML documents extracted from Sea Rescue system. Source collections are joined using fuzzy matching of ship names (names may contain errors); *compareShipName* function is applied for that.

The right column of the Table 4 contains an implementation of the rule for integration of data on ships in HIL.

Preliminary matching of elements from source collections is done with the help of *create link* entity resolution statement. The *from* clause of the statement refers to collections with entities to be matched (*ERRTableShips, Ship*). The *select* clause

Table 4. A specification and implementation of a rule for integration of data on ships

RIF-BLD	HIL

Forall ?Id ?name ?call
 ?beginDuty ?endDuty ?country
 ?ownerName ?imoNumber
(Exists ?owner (
And(
SARUnit(beginDuty->?beginDuty
 endDuty->?endDuty
 vehicle->?Id)
Vessel(vehicleId->?Id
 name->
  External(normalize(?nazv))
 call->?call Country->?country
 owner->?owner
 imoNumber->?imoNumber))
LegalEntity(entityId->?owner
 name->?ownerName)))
:-
And(
?ERRTableShips#ERRTableShips
?ERRTableShips[Id->?Id
 Name->?name Callsing->?call
 Dates->?Dates]
?Dates#Dates
?Dates[StartDate->?beginDuty
 EndDate->?endDuty]
Ship(
 PlatformName->?nazv
 Designator->?call
 CountryName->
  ?country
 OrganizationName->
  ?ownerName
 ImoNumber->?imoNumber)
External(
 compareShipName(?name, ?nazv))
))
	create link ShipLink as
	select [
	 Callsing_Name:
	 [Callsing: es.Callsing,
	  Name: es.Name],
	 Designator_PlatformName:
	 [Designator: s.Designator,
	  PlatformName: s.PlatformName]]
	from ERRTableShips es, Ship s
	match using
	 rule1: es.Callsing = s.Designator and
	   compareShipName(es.Name,
	     s.PlatformName);
	
	insert into SARUnit
	select [vehicle: s.Id,
	         beginDuty: s.Dates.StartDate,
	         endDuty: s.Dates.EndDate
	       ]
	from ShipLink sl, ERRTableShips s
	where sl.Callsing_Name.Name = s.Name
	 and
	 sl.Callsing_Name.Callsing =
	  s.Callsing;
	
	insert into Vessel
	select [vehicleId: s.Id,
	 name:
	  normalize(s.PlatformName),
	 call: s.Designator,
	 country: s.CountryName,
	 owner:
	  get_id(s.OrganizationName),
	 imoNumber: s.ImoNumber]
	from ShipLink sl, Ship s
	where sl.Callsing_Name.Callsing =
	       s.Designator
	  and sl.Callsing_Name.Name =
	       s.PlatformName;
	
	insert into LegalEntity
	select [entityId:
	  get_id(s.OrganizationName),
	 name: s.OrganizationName]
	from ShipLink sl, Ship s
	where sl.Callsing_Name.Name =
	       s.PlatformName and
	  sl.Callsing_Name.Callsing =
	   s.Designator;

defines compound keys (*Callsing_Name, Designator_PlatformName*) for source entities. *Match using* clause contains an entity matching rule *rule1*.

For every target relation a separate *insert* statement is declared. Target relations are populated only on the basis of entities that are matched by *create link* statement.

Existential quantifier is implemented with the help of *get_id* function. The function generates unique identifiers thus producing a value for existential variable *?owner*. Generated value is assigned to *LegalEntity.entityId* primary key and *Vessel.owner* foreign key.

3 Specification and Implementation of Document Data Integration Rules

This section considers an example of integration rule for document data. Source collection contains messages extracted from social networks on accidents in Arctic region as well as data on entities like persons, ships or geographic locations extracted from messages [8]. Data are stored in MongoDB and exported as JSON documents for further integration.

The left column of the Table 5 considers an example of data on accidents. *Entities* are extracted from *Messages*. Entities are bounded with messages via values of compound identifier *id*. The right column contains the respective entities of the target schema.

Table 5. Data on accidents and the respective elements of the target schema

Source data example (document model)	Target schema elements (relational model)

{"Messages": [{
 "id": {"coll_id": "8002",
 "res_id": {
 "site_id": "9b290c9f3bda",
 "doc_id": "3649a5559a62"}},
 "annotation": "Chinese seismic vessel aimed for
   Russian #Barents Sea oil at logistics port
   #Kirkenes",
 "metafields": {
  "mf203": "2016-4-30", "mf205": "eng",
  "mf200": "iceblogger"}
}]}
 	
{"Entities": [{
 "id": {"coll_id": "8002",
 "res_id": {
 "site_id": "9b290c9f3bda",
 "doc_id": "3649a5559a62"}},
"entities": [
  { "s_token": "Barents Sea", "s_tag": "I-ALOC",
    "s_end": 53, "s_begin": 42},
  { "s_token": "Kirkenes", "s_tag": "I-ALOC",
    "s_end": 85, "s_begin": 77}
]}] }
	Document(
	 documentId,
	 collection,
	 source,
	 content,
	 time,
	 language,
	 author)
	
	ExtractedEntity(
	 entityId,
	 document,
	 token,
	 tag,
	 begin
	 end)

Table 6 considers a specification and implementation of a rule for integration of data on accidents extracted from social networks.

Table 6. A specification and implementation of a rule for integration of data on accidents extracted from social networks

RIF-BLD	HIL

Forall ?m ?ext ?doc ?coll ?src
 ?cont ?time ?lang ?auth ?mid
 ?mres ?mf ?eid ?eres
(Exists ?ent(And(
Document(documentId->?doc
 collection->?coll source->?src
 content->?cont time->?time
 language->?lang author->?auth)
ExtractedEntity(entityId->?ent
 document->?doc token->?tok
 tag->?tag begin->?beg
 end->?end)))
:-
And(?m#Messages ?ext#Entities
?m[id->?mid annotation->?cont
  metafields->?mf]
?mid[coll_id->?coll
  res_id->?mres]
?mres[site_id->?src doc_id->?doc]
?mf[mf203->?time mf205->?lang
 mf200->?auth]
?ext[id->?eid entities->?ents]
?eid[coll_id->?coll
 res_id->?eres]
?eres[site_id->?src doc_id->?doc]
?ext#?ents
?ext[s_token->?tok s_tag->?tag
  s_begin->?beg s_end->?end]))
	insert into Document
	select [
	 documentId: m.id.res_id.doc_id,
	 collection: m.id.coll_id,
	 source: m.id.res_id.site_id,
	 content: m.annotation,
	 time: m.metafields.mf203,
	 language: m.metafields.mf205,
	 author: m.metafields.mf200]
	from Message m;
	
	insert into ExtractedEntity
	select [
	 entityId: get_id(strcat(
	  ext.id.res_id.site_id,
	  ext.id.res_id.doc_id)),
	 document: ext.id.res_id.doc_id,
	 token: e.s_token,
	 tag: e.s_tag,
	 begin: e.s_begin,
	 end: e.s_end]
	from Entities ext, ext.entities e;

Similarly to examples provided in the previous section the head of the RIF-BLD rule combines predicates that correspond to entities of target schema. Membership and frame predicates from rule body correspond to structure of documents and extracted entities stored in the source database.

4 General Principles for Specification of Data Integration Rules Using RIF-BLD and Their Implementation Using HIL

Basic principles for specification of data integration rules using RIF-BLD applied in this work can be generalized in the following way:

- integration rules are logical implications; the premise of an implication is called the *body* of a rule, the consequence of an implication is called the *head* of a rule;
- heads of rules are formulas binding predicates with conjunction;

- bodies of rules are formulas binding predicates with conjunction or disjunction;
- head predicates can refer only target schema entities;
- predicates from bodies can refer only source schemas entities;
- properties of tuples belonging to relations are described using predicates with named arguments [5];
- properties of data structures of arbitrary nesting (for non-relational models like XML, document, graph-based) are described using frame and membership predicates [5];
- values of attributes of source and target entities are bound using variables;
- free variables of the body of a rule are bound by outermost universal quantifier *Forall*;
- free variables of the head of a rule not presented in the body of the rule are bound by existential quantifier *Exists* in the head. In fact, these variables correspond with data that are not defined explicitly in source collections;
- nontrivial predicate conditions (that are not equality predicates) and functions resolving conflicts are described as external terms [5].

Basic principles for implementation of RIF-BLD data integration rules using HIL applied in this work can be generalized in the following way:

- source and target schemas entities, functions for conflict resolution and nontrivial predicate conditions are declared using *declare* directive; for functions input and output types are also declared;
- functions are implemented using Jaql language in separate sections of a HIL program or using Java language in external files;
- for every entity or relation in the head of a rule an *insert* statement is defined that creates tuples for this entity or relation:
  - source entities are declared in the *from* clause of the statement; every predicate with named variables and membership predicate in the body of a rule corresponds with a declaration;
  - attributes of target relations and expressions for their values are declared in the *select* clause of the statement. Names of attributes are taken from predicates in the head. Expressions for attribute values are constructed in accordance with terms in the head predicates or with terms in the body frame predicates;
  - predicates that reflect ways of joining the source entities and their selection are described in the *where* clause. Joining predicates are constructed on the basis of coincidence of variable names in the body frame predicates. Selection predicates are constructed on the basis of body predicates that are conditions over separate source collections;
- if join of source collections in the body of a rule is done using nontrivial predicates then preliminary join is done with a *create link* entity resolution statement:
  - joined collections are declared in the *from* clause;
  - keys (simple or compound) identifying source entities are declared in the *select* clause;
  - rules for source entity matching are described in the *match* clause;
  - collection of pairs of matched entities is used as source in *insert* statements producing tuples of target relations.

Implementation principles described above can be formalized further with the following implementation patterns.

Pattern 1. Simple attribute value binding.

RIF-BLD	HIL

`Forall ?sst ?v1 ?v2 ?v3 (`
`AND(`
` TargetRel1(a1->?v1  a3->?v3)`
` TargetRel2(a2->External(f(?v2)))`
`)`
`:-`
`And(`
` SourceRel(b1->?v1)`
` ?sst#SourceStruct`
` ?sst[b2->?v2  b3->?v3]`
`))`
	`declare TargetRel1: ?`
	`declare TargetRel2: ?`
	`declare SourceRel: ?`
	`declare SourceStruct: ?`
	
	`declare f: function Type1 to Type2;`
	
	`insert into TargetRel1`
	`select [a1: sr.b1, a3: sst.b3]`
	`from SourceRel sr,`
	`     SourceStruct sst;`
	
	`insert into TargetRel2`
	`select [a3: f(sr2.b2)]`
	`from SourceStruct sst;`

Several general principles of implementation are revealed by this pattern: source entities (relation *SourceRel* and structure *SourceStruct*), target entities (relations *TargetRel1* and *TargetRel2*) and a function *f* are declared by *declare* directive; *insert* statements are constructed for both target relations; source entities are associated with variables in *from* clauses of *insert* statements. Source and target attribute values bound by variables in RIF-BLD are represented by expressions in *select* clauses. For instance, values of *SourceRel.b1* and *TargetRel1.a1* bound by *?v1* variable are represented by *a1: sr.b1* expression. External functions like *f* can be used to bind attribute values.

Pattern 2. Unnesting a set attribute.

RIF-BLD	HIL

`Forall ?sst ?v1 ?v2 ?v3 (`
`TargetRel(a->?v3)`
`:-`
`And(`
` ?sst#SourceStruct  ?sst[b1->?v1]`
` ?v2#?v1  ?v2[b2->?v3]`
`))`
	`insert into TargetRel`
	`select [a: v2.b2]`
	`from SourceStruct sst, sst.b1 v2;`

If a source structure (for instance, *SourceStruct*) includes an attribute of a set type (for instance, *b1*) and internal substructures of the attribute values are bound with target attributes (for instance, attribute *b2* is bound with *TargetRel.a* by a variable *?v3*) then the attribute set value is declared as a source in the *from* clause of the *insert* statement (*sst.b1 v2*).

Pattern 3. Structure path expression.

<table>
<tr><th>RIF-BLD</th><th>HIL</th></tr>
<tr><td><pre>Forall ?sst ?v1 ?v2 ?v3 (
TargetRel(a->?v3)
:-
And(
 ?sst#SourceStruct1 ?sst[b1->?v1]
 ?v1#SourceStruct2 ?v1[b2->?v2]
 ?v2#SourceStruct3 ?v2[b3->?v3]
))</pre></td><td><pre>insert into TargetRel
select [a: sst.b1.b2.b3]
from SourceStruct sst;</pre></td></tr>
</table>

Source structures are allowed to have an arbitrary number of nested substructure levels. These levels can be referenced in RIF-BLD via membership predicates (for instance, *?v1#SourceStruct2*) and frame predicates (for instance, *?v1[b2-> ?v2]*). In HIL such groups of predicates are represented by path expressions in *select* clauses of statements (for instance, *sst.b1.b2.b3*).

Pattern 4. Existential variable in the head of a rule.

<table>
<tr><th>RIF-BLD</th><th>HIL</th></tr>
<tr><td><pre>Forall ?v1 v2 (
Exists ?v3 (
 TargetRel(a1->?v1 a2->?v3)
)
:-
And(
 SourceRel(b1->?v1 b2->?v2)
))</pre></td><td><pre>insert into TargetRel
select [a1: sr.b1,
        a2: get_id(sr.b2)]
from SourceRel sr;</pre></td></tr>
</table>

Data required to initialize a target attribute (for instance, *a2*) may not be explicitly defined in source entities (for instance, in the *SourceRel* relation). In this case a variable denoting the value of the attribute (for instance, *v3*) have to be existentially quantified. In HIL existential quantification have to be implemented explicitly by some concrete value (generated, for instance, by the *get_id* function).

Pattern 5. Simple join and selection of sources.

<table>
<tr><th>RIF-BLD</th><th>HIL</th></tr>
<tr><td><pre>Forall ?sst ?vb1 ?vb2 ?vc2 (
TargetRel(a1->?vb2 a2->?vc2)
:-
And(
 SourceRel(b1->?vb1 b2->?vb2)
 ?sst#SourceStruct
 ?sst[c1->?vb1 c2->?vc2]
 External(pred(?vc2))
))</pre></td><td><pre>insert into TargetRel
select [a1: sr.b2, a2: sst.b3]
from SourceRel sr,
     SourceStruct sst
where sr.b1 = sst.c1 and
      pred(sst.c2);</pre></td></tr>
</table>

Selection predicates for attribute values of source collections (for instance, *pred(?vc2)*) are represented by the respective predicates in the *where* clause of HIL statements. Simple join conditions for source collections (for instance, attributes *SourceRel.b1* and *SourceStruct.c1* have to have the same value denoted by *?vb1*) can also be represented by predicates in the *where* clause (for instance, *sr.b1 = sst.c1*).

Pattern 6. Nontrivial matching of source entities.

<table>
<tr><th>RIF-BLD</th><th>HIL</th></tr>
<tr><td><pre>
Forall ?sst ?vb1 ?vb2 ?vc (
TargetRel(a1->?vb1 a2->?vc)
:-
And(
 ?sst#SourceStruct
 ?sst[b1->?vb1 c1->?vc]
 SourceRel(b2->?vb2 c2->?vc)
 Or(External(pred1(?vb1 ?vb2))
  External(pred2(?vb1 ?vb2 ?vc)))
))
</pre></td><td><pre>
create link
 SourceStruct_SourceRel_link as
select[
 b1_c1: [b1: sst.b1, c1: sst.c1]
 b2_c2: [b2: sr.b2, c2: sr.c2]
]
from SourceStruct sst,
     SourceRel sr
match using
 rule1: sst.c1 = sr.c2 and
        pred1(sst.b1, sr.b2),
 rule2: sst.c1 = sr.c2 and
      pred2(sst.b1, sr.b2, sst.c1)

insert into TargetRel
select [a1: sst.b1, a2: sst.b2]
from SourceStruct sst,
  SourceStruct_SourceRel_link lnk
where sst.b1 = lnk.b1_c1.b1 and
      sst.c1 = lnk.b1_c1.c1
</pre></td></tr>
</table>

Source entities can be matched using complicated conditions. For instance, entities from *SourceStruct* and *SourceRel* are matched using equality of *SourceStruct.c1* and *SourceRel.c2* attributes as well as using a disjunction of two predicates: *pred1(?vb1 ?vb2)* and *pred2(?vb1 ?vb2 ?vc)*. Such matching is implemented via *create link* statement. Attributes *SourceStruct.b1* and *SourceStruct.b1* are identified as the compound key for *SourceStruct*; attributes *SourceRel.b2* and *SourceRel.c2* are identified as the key for *SourceRel*. Matching predicates are implemented via a couple of rules in the *match* clause of the *create link* statement. A collection of matched entities *SourceStruct_SourceRel_link* is used as a source for the *insert* statement producing tuples of the *TargetRel* relation.

5 Conclusions

This work considers an approach for specification of data integration rules using RIF-BLD logic dialect that is a recommendation of W3C. Wide range of RIF-BLD specification capabilities allows combining entities defined in different sources represented in different data models in the same rule. Rules considered in the paper can contain target entity predicates only in the heads and source entity predicates only in the bodies. Logical semantics of RIF-BLD provides for unambiguous interpretation of data integration rules and allows their implementation using various languages. The paper proposes an approach for implementation of RIF-BLD rules using IBM High-level integration language. Distributed execution of HIL programs over Hadoop is achieved by their compilation into MapReduce programs.

Further research is required to solve issues of interpretation and implementation of RIF-BLD programs that allow target entities to be referenced in rule bodies. An adaptation of data exchange chase procedure [9] can be applied for that. Other direction of further research is an implementation of automatic mapping of RIF-BLD specifications into HIL programs.

Acknowledgement. The research is partially supported by Russian Foundation for Basic Research, projects 15-29-06045, 18-07-01434.

References

1. Apache Hadoop Project (2017). http://hadoop.apache.org/
2. Ballard, C., Alon, T., Dronavalli, N., Jennings, S., Lee, M., Toratani, S.: IBM InfoSphere Information Server Deployment Architectures (2012). ibm.com/redbooks
3. Bar-Or, A., Choudhary, S.: Transform XML using the DataStage XML stage. IBM developerWorks (2011)
4. Beyer, K.S., Ercegovac, V., Gemulla, R., Balmin, A., Eltabakh, M., Kanne, C.-C., Ozcan, F., Shekita, E.J.: Jaql: a scripting language for large scale semistructured data analysis. In: 37th International conference on very large data bases VLDB, pp. 1272–1283. Curran Associates, New York (2011)
5. Boley, H., Kifer, M. (eds.): RIF Framework for Logic Dialects. W3C Recommendation, 2nd edn., 5 February 2013
6. Boley, H., Kifer, M. (eds.): RIF Basic Logic Dialect. W3C Recommendation, 2nd edn., 5 February 2013
7. Burdick, D., Hernández, M.A., Ho, H., Koutrika, G., Krishnamurthy, R., Popa, L., Stanoi, I. R., Vaithyanathan, S., Das, S.: Extracting, linking and integrating data from public sources: a financial case study. IEEE Data Eng. Bull. **34**(3), 60–67 (2011)
8. Devyatkin, D., Shelmanov, A.: Text processing framework for emergency event detection in the Arctic zone. In: Kalinichenko, L., Kuznetsov, Sergei O., Manolopoulos, Y. (eds.) DAMDID/RCDL 2016. CCIS, vol. 706, pp. 74–88. Springer, Cham (2017). https://doi.org/10.1007/978-3-319-57135-5_6
9. Fagin, R., Kolaitis, P., Miller, R., Popa, L.: Data exchange: semantics and query answering. Theoret. Comput. Sci. **336**(1), 89–124 (2005)
10. Hernandez, M., Koutrika, G., Krishnamurthy, R., Popa, L., Wisnesky, R.: HIL: A high-level scripting language for entity integration. In: 16th Conference (International) on Extending Database Technology Proceedings EDBT 2013, pp. 549–560 (2013)
11. IBM InfoSphere BigInsights Version 3.0 Information Center. https://goo.gl/lZpEQd
12. InfoSphere Big Match for Hadoop. Technical Overview. https://goo.gl/0TMqvw
13. Introducing JSON. http://www.json.org/
14. Miner, D.: MapReduce Design Patterns: Building Effective Algorithms and Analytics for Hadoop and Other Systems. O'Reilly Media, Newton (2012)
15. The Apache Hive data warehouse software. http://hive.apache.org/
16. Briukhov, D.O., Skvortsov, N.A., Stupnikov, S.A.: Methods of integration of multistructured data on Arctic zone for extraction of information aimed at support of search and rescue operations. Highly Available Syst. **13**(2), 3–19 (2017)
17. The Unified State System of Information on the Global Ocean. http://portal.esimo.ru/portal
18. Complex integrated information system MoRe. http://www.marsat.ru/ciis-more
19. Sea Rescue (Poisk-More) Software Suite. http://map.geopallada.ru/
20. Skvortsov, N.A., Briukhov, D.O.: Development of information warehouse schema for support of search and rescue activities in the Arctic region. Highly Available Syst. **13**(2), 20–44 (2017)

21. Stupnikov, S.: Specification and implementation of multimodel data integration rules. In: Selected Papers of the XIX International Conference on Data Analytics and Management in Data Intensive Domains (DAMDID/RCDL 2017), CEUR Workshop Proceedings, vol. 2022, pp. 197–205 (2017)
22. Steinberg, D., Budinsky, F., Paternostro, M., Merks, E.: EMF: Eclipse Modeling Framework, 2nd edn. Addison-Wesley Professional, Boston (2008)
23. EMFText Concrete Syntax Mapper. http://www.emftext.org/index.php/EMFText
24. Abstract and concrete syntax of RIF-FLD. GitHub Repository (2018). https://github.com/sstupnikov/ModelTransformation/tree/master/RIF_FLD/

Approach to Forecasting the Development of Situations Based on Event Detection in Heterogeneous Data Streams

Ark Andreev, Dmitry Berezkin, and Ilya Kozlov(✉)

Bauman Moscow State Technical University, Moscow, Russia
arkandreev@gmail.com, berezkind@bmstu.ru,
kozlovilya89@gmail.com

Abstract. The article deals with the problem of automated forecasting of situations development based on analysis of heterogeneous data streams. Existing approaches to solving this problem are analyzed and their advantages and disadvantages are determined. The authors propose a novel approach to forecasting of situations development based on event detection in data streams. The article analyzes various models of events and situations and event detection methods that are used in the processing of heterogeneous data. A method for generating possible scenarios of situations development is described. The method generates scenarios using the principle of historical analogy, taking into account the dynamics of the current situation's development. The probability of the generated scenarios' implementation is estimated via logistic regression. The generated set of scenarios is analyzed using Analytic Hierarchy Process to identify the optimistic and the pessimistic scenario. The authors describe a way to supplement scenarios with recommendations for decision makers. The results of experimental evaluation of the quality of the proposed approach are presented.

Keywords: Analogy · Decision support system · Event detection
Forecasting · Scenario analysis · Situational analysis · Text stream analysis

1 Introduction

Modern information systems solve a wide variety of tasks of processing data that is downloaded from the Internet or from specialized sources. Many of such systems have to deal with the data's heterogeneity and constant growth of its volumes. Data may be presented in numerical or textual form, may be unstructured or transformed into a structured form using different knowledge representation models. In the Internet, data is usually represented in unstructured form, which corresponds to a natural way for people to express their thoughts and share their knowledge.

Data that is regularly received from sources (such as news, analytical articles, legal documents, stock index values, sensor readings) is dynamic and, therefore, reflects the development of various situations. Hence it can be used to support decision making in real time. In order to make the best possible managerial decisions one should be able not only to track the current development of the situation but also to predict its further development. This applies to various subject areas: forecasting further development of

L. Kalinichenko et al. (Eds.): DAMDID/RCDL 2017, CCIS 822, pp. 213–229, 2018.
https://doi.org/10.1007/978-3-319-96553-6_16

the industry allows companies to make the most profitable deals, and predicting the development of an emergency situation at an industrial facility allows taking timely actions to minimize the negative consequences. Due to the large amount of data generated by the sources, analysis and forecasting should be performed automatically.

In [1], the authors proposed a method for forecasting the development of situations based on the analysis of a stream of text documents. This work is dedicated to the evolvement of the research presented in [1]. It describes a hybrid approach to predicting the further development of situations based on the processing of heterogeneous data streams. The approach is presented in Sect. 3. The article also expands [1] by describing various ways to represent events and situations in Sect. 4. The method for generating scenarios of situations development, presented in Sect. 5, has been refined to take into account the diversity of event models.

2 Related Work

Many different approaches have been proposed to the analysis and prediction of situations. Some articles describe situations by sets of certain numerical parameters [2]. In this case forecasting is performed by means of time series analysis and regression analysis methods. While such approaches are applicable for analyzing numerical data, more expressive models are needed to represent situations described in text documents in natural language.

Some works propose approaches to forecasting based on cognitive maps and signed digraphs [3]. They represent a situation as a graph where nodes correspond to the factors of the situation, and the edges reflect the influence of the factors on each other. In this case forecasting consists in estimating the future values of the factors by modeling the development of the situation under influence of various control actions. Representation of the current situation in a form of a cognitive map is created by an expert. Therefore such approaches are not applicable for automatic processing of data streams.

Many researchers use the principle of analogy: forecasting of the further development of a situation is performed by searching similar situations that occurred in the past. Works based on the principle of analogy use various approaches to representing situations.

In [4] a situation is represented as a fragment of a semantic network containing objects relevant to the situation and relations between them. Such a model can be generated automatically only for certain domains which makes the approach inapplicable for predicting the development of arbitrary situations.

Some articles describe situations by sets or vectors of parameters with certain values [5, 6]. In order to detect the analogy situations are compared by various similarity measures such as Euclidean distance, city block distance, Hamming distance, cosine similarity and others. The problem of these approaches is static description of situations: they don't take into account dynamics of situations' development when determining the similarity between them.

Dynamics can be taken into account by representing all possible variants of the situation's development as paths in a graph or automaton [7, 8]. Such approaches can detect strict analogy only: the current situation has to match one of the paths in the graph precisely. However, each situation described by texts in natural language has a number of distinctive features. Thus it is necessary to be able to detect an analogy between situations that have a certain difference in structure and content.

An approach based on a non-strict analogy was proposed in [9]. Situations are represented by sequences of events, the similarity between them is determined using a modified Levenshtein distance. But this approach assumes that an object and a subject are determined for each event, which cannot be done automatically for arbitrary text messages.

3 Hybrid Approach to Forecasting the Development of Situations

Events may be considered as changes occurring in the data stream and corresponding to different stages in the development of the situation [10]. Thus presenting a situation as a sequence of interrelated events reflects dynamics of its development, which is significant for forecasting. In this case forecasting consists in generating possible scenarios of the situation's further development. Each scenario is a hypothetical chain of events that may occur in the future. Also it is necessary to supplement each scenario with recommendations regarding actions that should be taken to promote or counteract the development of the situation according to this scenario.

Models of events differ significantly in various subject areas, and different methods are used to extract such events from heterogeneous data streams. In this regard, the hybrid approach to situation analysis should be able to handle different data models. On the other side, regardless of the domain, the development of a situation is considered as a sequence of events. Therefore, the approach must analyze situations and generate scenarios in a uniform way.

To meet these requirements, the authors propose an approach to forecasting the development of situations that consists of the following stages:

1. Periodic downloading of heterogeneous (numerical, tabular, textual) data from various sources, such as news portals on the Internet, document databases, relational databases.
2. Primary processing and cleansing of the data.
3. Detection of events ε_i in the data stream. Event model and event detection method are specific for each data type.
4. Constructing of situations that are chains of interrelated events: $s = (\varepsilon_s^1, \varepsilon_s^2, \ldots, \varepsilon_s^n)$. The way of combining events into a situation also depends on the characteristics of the data.
5. Generation of possible scenarios of the current situation's further development. Each scenario $\xi = (\varepsilon_\xi^1, \varepsilon_\xi^2, \ldots, \varepsilon_\xi^n)$ is a potential continuation of the situation. Scenarios are generated uniformly for all types of data. The generated set of scenarios is analyzed to identify three scenarios of particular interest for the decision maker (DM): the optimistic one, the pessimistic one and most probable one.

6. Preparation of recommendations for the DM. Each of the generated scenarios ξ is supplemented with a recommendation rec_{ξ}.

The stages of the proposed approach are demonstrated in Fig. 1. Experts tune the parameters of the methods that are used at various stages. After training, data streams are analyzed automatically. The analysis results in the set of generated scenarios and recommendations that are provided to the DM.

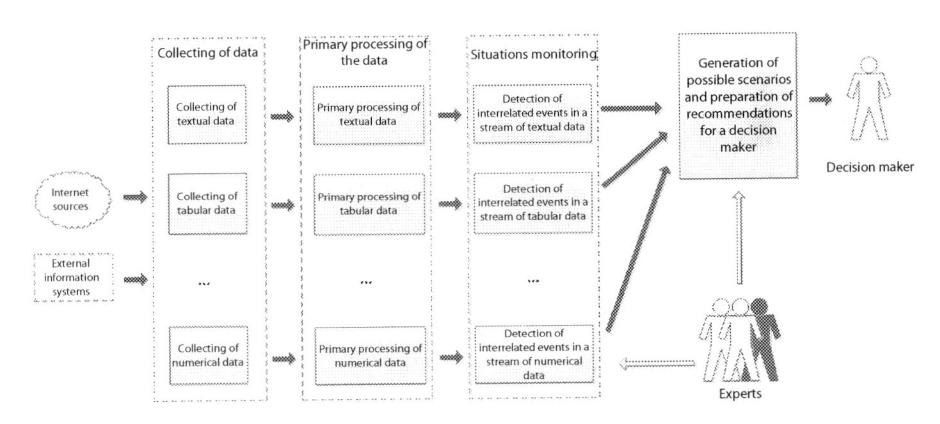

Fig. 1. Stages of the hybrid approach to forecasting the development of situations.

Primary processing consists in cleansing the target data of irrelevant and service information from the source using extraction rules that are specific to each source site. Whenever the structure of the site changes, it is necessary to promptly detect such a change and edit the settings of the extraction rules in order to avoid downloading incorrect data. To ensure high quality of the data, it is also important to detect and discard duplicates, that is, messages generated from other documents by slightly modifying their structure and content. The authors proposed solutions to these problems in [11, 12].

The prepared data is subjected to event detection procedure. The following section describes various ways to represent events that are used when working with different types of data.

4 Models of Events and Situations

A lot of various models have been suggested to represent events reflected in a stream of text documents. In [13], an ontological representation of an event is considered, however in practice researchers use other models that are more convenient for automatic analysis of a text stream.

In [5] an event is considered as a change in properties of the text stream, which is characterized by a significant change in topical composition of the documents. This approach consists in splitting all documents of the stream into clusters, and the current state of the stream is characterized by distribution of recently downloaded documents over the clusters. The occurrence of a new event is detected when the distribution changes significantly.

In some papers [14, 15], each event is characterized by a burst of intensity of appearance of certain words in the text stream. Thus, the event is characterized by the moment of its occurrence and a set of words that had a burst of intensity at that moment: $\varepsilon_i = \left(\left\{l_i^1, l_i^2, \ldots, l_i^n\right\}, t_i\right)$, where l_i^j is a word and t_i is a point of time. Events are detected by analyzing signals constructed for each word that is present in the text stream. Words are grouped into clusters based on the correlation between their signals. The moment when those words' intensity experienced a surge is considered as the time of the event's occurrence.

The above-described ways of representing events are not applicable for monitoring the development of situations, since they do not provide the possibility to detect interrelated events and build situational chains. This problem can be solved by using approaches that represent an event by a single text document: $\varepsilon_i = d_i$. Such approaches are based on the detection of text messages containing information about events not described in previously downloaded documents. There are various methods to detect such messages.

In [16], the documents of the text stream are ranked in such a way that the biggest weight is assigned to messages with the highest similarity to recently downloaded documents and the lowest similarity to messages downloaded over a considerable period of time. The heaviest documents represent new events.

In [17], event detection is regarded as a task of binary classification of new documents. Each document is assigned to one of two classes depending on the presence or absence of information about a new event in the document. If similarity between the document and each of previously downloaded messages does not exceed a threshold value, it is considered that this document describes a new event. To perform the comparison each document is represented by a vector model: $\varepsilon_i = \left(w_i^1, w_i^2, \ldots, w_i^n\right)$, where w_i^j is a weight that reflects the importance of the j th word for the i th event. Weights are calculated via TF-IDF method. Models of documents are compared using cosine distance.

Events represented by single documents can be used to create situational chains. Two events are assigned to the same situation if similarity between their vector models exceeds a threshold value.

However, in many subject areas events are described not by a single document, but by a set of messages (for example, a group of news published by different agencies). Each of these messages can contain important information about the event, that is absent in other documents. In this regard, models based on groups of documents can represent information about the event more comprehensively: $\varepsilon_i = \left\{d_i^1, d_i^2, \ldots, d_i^k\right\}$.

Approaches based on such models consider event detection as a task of clustering the stream of text messages. For instance, in [18] events are identified via hierarchical clustering of a static collection of documents. A shortcoming of this approach is the need to redo the clustering completely whenever the collection is replenished with new documents. The task of event detection in a collection constantly receiving new documents is usually solved by using incremental clustering [19]. Every new document is compared to previously formed clusters, each of which corresponds to a certain event.

Depending on the similarity between the document and the events, the document is either assigned to the closest cluster or used to create a new cluster. To perform such a comparison, the cluster is usually represented by a vector of words. In [20] the vector is extended with seven types of named entities automatically extracted from texts (such as "Persons", "Organizations", "Locations", etc.). The similarity between the event and the document is determined by cosine distance between their vector models multiplied by a decay coefficient that depends on the interval between the time of the event's occurrence and the moment of the document's publication.

In [21] the authors proposed a clustering-based approach to event detection which can be flexibly adjusted to different subject areas. Each document is represented by a complex model that contains components describing content, structure and metadata of the document: $d_i = \left(d_i^w, d_i^{tw}, d_i^c, d_i^p, d_i^n, d_i^{dt}, d_i^e, d_i^g, d_i^t\right)$. In particular, some of the model's components are: a vector of keywords, sets of named entities related to the document, a set of numerical values extracted from the text, publication time of the message. Each event ε_j is represented by a similar multi-component model that reflects features of a group of documents related to the event. In order to determine the similarity between the document and the event, a pairwise comparison of their models' components is performed. Components are compared via proximity measures such as cosine distance, Levenshtein distance, Jaccard index. The results of pairwise comparisons of the components are assembled into a vector $\rho_{i,j} = \left(\rho_{i,j}^w, \rho_{i,j}^{tw}, \rho_{i,j}^c, \rho_{i,j}^p, \rho_{i,j}^n, \rho_{i,j}^{dt}, \rho_{i,j}^e, \rho_{i,j}^g, \rho_{i,j}^t\right)$ that is used to compute the similarity between d_i and ε_j via Support Vector Machine. Depending on the similarity it is determined whether the document relates to the event or not.

More detailed description of the document and event models, as well as the event detection method, is given in [21]. The method can be used to analyze documents in any language but requires topical queries prepared by experts, as well as dictionaries of names of persons and organizations and geographical names in the desired language. The higher quality of event detection may be achieved by utilizing methods of morphological, syntactic and semantic analysis of texts.

The detected events are analyzed to identify pairs $p_{ij} = \left(\varepsilon_i, \varepsilon_j\right)$ of events that potentially belong to the same situation. These pairs are used to build a situational graph $G = (E, P)$ where nodes $E = \{\varepsilon_i\}$ correspond to events and edges $P = \{p_{ij}\}$ correspond to the identified pairs (each edge is directed from the earlier event in the pair towards the later one). Each path in G is a potential situation: $s = \left(\varepsilon_s^1, \varepsilon_s^2, \ldots, \varepsilon_s^n\right)$. The graph model reflects the non-linearity of the analyzed processes: any event can belong to multiple situations.

Figure 2 demonstrates an example of detected situation that consists of four events related to testing of driverless car service by Uber.

Some researchers consider event detection in a stream of text documents as an Information Extraction task [22]. In this case, each event is represented as a frame that contains slots holding information about the participants, conditions and time scope of the event [23]. Events are extracted from texts by means of lexical-syntactic or lexical-semantic patterns. One has to prepare separate patterns for each type of events that should be extracted.

Situation: Uber's driverless car service testing

Start date and time: 09/14/2016 06:00:00
End date and time: 12/22/2016 16:28:00
Geographic names: USA, California, Pennsylvania, Arizona
Names of related organizations: Uber, Ford, Google, Tesla
Names of related persons: T. Kalanick, A. Levandowski, D. Ducey
Topics: Driverless taxi, autonomous vehicles, transport

Uber launches driverless car service in landmark US trial

Document name	Publication date	Publication time	Source
We Take a Ride in the Self-Driving Uber Now Roaming Pittsburgh	09/14/2016	06:00:00	WIRED
Uber launches driverless car service in landmark US trial	09/14/2016	16:16:00	ABC Online
Watch: Uber launches driverless car service	09/14/2016	16:27:00	The Telegraph

State regulators demand Uber halt self-driving car program, threaten legal action

Document name	Publication date	Publication time	Source
State regulators demand Uber halt self-driving car program, threaten legal action	12/14/2016	12:25:00	The Mercury News
If Uber doesn't like California's rules, it can test its driverless cars elsewhere	12/17/2016	05:00:00	Los Angeles Times
Uber threatened with legal action over safety of self-driving cars in California	12/17/2016	09:28:00	The Sun

Uber halts S.F. self-driving pilot in showdown with DMV

Document name	Publication date	Publication time	Source
Uber Halts San Francisco Driverless Cars As DMV Revokes Test Fleet Registration	12/21/2016	21:18:00	Forbes
Uber halts S.F. self-driving pilot in showdown with DMV	12/21/2016	22:17:00	SFGate
Roadblock: Uber's driverless fleet stops San Francisco experiment	12/22/2016	06:04:00	TechRepublic

Uber dispatches driverless cars to Arizona

Document name	Publication date	Publication time	Source
Uber sends self-driving cars to Arizona after failed San Francisco pilot	12/22/2016	12:35:00	The Mercury News
Uber dispatches driverless cars to Arizona	12/22/2016	16:28:00	USA TODAY

Fig. 2. An example of event detection and situation construction.

Events represented by frames are usually combined into a situational chain if they have identical values of certain slots. For instance, in [9] chains are composed of events with the same "subject-object" pair.

Similar representation of events (namely, as records with a set of fields) is used in process mining technique [7]. Sequences of such events describe the development of the analyzed processes in a system or organization. In this case events are combined into a chain if they have the same process instance ID.

Some approaches to event detection consider not the content of documents, but their amount. They analyze time series, each value of which is the number of messages received during the corresponding time interval [24]. Events are detected by means of generative models that assume that each message can relate to each of the events with different probabilities, and these probabilities depend on the document publication time. Events in this case are characterized by their probability density functions [25].

Researchers also have to deal with the task of detecting events in time series when processing numeric data that is dynamically received from sources (for example, sequences of exchange rates values or measuring instruments readings). In this task an event is determined by a moment of significant change in the time series characteristics [26]. Such moments are detected by means of statistical methods, in particular, maximum likelihood method. The events detected in the time series are combined into chains that are subjected to further analysis.

Another technique based on event detection in data streams is complex event processing (CEP). CEP assumes that multiple basic events are dynamically detected in data streams, such as sensor triggering or certain changes in the characteristics of numerical or text data. Basic events are atomic and occur at certain time moments. A complex event is aggregated from basic events and other complex events by means of logical rules: $\varepsilon^c = E(a, c, t_b, t_e)$, where $a = \{a_1, a_2, \ldots, a_m\}$ is a set of complex event's attributes, $c = \{\varepsilon_1, \varepsilon_2, \ldots, \varepsilon_n\}$ is a set of basic and complex events that cause event ε^c to happen, t_b and t_e are moments of the event's beginning and ending. Researchers have proposed various languages based on temporal logics to create the rules for constructing complex events [27, 28]. CEP assumes that events are connected by hierarchical relations, thus the situational chain may be represented as a complex event of a higher level.

5 Method for Generating Scenarios of Situations Development

Approaches that consider situations as sequences of events usually perform forecasting using the principle of historical analogy. These approaches search for analogues of the current situation to assign this situation to one of the classes that describe various possible outcomes of the situation development [9]. However, detecting an analogue chain s_e for the current sequence s_c can not only show the possible result of s_c but also explain which events may lead to this result. Such a forecast can be made if the entire current sequence is similar to the initial part $st(s_e, s_c)$ of the analogous chain. In this case, one can predict an occurrence of a sequence of events that will resemble the final part of the analogue chain $fin(s_e, s_c)$. This final part can be considered as a possible scenario of the further development of the current situation s_c.

Implementation of this approach to scenario generation requires a collection of sample situations $S_e = \{s_e^i\}$. The samples are prepared by experts in accordance with the subject area and specifics of the task. For instance, to analyze the situation related to driverless taxi testing (Fig. 2) the authors used a set of sample situations reflecting the development of various technologies in the past.

If the current situations are considered as possible paths in the situational graph G, the process of forecasting situation development consists of the following steps:

1. Detecting a pair of analogous events $(\varepsilon_c, \varepsilon_e)$ where ε_c is a new or recently changed event from the situational graph and ε_e belongs to one of the sample situations s_e.
2. Searching for a chain s_c (current situation) in the situational graph that contains ε_c and has the highest similarity with the initial part $st(s_e, s_c)$ of the sample situation s_e. If the similarity between $st(s_e, s_c)$ and s_c is high enough, s_e is considered as an analogue of s_c and its final part $fin(s_e, s_c)$ is regarded as a possible scenario of the further development of s_c.
3. Identifying three scenarios of particular interest for the DM: the most probable one, the optimistic one and the pessimistic one.
4. Preparing recommendations regarding actions that should be taken to promote the development of the current situation according to the most favorable scenarios.

A method for detecting analogous events depends on the model of an event and thus is chosen individually for every subject area and data type. In [1], the authors proposed a method for determining the analogy between events extracted from a stream of text documents.

The authors suggest using the situational graph model when processing streams of text documents in natural language. However, when working with other types of data (for example, when analyzing event logs by means of Process Mining), current situation can be unambiguously constructed at the stage of event detection. In this case, one just needs to compare the current chain s_c to the initial parts $st(s_e, s_c)$ of sample situations $s_e \in S_e$ and determine which ones are analogues of s_c.

In both cases, it is necessary to solve the task of calculating the similarity between the current and sample situations in order to determine whether they are analogous.

5.1 Calculating the Similarity Between Situations

The authors use the following restriction when comparing two situations: the order of events in the first situation must be identical to the order of their analogues in the second one. Events order reflects the causal relations between the events and determines the logical structure of the situation. Hence situations with different order of the corresponding events cannot be considered as analogues.

Thus, the compared situations may contain several pairs of analogous events that are located in the situations in the identical order. They are denoted by grey circles in Fig. 3. Analogy between events is illustrated by dashed lines. Besides, each of the situations may contain events without analogues in the other chain. In Fig. 3 additional events of the current situation are denoted by horizontally hatched circles. Circles denoting additional events of the sample situation are vertically hatched.

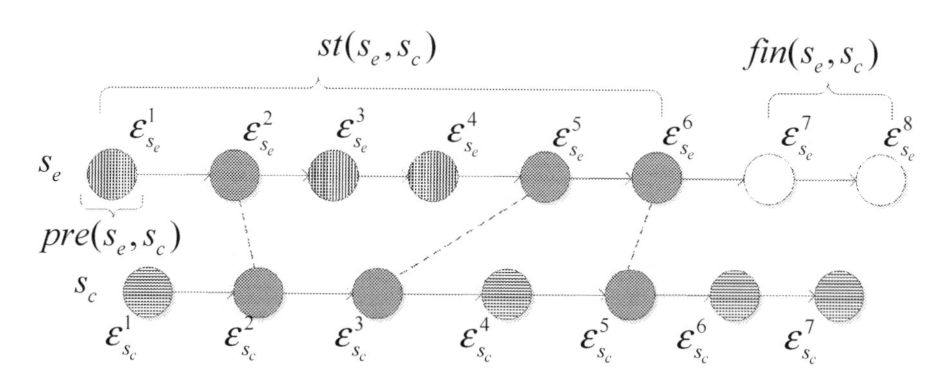

Fig. 3. Comparison of event chains.

The goal of the situations comparison is to determine whether the current situation is analogous to the initial part $st(s_e, s_c)$ of the sample. Thus events that compose the final part $fin(s_e, s_c)$ of the sample situation (events that follow the last event that has an analogue in the current situation) don't affect the similarity between the chains. These events are denoted with white circles.

Table 1 demonstrates an example of comparison of the current and the sample situation. In this case, (ε_1^1,ε_2^1), (ε_1^2,ε_2^2) and (ε_1^3,ε_2^3) are pairs of analogous events, ε_1^4 is an additional event of the current situation, and ε_2^4 is the only event of the final part of the sample situation.

Table 1. Comparison of the current and the sample situation.

Current situation	Sample situation
ε_1^1 : Uber launches driverless car service in landmark US trial	ε_2^1 : Samsung Galaxy Note 7 unveiled
ε_1^2 : State regulators demand Uber halt self-driving car program, threaten legal action	ε_2^2 : US government urges Samsung Galaxy Note 7 owners to stop using the smartphone
ε_1^3 : Uber Halts San Francisco Driverless Cars As California Department of Motor Vehicles Revokes Test Fleet Registration	ε_2^3 : US Government announces official Samsung Galaxy Note 7 recall
ε_1^4 : Uber sends self-driving cars to Arizona after failed San Francisco pilot	
	ε_2^4 : Samsung ends Galaxy Note 7 production, sales amid safety fears

Similarity between the situations is measured via modified Levenshtein distance: the distance between chains is determined by the summarized weight of the operations that are needed to transform one chain into another. The initial part $st(s_e, s_c)$ of the sample situation can be transformed into the current situation s_c using the following operations:

- Deleting an event $\varepsilon_{s_e}^i$ that doesn't have an analogue in s_c. The weight of the operation $w_{del}\left(\varepsilon_{s_e}^i\right)$ reflects the significance of the deleted event. The method of calculating the weight depends on the event model. The summarized weight of operations of this type is calculated as $W_{del} = \sum_{\varepsilon_{s_e} \in E_{del}} w_{del}(\varepsilon_{s_e})$ where E_{del} denotes a set of deleted events.
- Adding an event $\varepsilon_{s_c}^i$ that doesn't have an analogue in $st(s_e, s_c)$. The weight of the operation $w_{add}\left(\varepsilon_{s_c}^i\right)$ is determined similarly to $w_{del}\left(\varepsilon_{s_e}^i\right)$. The summarized weight of operations of this type is calculated as $W_{add} = \sum_{\varepsilon_{s_c} \in E_{add}} w_{add}(\varepsilon_{s_c})$ where E_{add} denotes a set of added events.
- Replacing an event $\varepsilon_{s_e}^i$ from the sample chain to its analogue $\varepsilon_{s_c}^j$ from the current situation. The weight of the operation $w_{rep}\left(\varepsilon_{s_e}^i, \varepsilon_{s_c}^j\right)$ is determined by the distance between events $\varepsilon_{s_e}^i$ and $\varepsilon_{s_c}^j$. The method of calculating this distance depends on the event model. The summarized weight of operations of this type is calculated as $W_{rep} = \sum_{\left(\varepsilon_{s_e}^i, \varepsilon_{s_c}^j\right) \in P_{rep}} w_{rep}\left(\varepsilon_{s_e}^i, \varepsilon_{s_c}^j\right)$ where P_{rep} denotes a set of pairs $\left(\varepsilon_{s_e}^i, \varepsilon_{s_c}^j\right)$ of analogous events.

- Modifying (shortening or lengthening) of the time interval $t_{s_e}^{i,j}$ between events. The interval $t_{s_e}^{i,j}$ in the sample situation corresponds to the interval $t_{s_c}^{k,l}$ in the current situation where $\varepsilon_{s_c}^{k}$ is the analogue of $\varepsilon_{s_e}^{i}$, and $\varepsilon_{s_c}^{l}$ is the analogue $\varepsilon_{s_e}^{j}$. The weight of the operation $w_{trep}\left(t_{s_e}^{i,j}, t_{s_c}^{k,l}\right)$ is determined by relative difference between the interval lengths: $w_{trep}\left(t_{s_e}^{i,j}, t_{s_c}^{k,l}\right) = \left|t_{s_e}^{i,j} - t_{s_c}^{k,l}\right| / t_{s_e}^{i,j}$. The summarized weight of operations of this type is calculated as $W_{trep} = \sum_{\left(t_{s_e}^{i,j}, t_{s_c}^{k,l}\right) \in T_{trep}} w_{trep}\left(t_{s_e}^{i,j}, t_{s_c}^{k,l}\right)$ where T_{trep} denotes a set of pairs $\left(t_{s_e}^{i,j}, t_{s_c}^{k,l}\right)$ of intervals between events.

Weights of operations are calculated via various methods and hence they should be included into the integral distance between situations with different coefficients. Furthermore, the distance must be normalized, since the shorter the compared chains are, the less modifying operations can be performed without breaking the analogy between situations. Therefore, the distance between the sample and the current situations is determined as

$$\rho(s_e, s_c) = \frac{\theta^T W}{len(st(s_e, s_c))} = \frac{\left(\theta_{del} W_{del} + \theta_{add} W_{add} + \theta_{rep} W_{rep} + \theta_{trep} W_{trep}\right)}{len(st(s_e, s_c))} \quad (1)$$

$len(st(s_e, s_c))$ is the number of events in the initial part $st(s_e, s_c)$ of the sample situation s_e. θ_{del}, θ_{add}, θ_{rep} and θ_{trep} are coefficients that determine the contribution of various types of operations to the integral distance value.

5.2 Determining the Probability of Analogy Between Situations

Distance $\rho(s_e, s_c)$ should be used to determine whether the situations are analogous and, if they are, to estimate the probability that the current situation s_c will develop according to the scenario based on s_e. Due to these requirements comparison of the situations was considered as a logistic regression task. For this purpose, a variable y is introduced that assumes the value 1 if the situations are not analogous and the value 0 otherwise.

It is assumed that the probability of event $y = 0$ (that is, probability that the situations are analogous) is defined by logistic function:

$$\mathrm{P}(y = 0 | s_e, s_c) = 1 - \frac{1}{1 + e^{-\frac{\theta^T W}{len(st(s_e, s_c))}}} \quad (2)$$

The values of the parameters θ are selected by the maximum likelihood method using a training set consisting of labeled pairs of analogous and non-analogous situations.

The result of logistic regression can also be used to perform binary classification of the situation pair: the chains s_e and s_c are considered to be analogues if $\mathrm{P}(y = 0 | s_e, s_c) > 0.5$.

5.3 Generating Scenarios for the Current Situation

If the set of events is represented by the situational graph G, a certain path in this graph has to be chosen to be considered as the current situation. The construction of this situation begins with a new or recently modified event ε_c that has an analogue in the sample situation s_e. Then, at each step, situation is extended by adding one of the neighbors of the event that is currently the first or the last in the chain. The added event may be regarded as an analogue of some event from s_e, or as an additional event that has no analogues in the sample chain. Since there are multiple ways of extending the situation at each step and all of them have to be analyzed, the result of the construction is a tree of possible variants of the current situation. This tree is traversed to find a chain s_c^{max} that has the highest similarity with the sample situation. This chain is considered as a completed current situation.

If $\mathrm{P}(y = 0|s_e, s_c^{max}) > 0.5$, the current situation is recognized as an analogue of s_e. In this case $fin(s_e, s_c^{max})$ is regarded as a possible scenario of the further development of the current situation, and $\mathrm{P}(y = 0|s_e, s_c^{max})$ is the estimation of probability of the current situation's development in accordance with this scenario.

A group of sample situations that are analogous to the current one are used to generate a set of possible scenarios of its further development. The final part of the chain that has the highest probability of the analogy with the current situation ($s_e^{prob} = \arg\max_{s_e}\left[\mathrm{P}(y = 0|s_e, s_c^{max})\right]$), is designated as the most probable scenario.

5.4 Identifying the Optimistic and the Pessimistic Scenarios

Aside from the most probable scenario the DM is particularly interested in two other scenarios - the optimistic one and the pessimistic one. They may be considered as scenarios with the highest and the lowest optimality. Optimality of scenarios is estimated by means of analytic hierarchy process (AHP) - a technique used to determine priorities of several alternatives that are estimated via a group of criteria [29]. The technique is based on performing pairwise comparisons of criteria regarding the goal and pairwise comparisons of alternatives regarding every criterion. In this case the alternatives are the generated scenarios and the goal is to select the optimal one. The set of criteria depends on the domain. In case of analyzing scenarios for driverless taxi testing possible criteria are "Safety", "Economical effectiveness" and "Duration".

When preparing a set of sample situations S_e experts determine priorities of the criteria regarding the goal by performing pairwise comparisons. They also determine criteria values for every situation, which allows any pair of scenarios to be automatically compared regarding any criterion based on criteria values of corresponding sample situations. Therefore, when a set of scenarios is generated for the current situation, it is possible to automatically calculate their priorities regarding the goal and identify the optimistic and the pessimistic scenarios.

Figure 4 demonstrates the optimistic and the pessimistic scenarios generated for the situation related to the driverless car service testing. It also illustrates the most probable scenario identified via logistic regression.

a) The optimistic scenario (Permission for testing the technology)

Event of an analogous situation	Recommendation	
Amazon was given permission to test drone delivery by the UK government	*actor*	Company management
	action	Initiate receiving a special permission
	period	1 month

b) The most probable scenario (Suspension of the technology usage before evidence of safety is provided)

Event of an analogous situation	Recommendation	
Suspension of the unconventional gas and fracking industry in Scotland until a full examination of health and environmental impacts is carried out	*actor*	Company management
	action	Initiate preparing the evidence of the technology's safety
	period	3 months

c) The pessimistic scenario (Termination of the technology usage due to safety problems)

Event of an analogous situation	Recommendation	
Samsung permanently stops Galaxy Note 7 production for the benefit of consumers' safety	*actor*	Company management
	action	Reallocate resources to the development of other technologies
	period	3 months

Fig. 4. An example of generating scenarios and recommendations.

5.5 Generating Recommendations for the Decision Maker

When preparing the collection of sample situations experts supplement every event ε_e of every situation s_e with recommendations regarding actions that should be taken if a similar event occurs in the future. Recommendation $rec_{\varepsilon_e} = \langle action_{\varepsilon_e}, actor_{\varepsilon_e}, period_{\varepsilon_e} \rangle$ designates an $action_{\varepsilon_e}$ that must be taken by an $actor_{\varepsilon_e}$ within a $period_{\varepsilon_e}$ to facilitate or counteract the development of the current situation in accordance with the scenario based on s_e.

Figure 4 demonstrates recommendations generated for the situation related to the driverless car service testing.

6 Experimental Results

The authors have developed a system for automated monitoring and forecasting of situations development that implements the proposed approach. The training of the system is carried out by experts based on collections of sample events and situations. The trained system automatically analyzes streams of heterogeneous data, detects events, constructs situational chains, generates possible scenarios of the current situation's development and prepares recommendations. Results of the analysis and forecast are provided to the DM.

The authors have carried out several experiments to estimate the quality of the system's operation in application to the analysis of a stream of text documents. Analysis of the quality of event detection subsystem is given in [21]. The experiments showed that the value of F-measure may reach 79.8% when using 1300 pairs of documents and events for training. Values of precision and recall under similar conditions were 85.2% and 76% respectively.

Another experiment has been carried out to analyze the quality of scenario generation subsystem. The experiment's goal was to evaluate the ability of the subsystem to detect pairs of situations marked by experts as analogous. The quality of the

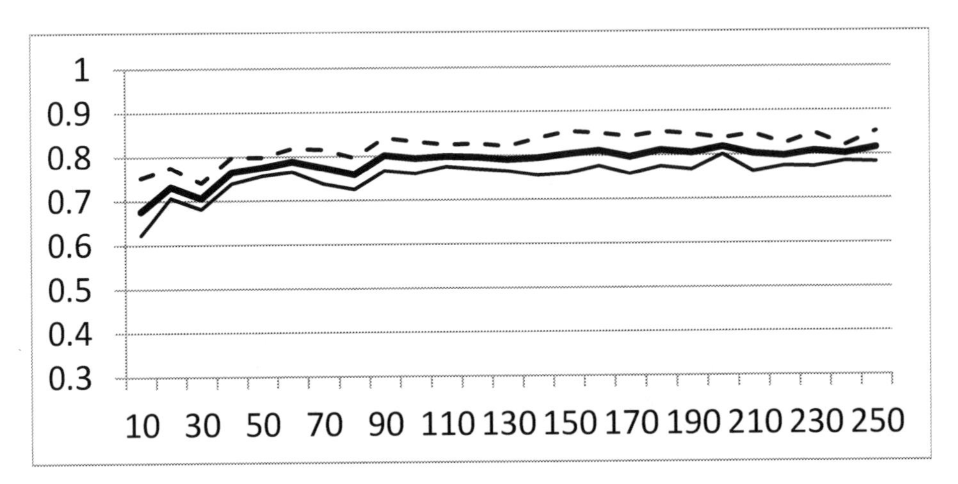

Fig. 5. Dependency between an amount of training examples and precision (thin solid line), recall (dashed line), and F-measure (thick solid line) of analogous situations detection method.

subsystem's operation was evaluated by traditional machine learning measures, including precision, recall and F-measure. Figure 5 illustrates the dependency between an amount of labeled situation pairs used for training and values of the quality measures.

The experiment demonstrated that 90 situation pairs were enough for the training. When using such amount of examples the value of F-measure reaches 0.8. Further increase of the training set did not impact the quality measures' values.

7 Future Research Directions

While providing the DM with possible scenarios and recommendations, the proposed approach does not support further management of the situation development. The DM needs to know if the actual development of the situation fits the previously generated scenario. In case of divergence the DM should be provided with recommendations regarding corrections that should be applied to the chosen plan of actions.

Hence, a possible direction for enhancement of the approach is the development of more complex graph models of sample situations. Such models should be able to reflect various ways of possible development of the current situation, taking into account possible actions of the DM at each stage of the management of the situation.

The article describes an experiment to estimate the quality of analogous situations detection, but the quality of forecasting also has to be estimated. In this regard, the authors plan to develop criteria of the quality of the situational forecast and to analyze the forecasting results generated via the proposed approach according to these criteria.

It is also planned to carry out experiments to analyze the applicability of the proposed approach for processing various types of data.

8 Conclusion

The article describes a hybrid approach to forecasting the development of situations based on event detection in heterogeneous data streams. It is shown that situations in various domains can be represented by chains of interrelated events. Forecasting consists in the generation of scenarios that are potential continuations of the current situation. Scenarios are generated using the principle of historical analogy: if a certain chain of events detected in the data stream is close to one of the sample situations, this sample situation is considered as an analogue of the current situation. Its final part is regarded as a possible scenario of the current situation's further development. The proposed scenarios generation method takes into account the dynamics of the situations development and is able to detect non-strict analogy between situations.

The optimistic and the pessimistic scenarios are determined via analytic hierarchy process. The most probable scenario is identified by means of logistic regression. The generated scenarios are supplemented with recommendations regarding actions that should be taken to promote the situation's development according to the optimal scenario.

The proposed approach provides uniform generation of scenarios in various domains while being able to handle different models of events that are used when processing heterogeneous data.

References

1. Andreev, A.M., Berezkin, D.V., Kozlov, I.A.: Method for forecasting of situations development based on event detection in text stream. In: Selected Papers of the XIX International Conference on Data Analytics and Management in Data Intensive Domains (DAMDID/RCDL 2017), pp. 367–374. CEUR-WS.org, Aachen, Germany (2017). (In Russian)
2. Zacarinny, A.A., Suchkov, A.P.: Some approaches to situational analysis of event flows. Open Educ. **1**, 39–46 (2012). (In Russian)
3. Kulinich, A.A.: Computer systems of cognitive cards modeling: approaches and methods. Control Sci. **3**, 2–16 (2010). (In Russian)
4. Eremeev, A., Varshavsky, P.: Analogous reasoning and case-based reasoning for intelligent decision support systems. Inf. Theories Appl. **13**(4), 316–324 (2006)
5. Aggarwal, C.C., Subbian, K.: Event detection in social streams. In: Proceedings of the 2012 SIAM international conference on data mining, pp. 624–635. Society for Industrial and Applied Mathematics, Philadelphia, Pennsylvania, USA (2012). https://doi.org/10.1137/1.9781611972825.54
6. Eremeev, A.P., Varshavsky, P.R.: Modeling of case-based reasoning in intelligent decision support systems. Artif. Intell. Decis. Making **2**, 45–57 (2009). (In Russian)
7. van der Aalst, W.M., Reijers, H.A., Weijters, A.J., van Dongen, B.F., De Medeiros, A.A., Song, M., Verbeek, H.M.W.: Business process mining: An industrial application. Inf. Syst. **32**(5), 713–732 (2007). https://doi.org/10.1145/775047.775150
8. Borisov, V.V., Zernov, M.M.: An implementation of a situational approach based on a fuzzy hierarchical situation-event network. Artificial Intelligence and Decision Making **1**, 18–30 (2009). (In Russian)

9. Ahremenko, A.S.: Political analysis and forecasting. Gardariki, Moscow, Russia (2006). (In Russian)
10. Allen, J.F., Ferguson, G.: Actions and events in interval temporal logic. J. Logic Comput. **4** (5), 531–579 (1994). https://doi.org/10.1093/logcom/4.5.531
11. Andreev, A.M., Berezkin, D.V., Kozlov, I.A., Simakov, K.V.: Unsupervised approach to web wrapper maintenance. Informatics and Applications **7**(3), 2–13 (2013). https://doi.org/10.14357/19922264130301. (In Russian)
12. Andreev, A.M., Berezkin, D.V., Kozlov, I.A., Simakov, K.V.: Multi-criteria method for detecting near-duplicates in a stream of text messages. Systems and Means of Informatics **25** (1), 34–53 (2015). https://doi.org/10.14357/08696527150103. (In Russian)
13. Raimond, Y., Abdallah, S.: The event ontology. http://motools.sourceforge.net/event/event.html. Accessed 21 Jan 2018
14. Weng, J., Lee, B.S.: Event detection in Twitter. In: Fifth International AAAI Conference on Weblogs and Social Media, pp. 401–408. AAAI Press, Palo Alto (2011)
15. Fung, G.P.C., Yu, J.X., Yu, P.S., Lu, H.: Parameter free bursty events detection in text streams. In: Proceedings of the 31st International Conference on Very Large Data Bases VLDB 2005, pp. 181–192. VLDB Endowment (2005)
16. Lande, D.V., Braichevsky, S.M., Grigoryev, A.N., Darmokhval, A.T., Radetzky, A.B.: Detection of new events from news flow. In: Proceedings of the International Conference « Dialog–2007 » on Computer Linguistics and Intelligent Technologies, pp. 349–352. Nauka, Moscow, Russia (2007). (In Russian)
17. Yang, Y., Carbonell, J.G., Brown, R.D., Pierce, T., Archibald, B.T., Liu, X.: Learning approaches for detecting and tracking news events. IEEE Intelligent Systems **14**(4), 32–43 (1999). https://doi.org/10.1109/5254.784083
18. Yang, Y., Pierce, T., Carbonell, J.: A study on retrospective and on-line event detection. In: Proceedings of the 21st Annual International ACM SIGIR Conference on Research and Development in Information Retrieval, pp. 28–36. ACM, New York (1998). https://doi.org/10.1145/290941.290953
19. Aggarwal, C.C., Yu, P.S.: On clustering massive text and categorical data streams. Knowl. Inf. Syst. **24**(2), 171–196 (2010). https://doi.org/10.1007/s10115-009-0241-z
20. Yang, Y., Zhang, J., Carbonell, J., Jin, C.: Topic-conditioned novelty detection. In: Proceedings of the Eighth ACM SIGKDD International Conference on Knowledge Discovery and Data Mining, pp. 688–693. ACM, New York (2002). https://doi.org/10.1145/775047.775150
21. Andreev, A.M., Berezkin, D.V., Kozlov, I.A.: Automated topic monitoring based on event detection in text stream. Inf.-measur. Control Syst. **15**(3), 49–60 (2017). (In Russian)
22. Hogenboom, F., Frasincar, F., Kaymak, U., de Jong, F.: An overview of event extraction from text. In: CEUR Conference Proceedings, Workshop on Detection, Representation, and Exploitation of Events in the Semantic Web (DeRiVE 2011) at Tenth International Semantic Web Conference (ISWC 2011), pp. 48–57. CEUR-WS.org, Aachen, Germany (2011)
23. Yakushiji, A., Tateisi, Y., Miyao, Y., Tsujii, J.I.: Event extraction from biomedical papers using a full parser. In: Pacific Symposium on Biocomputing, pp. 408–419 (2001)
24. Zhao, Q., Mitra, P., Chen, B.: Temporal and information flow based event detection from social text streams. In: Proceedings of the Twenty-Second AAAI Conference on Artificial Intelligence, pp. 1501–1506. AAAI Press, Palo Alto (2007)
25. Sakaki, T., Okazaki, M., Matsuo, Y.: Earthquake shakes Twitter users: real-time event detection by social sensors. In: Proceedings of the 19th international conference on World wide web, pp. 851-860. ACM, New York (2010). https://doi.org/10.1145/1772690.1772777

26. Guralnik, V., Srivastava, J.: Event detection from time series data. In: Proceedings of the fifth ACM SIGKDD International Conference on Knowledge Discovery and Data Mining, pp. 33–42. ACM, New York (1999). https://doi.org/10.1145/312129.312190
27. Yao, W., Chu, C.H., Li, Z.: Leveraging complex event processing for smart hospitals using RFID. J. Netw. Comput. Appl. **34**(3), 799–810 (2011). https://doi.org/10.1016/j.jnca.2010.04.020
28. Gyllstrom, D., Wu, E., Chae, H.J., Diao, Y., Stahlberg, P., Anderson, G.: SASE: Complex event processing over streams. In: Proceedings of the International Conference on Innovative Data Systems Research (CIDR), pp. 407–411 (2007)
29. Saaty, T.L.: The Analytic Hierarchy Process. McGraw-Hill, New York (1980)

Integrating DBMS and Parallel Data Mining Algorithms for Modern Many-Core Processors

Timofey Rechkalov and Mikhail Zymbler(✉)

South Ural State University, Chelyabinsk, Russia
trechkalov@yandex.ru, mzym@susu.ru

Abstract. Relational DBMSs (RDBMSs) remain the most popular tool for processing structured data in data intensive domains. However, most of stand-alone data mining packages process flat files outside a RDBMS. In-database data mining avoids export-import data/results bottleneck as opposed to use stand-alone mining packages and keeps all the benefits provided by a RDBMS. The paper presents an approach to data mining inside a RDBMS based on a parallel implementation of user-defined functions (UDFs). Such an approach is implemented for PostgreSQL and modern Intel MIC (Many Integrated Core) architecture. The UDF performs a single mining task on data from the specified table and produces a resulting table. The UDF is organized as a wrapper of an appropriate mining algorithm, which is implemented in C language and is parallelized by the OpenMP technology and thread-level parallelism. The heavy-weight parts of the algorithm are additionally parallelized by intrinsic functions for MIC platforms to reach the optimal loop vectorization manually. The library of such UDFs supports a cache of precomputed mining structures to reduce costs of further computations. In the experiments, the proposed approach shows good scalability and overtakes *R* data mining package.

Keywords: Data mining · In-database analytics · PostgreSQL
Clustering · Partition Around Medoids (PAM) · Thread-level parallelism
OpenMP · Intel Xeon Phi

1 Introduction

Nowadays relational DBMSs (RDBMSs) remain the most widely used tool for processing structured data in data intensive domains (e.g. finance, medicine, physics, etc.). Meanwhile, most of data mining algorithms deal with flat files. In data intensive domains, exporting of data sets and importing of mining results inhibit analysis of large databases outside a RDBMS [18]. Data mining inside a RDBMS avoids export-import bottleneck and provides the end-user with all the built-in RDBMS's services (query optimization, data consistency and security, etc.).

Approaches to integrating data mining with RDBMSs include special data mining languages and SQL extensions, SQL implementation of mining algorithms and user-defined functions (UDFs) implemented in high-level programming language. In order to increase performance of data analysis, the latter could serve as a subject of applying parallel processing on modern many-core platforms.

L. Kalinichenko et al. (Eds.): DAMDID/RCDL 2017, CCIS 822, pp. 230–245, 2018.
https://doi.org/10.1007/978-3-319-96553-6_17

In [25], we presented an approach to data mining inside the PostgreSQL open-source DBMS exploiting capabilities of modern Intel MIC (Many Integrated Core) [1] platform. The mining UDF is organized as a wrapper of an appropriate algorithm, which is implemented in C language and is parallelized for Intel MIC platform by OpenMP technology and thread-level parallelism. We took Partition Around Medoids (PAM) clustering algorithm [9] and wrapped its parallel implementation for Intel MIC proposed in [24]. Our experiments on the platforms of Intel Xeon CPU and Intel Xeon Phi, Knights Corner (KNC) generation, showed an efficiency of the proposed approach.

In this paper, we give a more detailed description of our approach and extend the study mentioned above as follows. We enhance parallel PAM by accelerating its step of distance matrix computation and conduct additional experiments on Intel Xeon Phi, Knights Landing (KNL), which is the second-generation MIC architecture product from Intel.

The rest of the paper is organized as follows. In Sect. 2, we discuss related works. The proposed approach is described in Sect. 3. We give the results of experimental evaluation of our approach in Sect. 4. Section 5 concludes the paper.

2 Related Work

The problem of integrating data analytics with relational DBMSs has been studied since data mining research originates. Early developments considered data mining query languages [5, 7] and implementation of data mining functionality in SQL, e.g. clustering algorithms [15, 17], association rules [23, 27], classification [20, 26], and graph mining [4, 21].

In [18], authors proposed integration of correlation, linear regression, PCA and clustering into the Teradata DBMS based on UDFs. In [19], it was shown that UDFs implementing common vector operations are as efficient as automatically generated SQL queries with arithmetic expressions, and queries calling scalar UDFs are significantly more efficient than equivalent queries using SQL aggregations. In [8], a technique for execution of aggregate UDFs based on data parallelism was proposed, which will be embodied later in Teradata DBMS.

The MADlib library [6] provides many methods for supervised learning, unsupervised learning and descriptive statistics for PostgreSQL. The MADlib exploits UDAs, UDFs, and a sparse matrix C library to provide efficient representations on disk and in memory. Since many statistical methods are iterative, authors wrote a driver UDF in Python to control iteration in such a way that all large data movement is done within the database engine and its buffer pool.

The Bismarck system [3] provides a unified architecture for in-database analytics, facilitating UDFs as a convenient interface for the analyst to describe their desired analytics models. This development is based on incremental gradient descent (IGD), which is a general technique to solve a large class of analytical models expressed as convex optimization problem (e.g. logistic regression, support vector machines, etc.). Authors showed that IGD has a data access pattern identical to the UDA access pattern and provided a UDA-based implementation for the analytical problems mentioned above.

The DAnA [12] system automatically maps a high-level specification of in-database analytics queries to the FPGA accelerator. The accelerator implementation is generated from an UDF, expressed as part of a SQL query in a Python-embedded Domain-Specific Language. In order to implement efficient in-database integration, DAnA-generated accelerators contain a special hardware structure, Striders, that directly interface with the buffer pool of the database. The Striders extract, cleanse, and process the training data tuples, which are consumed by a multi-threaded FPGA engine that executes the analytics algorithm.

In this paper, we suggest an approach to embedding data mining functions into PostgreSQL. As methods mentioned above, our approach exploits UDFs. The implementation of those systems, however, involves a combination of SQL, UDFs, and driver programs written in other languages, so the systems could become obscure and relatively difficult to maintain. Our approach assumes parallelization of UDFs for Intel many-core platform that current RDBMS is running on and hiding the parallelization details from the RDBMS. This make it possible to port our approach to some other open-source RDBMS with possible non-trivial but mechanical software development effort. Additionally, there is a special module in our approach, which provides a cache of precomputed mining structures and allows UDF to reuse these structures in order to reduce costs of computations.

3 Data Mining Inside PostgreSQL Using Intel MIC

3.1 Key Ideas

The goal of our approach is to provide a database application programmer with the library of data mining functions, which could be run inside a DBMS as it is shown in Fig. 1. In this example, the *pgPAM* function applies the Partition Around Medoids (PAM) clustering algorithm [9] to data points from the specified input table and saves results in the output table (for the specified number of the input table columns and number of clusters). An application programmer is not obliged to export input data

```
#include <libpq-fe.h> // API of PostgreSQL
#include "pgmining.h" // API of pgMining library

void main (void)
{
  char * inpTab = "points";
  char * outTab = "clusters";
  int dim = 3;
  int k = 5;
  char * conninfo="user=postgres port=5432 host=localhost";

  PGconn * conn = PQconnectdb(conninfo);
  pgPAM(conn, inpTab, dim, k, outTab);
  PQexec(conn, strcat("SELECT * FROM ", outTab));
  PQfinish(conn);
}
```

Fig. 1. An example of using data mining function inside PostgreSQL

from PostgreSQL and import mining results back. At the same time, PAM encapsulates parallel implementation [24] based on OpenMP and thread-level parallelism.

Implementation of such an approach is based on the following ideas. A data mining algorithm is implemented with C language and parallelized for Intel MIC by OpenMP technology. Next, we cover the parallel mining function by two wrappers, namely a system-level wrapper and a user-level wrapper. The *user-level wrapper* registers the system-level wrapper in the database schema, connects to the database server and calls system-level wrapper. The *system-level wrapper* is an UDF, which parses parameters of the user-level wrapper, calls the parallel mining function and saves the results in the table(s).

3.2 Component Structure

Figure 2 depicts the component structure of our approach being applied to PostgreSQL.

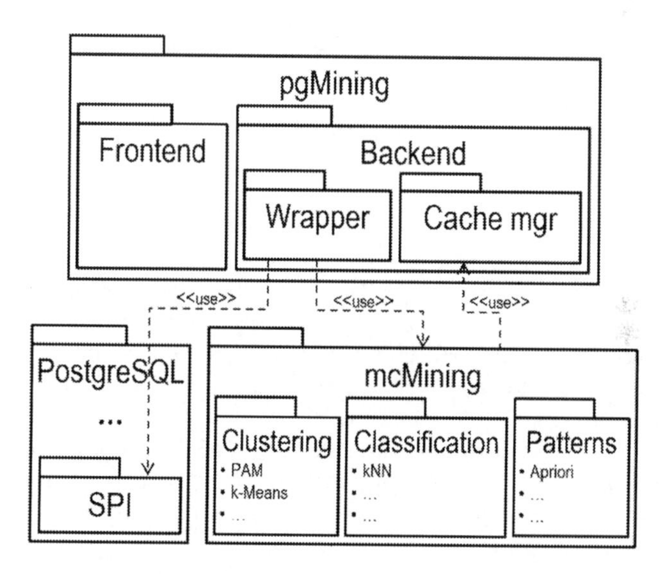

Fig. 2. Component structure of the proposed approach

The *pgMining* is a library of functions for data mining inside PostgreSQL. The *mcMining* is a library that exports data mining functions, which are parallelized for Intel MIC systems and are subject of wrapping by the respective functions from *pgMining* library. Implementation of the *pgMining* library uses the PostgreSQL SPI (Server Programming Interface), which provides the low-level functions for data access. The *pgMining* library consists of two following subsystems, namely *Frontend* and *Backend*, where the former provides presentation layer and the latter provides data access layer of concerns for an application programmer.

3.3 Frontend

The *Frontend* implements a user-level wrapper. A *Frontend* function wraps the respective UDF from *Backend*, which is loaded into PostgreSQL and executed as "INSERT INTO ... SELECT ..." query to save mining results in the specified table. An example of the *Frontend* function is given in Fig. 3.

```
// PAM clustering inside PostgreSQL
// Returns 0 in case of success or negative error code.
int pgPAM (
  PGconn * conn,  // ID of PostgreSQL connection
  char * inpTab,  // Name of input table
  int dim,        // Number of coordinates in data point
  int k,          // Number of clusters
  char * outTab)  // Name of output table
{
  PQexec(conn, "CREATE FUNCTION
    wrap_pgPAM(text, integer, integer, real) RETURNS text AS
    'pgmining', 'wrap_pgPAM' LANGUAGE C STRICT;");
  PQexec(conn, "CREATE %s TABLE (data text)", outTab);
  return PQexec(conn, "INSERT INTO %s
   SELECT wrap_pgPAM(%s, %d, %d);", outTab, inpTab, dim, k);
}
```

Fig. 3. Interface and implementation schema of function from *Frontend*

Such a function connects to PostgreSQL, carries out some mining task and returns exit code. The function mandatory parameters are PostgreSQL connection ID, names of input and output tables, and a number of first left columns in the input table containing data to be mined. The rest parameters are specific to the mining task. As a side effect, the function creates a table with mining results, which are stored as a text in JSON format. Notation of such a text allows to define various results specific to the mining algorithm, e.g. for a clustering algorithm this text could describe output data points with associated numbers of clusters, centroids of resulting clusters, etc. Application programmer is then in charge of parse and extract the results, and save it in relation table(s) if necessary.

3.4 Backend

The *Wrapper* of the *Backend* implements a system-level wrapper. Figure 4 depicts an example of *Wrapper*'s function. Such a function is an UDF, which wraps a parallelized mining function from *mcMining* and performs as follows. Firstly, the function parses its input to form the parameters to call the *mcMining* function. After that, the function checks if input table(s) and/or supplementary mining structure(s) are in the cache maintained by *Cache manager* and then loads them if not. Finally, a call of the *mcMining* function with appropriate parameters is performed. The *Cache manager* provides buffer pool to store precomputed mining structures.

```
// Wrapper for PAM clustering inside PostgreSQL
// Returns clustering result as JSON string.
Datum wrap_pgPAM(PG_FUNCTION_ARGS)
{
  // Extract parameters of the algorithm
  char * inpTab = text_to_cstring(PG_GETARG_TEXT_P(0));
  int dim = PG_GETARG_INT32(1);
  int k = PG_GETARG_INT32(2);
  int N;
  // Check if mining structure is in the cache
  void * distMatr = cache_getObject(
    strcat(inpTab, "_distMatr"));
  if (distMatr == NULL) {
    // Check if input table is in the cache
    void * inpData = cache_getObject(inpTab);
    if (inpData == NULL) {
      // Allocate memory and load input table to the cache
      inpData = (float *) malloc(dim * N *sizeof(float));
      wrap_tabRead(inpData, inpTab, dim, &N);
      cache_putObject(inpTab, inpData, sizeof(inpData));
    }
    distMatr = mcCalcMatrix(inpData, dim, N);
    cache_putObject(strcat(inpTab, "_distMatr"),
      distMatr, sizeof(distMatr));
  }
  // Perform clustering and save results to the output ta-
ble
  mcPAM_res * outData = mcPAM_resCreate();
  mcPAM(N, k, outData, distMatr);
  PG_RETURN_TEXT(data2String(outData));
}
```

Fig. 4. Interface and implementation schema of function from *Backend*

Distance matrix $d_{ij} = dist(a_i, a_j)$ stores distances between each pair of a_i and a_j elements of input data set and is a typical example of mining structure to be cached. Being precomputed once, distance matrix could be used many times for clustering or kNN-based classification with various parameters (e.g. the number of clusters, the number of neighbors, etc.).

The *Cache manager* exports two basic functions depicted in Fig. 5. The *putObject* function loads a mining structure specified by its ID, buffer pointer and size into the cache. The *getObject* searches in the cache for an object with the given ID. An ID of mining structure is a string, which is made as a concatenation of an input table name and object informational string (e.g. "_distMatr"). In order to handle a situation when there is not enough space in the buffer pool to put a new object, *Cache manager* implements one of the replacement strategies, e.g. LRU-K [16], LFU-K [28], etc.

```
// Load an object to the cache.
// Returns 0 in case of success or negative error code.
int cache_putObject(
  char * objID,// An object's ID
  void * data, // Pointer to data buffer
  int size);   // Data size
// Search for an object by the given ID in cache.
// Returns pointer to the object in case of success or NULL.
void * cache_getObject(char * objID);
```

Fig. 5. Interface of *Cache manager* module

3.5 Library of Parallel Algorithms for Intel MIC

Figure 6 gives an example of the *mcMining* library function. Such a function encapsulates parallel implementation for Intel many-core systems based on OpenMP. In this example, we use Partition Around Medoids (PAM) [9] clustering algorithm, which is applicable when minimal sensitivity to noise data is required. The PAM provides such a property since it represents cluster centers by points of the input data set (*medoids*). Firstly, PAM computes distance matrix for the given data points. Then, the algorithm carries out an initial clustering by the successive selection of medoids until the required number of clusters have been found. Finally, the algorithm iteratively improves clustering in accordance with an objective function.

```
// PAM clustering parallelized for Intel MIC platform.
// Returns 0 in case of success or negative error code.
int mcPAM(
    int N,              // Number of data points
    int k,              // Number of clusters
    void * outData,     // Array of output centroids
    void * distMatr); // Precomputed distance matrix
```

Fig. 6. Interface of function from *mcMining* library

In our previous work [24], a parallel version of the PAM algorithm exploits auto-vectorization of distance matrix computation. Auto-vectorization relies on a compiler's ability to transform the loops into sequences of vector operations and utilize vector processor units. Thus, auto-vectorization does not guarantee the optimal vectorization and, in turn, the best performance.

3.6 Advanced Vectorization of Parallel Algorithms for Intel MIC

In this study, we take a step forward and accelerate parallel PAM by speeding up its first phase, keeping in mind that according to experiments in our previous work, distance matrix computation takes up to 80% of overall runtime.

We implemented distance matrix computation phase by intrinsics instead of auto-vectorization. Intrinsics are assembly-coded functions that wrap processor specific instruction sets and allow us to use the C function calls and variables in place of assembly instructions. Intrinsics are expanded inline eliminating function call overhead. Thus, intrinsics allow make it possible to reach the optimal vectorization manually, at the expense of a programmer efforts and the code maintainability.

Moreover, we implemented sophisticated SoAoS (Structure of Arrays of Structures) data layout [22] to organize data points in memory as follows. Suppose, there are N data points where each point comprises of dim float coordinates and there is a platform-dependent parameter, namely $size_{vector}$. Then, data points are represented by an array comprising of N div $size_{vector}$ elements. Each element of the array is a structure comprising of dim attributes where each attribute is an array of $size_{vector}$ float coordinates. The $size_{vector}$ parameter is chosen with respect to a number of floats that could be processed in one vector operation for the given platform's instruction set (e.g. for Intel Xeon and its SSE instruction set we took $size_{vector}$= 4, and for Intel Xeon Phi and its AVX-512 instruction set we took $size_{vector}$= 16).

Figure 7 depicts implementation scheme of Euclidean distance matrix computation for Intel MIC platform based on intrinsics and the SoAoS data layout. The algorithm performs as follows.

```
// Computation of Euclidean distance matrix.
// Returns pointer to the matrix.
float * mcCalcMatrix (float * inpData, int dim, int N)
{
  const int vecSize = 16;
  float * distMatr = ALLOC_ALIGNED_FLOAT_ARRAY (N*N);
  float * SoAoS = ALLOC_ALIGNED_FLOAT_ARRAY (dim*N);
  SoAoS_permute(inpData, SoAoS, N, dim, vecSize);
  #pragma omp parallel for
  for (int i=0; i<N; i++) {
    for (int k=0; k<N; k+=vecSize) {
      VECTOR res = GET_ZERO_VEC ();
      for (int j=0; j < dim; j++) {
        VECTOR p1 = FILL_VEC (inpData + i*dim + j);
        VECTOR p2 = LOAD_VEC (SoAoS + k*dim + j* vecSize);
        VECTOR diff = SUB_VEC (p1, p2);
        res = FMADD_VEC (diff, diff, res);
      }
      STORE_VEC (distMatr + i*N + k, res);
    }
  }
  free(SoAoS);
  return distMatr;
}
```

Fig. 7. Parallel implementation of Euclidean distance matrix computation

Before the computation, we permute an array of input data points to represent them as a SoAoS layout. Computation is organized as three nested loops where the outer loop runs along the input data points and parallelized by the OpenMP `#pragma` compiler directive. The second inner loop runs along the SoAoS blocks of data points and initializes a vector register in order to store temporary results. The innermost loop runs along the coordinates of a data point and carries out the following actions by intrinsic functions. Vector register is filled by the j-th coordinate of an input data point.

Then, we read *j*-th coordinates of the sixteen data points from a SoAoS block. Next, we calculate the difference of the vectors mentioned above, square the difference and add the result to the vector register. At the end of the second loop, data from the vector register are moved to the resulting matrix. In the end, resulting matrix will comprise of squared Euclidean distances.

4 Experimental Evaluation

4.1 Background of the Experiments

In the experiments, we firstly evaluated how intrinsics and new data layout affect the performance of distance matrix computation. We also investigated the scalability of the modified *mcPAM* version on Intel MIC platforms depending on the number of threads employed. Here, we mean speedup and parallel efficiency as basic characteristics of a parallel algorithm, which are defined as follows. Speedup and parallel efficiency of a parallel algorithm being ran on k threads are calculated as

$$s(k) = \frac{t_1}{t_k},\ e(k) = \frac{t_1}{k \cdot t_k} \cdot 100\%$$

where t_1 and t_k are run times of the algorithm on one and k threads, respectively. A parallel algorithm with speedup closer to one and parallel efficiency at least 50% is considered to have good scalability.

We took the *pgPAM* function with the modified *mcPAM* function and compared the *pgPAM* performance with the PAM from the *R* package [14] as well.

We performed the evaluation on the Tornado SUSU supercomputer [10] and the RSC[1] cluster node. The former provides a node with two Intel MIC platforms, namely Intel Xeon and Intel Xeon Phi (KNC generation) and the latter provides a node with Intel Xeon Phi (KNL generation) platform (cf. Table 1 for the specifications).

Table 1. Specifications of hardware

Specifications	Host CPU	Coprocessor	CPU system
Model, Intel Xeon	2 × X5680	Phi (KNC), SE10X	Phi (KNL), 7250
Physical cores	2 × 6	61	68
Hyper threading factor	2	4	4
Logical cores	24	244	272
Frequency, GHz	3.33	1.1	1.4
Vector processing unit size, bit	128	512	512
Boot ability	Yes	No	Yes
Peak performance, TFLOPS	0.371	1.076	3.046

[1] http://www.rscgroup.ru/en/.

In the experiments, we used the datasets depicted in Table 2.

Table 2. Datasets used in experiments

Dataset	dim	# clusters	# points, $\times 2^{10}$	Semantic
FCS Human	423	10	18	Aggregated human gene information [2]
MixSim	5	10	35	Generator of synthetic datasets for evaluation of clustering algorithms [14]
Census	67	10	35	US Census Bureau population surveys [13]
Power	3	10	35	Household electricity consumption [11]

4.2 Results and Discussion

Figure 7 shows the results of the first series of experiments. We can see that auto-vectorization provides at least 1.5× and 5× faster run time of distance matrix computation for the Intel Xeon and Intel MIC platform, respectively, in comparison with scalar version. The use of intrinsics, in turn, provides up to 2× faster run time in comparison with auto-vectorization. Thus, we can conclude that intrinsics combined with proper data layout significantly increase performance of the most heavy-weight part of the parallel PAM algorithm.

The results of the experiments on the *mcPAM* scalability are depicted in Fig. 8. Both speedup and parallel efficiency are closer to linear when the number of threads matches the number of physical cores the algorithm is running on, for all the platforms (i.e. 12 cores for Intel Xeon, 60 cores for Intel Xeon Phi KNC and 68 cores for Intel Xeon Phi KNL, respectively). Speedup and parallel efficiency become sub-linear when the algorithm uses more than one thread per physical core. We can conclude that after acceleration of the phase of distance matrix computation, the algorithm still demonstrates good scalability for all the considered Intel MIC platforms.

Figure 9 shows the results of the experiments on the *pgPAM* performance on different datasets. We compared serial PAM from the *R* package with our both serial and parallel *pgPAM* where a distance matrix was precomputed or not. Parallel versions ran on the following Intel platforms, namely 2× Xeon CPU (24 threads), Xeon Phi KNC (240 threads), and Xeon Phi KNL (272 threads). We can see that the parallel and serial *pgPAM* versions outperform the *R* PAM for the given platforms and datasets. Next, caching the precomputed distance matrix, we improve the performance, especially in case of high-dimensional data. By using the scalable *mcPAM* algorithm, the *pgPAM* shows the better results on the Intel Xeon MIC systems than on the Intel Xeon CPU and performs best on Intel Xeon Phi KNL (Fig. 10).

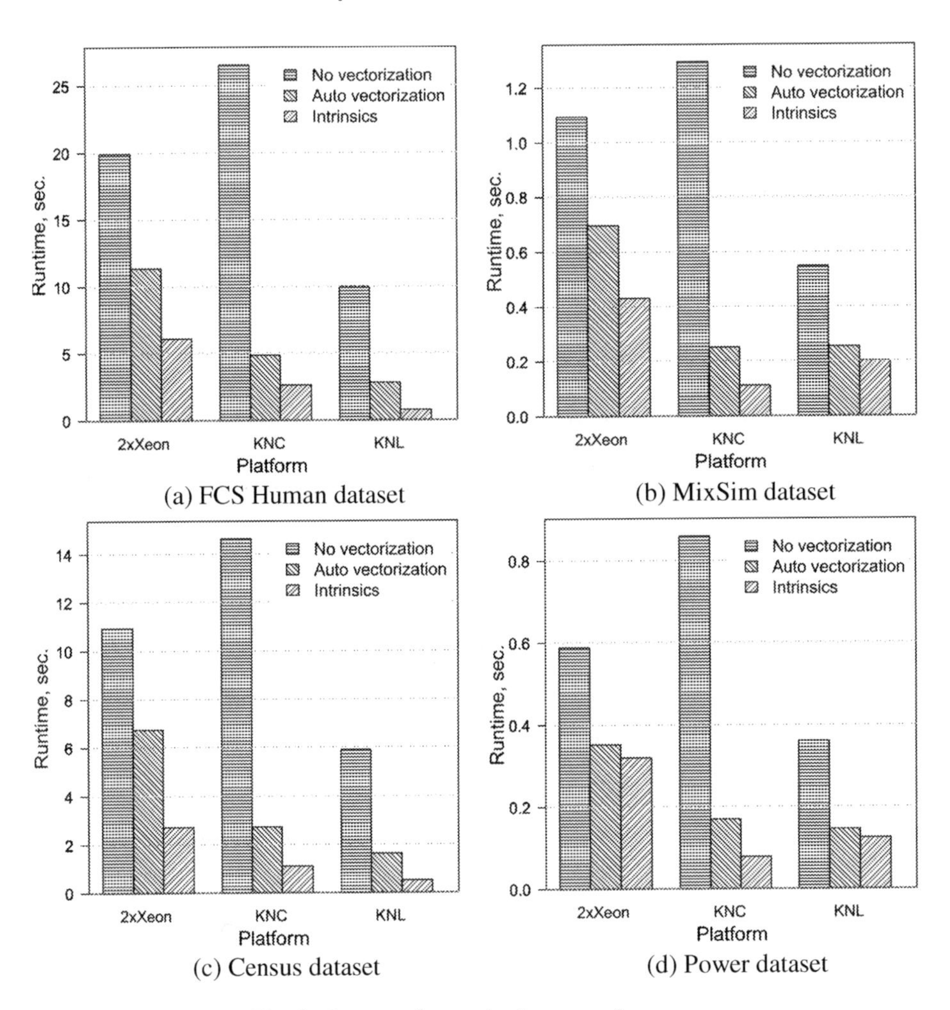

(a) FCS Human dataset

(b) MixSim dataset

(c) Census dataset

(d) Power dataset

Fig. 8. Impact of vectorization to performance

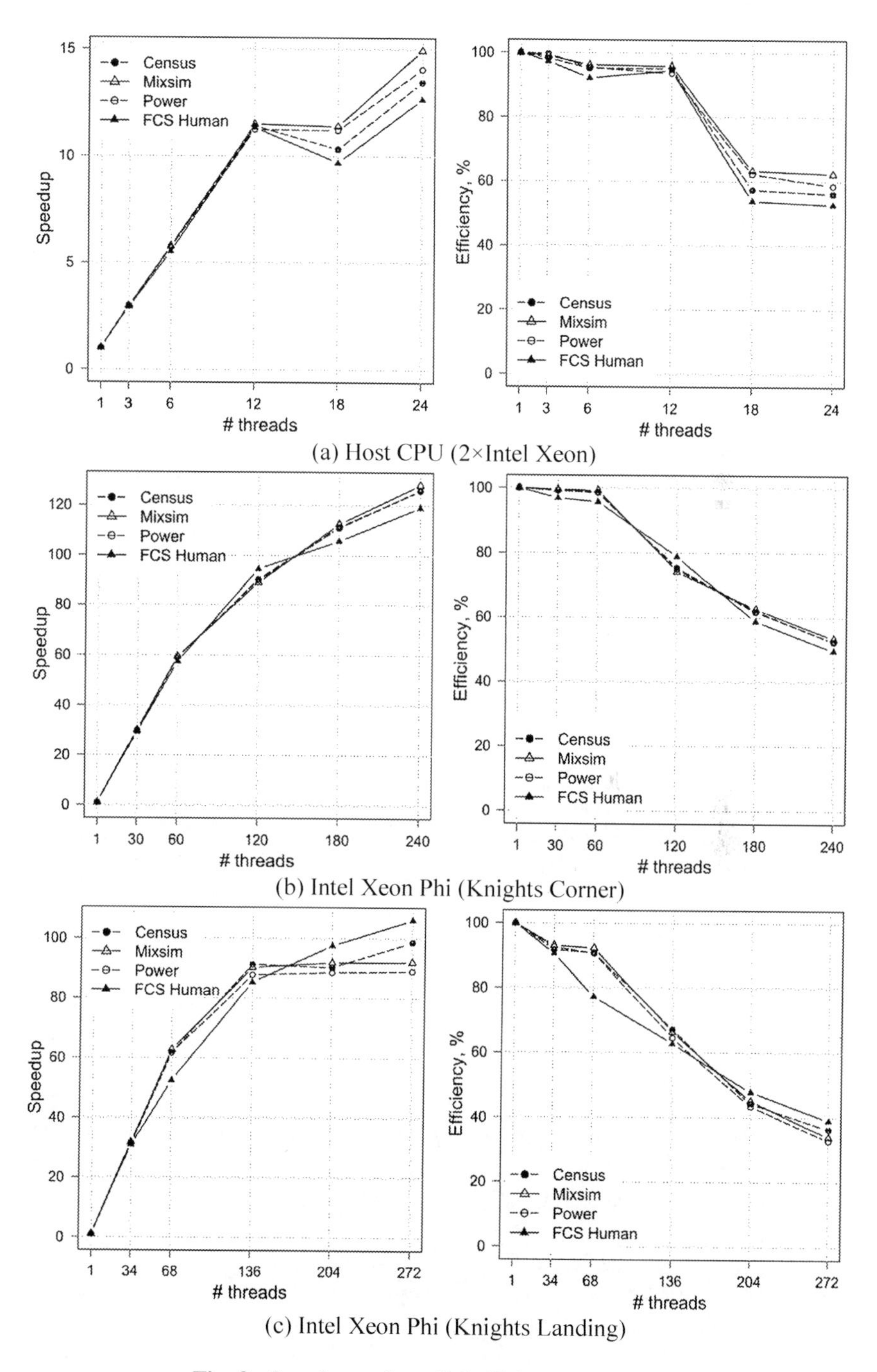

Fig. 9. Speedup and parallel efficiency of *mcPAM*

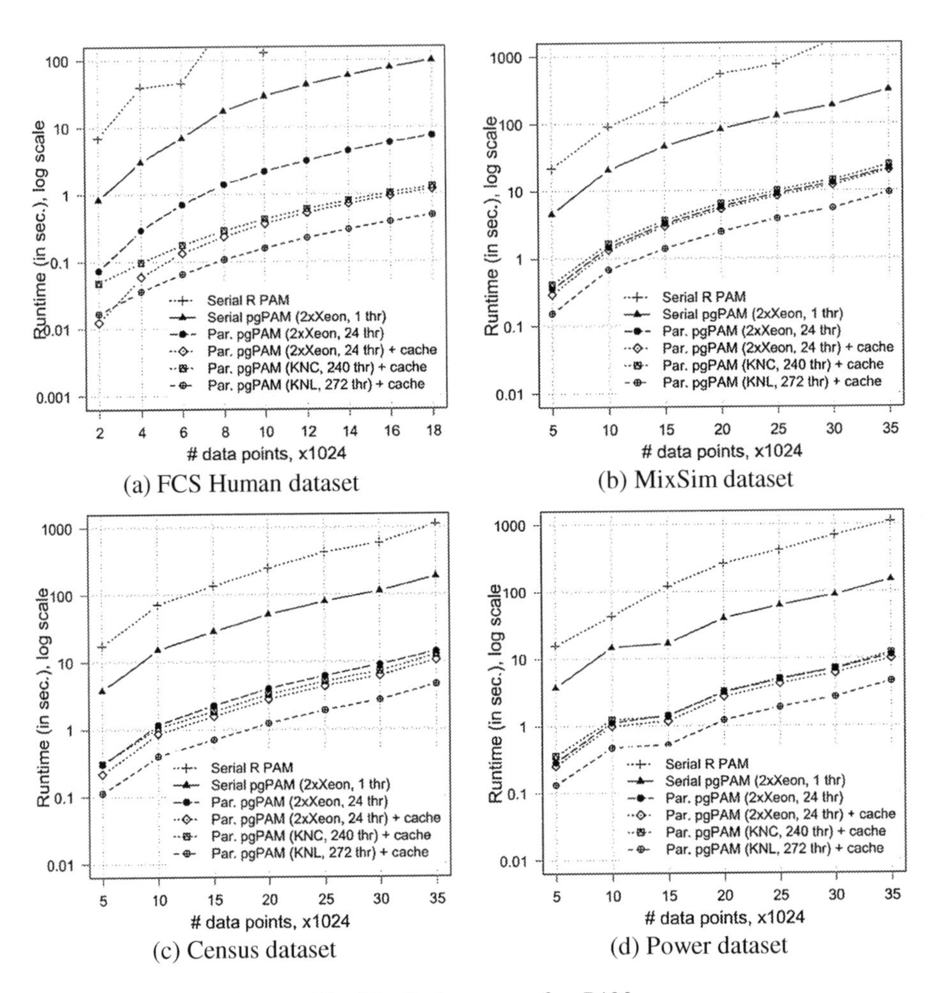

(a) FCS Human dataset

(b) MixSim dataset

(c) Census dataset

(d) Power dataset

Fig. 10. Performance of *pgPAM*

5 Conclusion

In this paper, we touch upon the problem of integrating data mining algorithms and a relational DBMS in data intensive domains. We presented an approach to implementation of in-database analytics that exploits capabilities of modern many-core platforms to improve performance of analytics. We implemented such an approach for PostgreSQL and Intel Many Integrated Core (MIC) architecture.

Our approach exploits the following key ideas. A data mining algorithm is implemented with C language and parallelized by the OpenMP technology for Intel MIC platforms. The parallel mining function is covered by two wrappers, namely a system-level wrapper and a user-level wrapper. The user-level wrapper registers the system-level wrapper in the database schema, connects to the database server and calls system-level wrapper. The system-level wrapper is an UDF, which parses parameters of the user-level wrapper, calls the parallel mining function and saves results in the table(s). The system-level wrapper is accompanied by a cache of precomputed mining structures (e.g. distance matrix) to reduce costs of computations. Since our approach assumes encapsulation of parallel implementation from PostgreSQL, it could be ported to some other open-source RDBMS, with possible non-trivial but mechanical software development effort.

In this study, in order to increase performance of in-database clustering, we additionally implemented distance matrix computation (which is the heaviest part of the clustering algorithm) using advanced data layout and intrinsic functions for MIC platforms. We evaluated our approach on modern Intel MIC platforms (Intel Xeon and Intel Xeon Phi with both Knights Corner and Knights Landing generations) using real datasets where our solution showed good scalability and performance and overtook the *R* data mining package.

Acknowledgments. This work was financially supported by the Russian Foundation for Basic Research (grant No. 17-07-00463), by Act 211 Government of the Russian Federation (contract No. 02.A03.21.0011) and by the Ministry of education and science of Russian Federation (government order 2.7905.2017/8.9). Authors thank RSC Group (Moscow, Russia) for the provided computational resources.

References

1. Duran, A., Klemm, M.: The Intel Many Integrated Core architecture. In: Smari, W.W., Zeljkovic, V. (eds.) HPCS, pp. 365–366. IEEE (2012)
2. Engreitz, J.M., Daigle Jr., B.J., Marshall, J.J., Altman, R.B.: Independent component analysis: mining microarray data for fundamental human gene expression modules. J. Biomed. Inform. **43**(6), 932–944 (2010)
3. Feng, X., Kumar, A., Recht, B., Re, C.: Towards a unified architecture for in-RDBMS analytics. In: Candan, K.S., Chen, Y., Snodgrass, R.T., Gravano, L., Fuxman, A. (eds.) Proceedings of the ACM SIGMOD International Conference on Management of Data, SIGMOD 2012, Scottsdale, AZ, USA, 20–24 May 2012, pp. 325–336. ACM (2012)
4. Garcia, W., Ordonez, C., Zhao, K., Chen, P.: Efficient algorithms based on relational queries to mine frequent graphs. In: Nica, A., Varde, A.S. (eds.) Proceedings of the Third Ph.D. Workshop on Information and Knowledge Management, PIKM 2010, Toronto, Ontario, Canada, pp. 17–24. ACM, 30 October 2010
5. Han, J., Fu, Y., Wang, W., Chiang, J., Gong, W., Koperski, K., Li, D., Lu, Y., Rajan, A., Stefanovic, N., Xia, B., Zaiane, O.R.: Dbminer: a system for mining knowledge in large relational databases. In: Simoudis, E., Han, J., Fayyad, U.M. (eds.) Proceedings of the Second International Conference on Knowledge Discovery and Data Mining (KDD-96), Portland, Oregon, USA, pp. 250–255. AAAI Press (1996)
6. Hellerstein, J.M., Re, C., Schoppmann, F., Wang, D.Z., Fratkin, E., Gorajek, A., Ng, K.S., Welton, C., Feng, X., Li, K., Kumar, A.: The MADlib analytics library or MAD skills, the SQL. PVLDB **5**(12), 1700–1711 (2012)
7. Imielinski, T., Virmani, A.: MSQL: a query language for database mining. Data Min. Knowl. Discov. **3**(4), 373–408 (1999)

8. Jaedicke, M., Mitschang, B.: On parallel processing of aggregate and scalar functions in object-relational DBMS. In: Haas, L.M., Tiwary, A. (eds.) SIGMOD 1998, Proceedings of the ACM SIGMOD International Conference on Management of Data, 2–4 June, 1998, Seattle, Washington, USA, pp. 379–389. ACM Press (1998)
9. Kaufman, L., Rousseeuw, P.J.: Finding Groups in Data: An Introduction to Cluster Analysis. Wiley, New York (1990)
10. Kostenetskiy, P., Safonov, A.: SUSU supercomputer resources. In: Sokolinsky, L., Starodubov, I. (eds.) PCT 2016, International Scientific Conference on Parallel Computational Technologies, Arkhangelsk, Russia, 29–31 March 2016, CEUR Workshop Proceedings, vol. 1576, pp. 561–573. CEUR-WS.org (2016)
11. Lichman, M.: UCI machine learning repository. Irvine, CA: University of California, School of Information and Computer Science (2013). http://archive.ics.uci.edu/ml/datasets/individual+household+electric+power+consumption
12. Mahajan, D., Kim, J.K., Sacks, J., Ardalan, A., Kumar, A., Esmaeilzadeh, H.: In-RDBMS Hardware Acceleration of Advanced Analytics. CoRR abs/1801.06027 (2018)
13. Meek, C., Thiesson, B., Heckerman, D.: The learning-curve sampling method applied to model-based clustering. J. Mach. Learn. Res. **2**, 397–418 (2002)
14. Melnykov, V., Chen, W.C., Maitra, R.: MixSim: an R package for simulating data to study performance of clustering algorithms. J. Stat. Softw. Artic. **51**(12), 1–25 (2012)
15. Miniakhmetov, R., Zymbler, M.: Integration of fuzzy c-means clustering algorithm with PostgreSQL database management system. Numer. Methods Programm. **13**(2(26)), 46–52 (2012)
16. O'Neil, E.J., O'Neil, P.E., Weikum, G.: The LRU-K page replacement algorithm for database disk buffering. In: Buneman, P., Jajodia, S. (eds.) Proceedings of the 1993 ACM SIGMOD International Conference on Management of Data, Washington, D.C., 26–28 May 1993, pp. 297–306. ACM Press (1993)
17. Ordonez, C.: Integrating k-means clustering with a relational DBMS using SQL. IEEE Trans. Knowl. Data Eng. **18**(2), 188–201 (2006)
18. Ordonez, C.: Building statistical models and scoring with UDFs. In: Chan, C.Y., Ooi, B.C., Zhou, A. (eds.) Proceedings of the ACM SIGMOD International Conference on Management of Data, Beijing, China, 12–14 June 2007, pp. 1005–1016. ACM (2007)
19. Ordonez, C., Garcia-Garcia, J.: Vector and matrix operations programmed with UDFs in a relational DBMS. In: Yu, P.S., Tsotras, V.J., Fox, E.A., Liu, B. (eds.) Proceedings of the 2006 ACM CIKM International Conference on Information and Knowledge Management, Arlington, Virginia, USA, 6–11 November 2006, pp. 503–512. ACM (2006)
20. Ordonez, C., Pitchaimalai, S.K.: Bayesian classifiers programmed in SQL. IEEE Trans. Knowl. Data Eng. **22**(1), 139–144 (2010)
21. Pan, C.S., Zymbler, M.L.: Very large graph partitioning by means of parallel DBMS. In: Catania, B., Guerrini, G., Pokorný, J. (eds.) ADBIS 2013. LNCS, vol. 8133, pp. 388–399. Springer, Heidelberg (2013). https://doi.org/10.1007/978-3-642-40683-6_29
22. Peng, Y., Grossman, M., Sarkar, V.: Static cost estimation for data layout selection on GPUs. In: 7th International Workshop on Performance Modeling, Benchmarking and Simulation of High Performance Computer Systems, PMBS@SC 2016, Salt Lake, UT, USA, 14 November 2016, pp. 76–86. IEEE (2016)
23. Rantzau, R.: Frequent itemset discovery with SQL using universal quantification. In: Meo, R., Lanzi, P.L., Klemettinen, M. (eds.) Database Support for Data Mining Applications. LNCS (LNAI), vol. 2682, pp. 194–213. Springer, Heidelberg (2004). https://doi.org/10.1007/978-3-540-44497-8_10

24. Rechkalov, T., Zymbler, M.: Accelerating medoids-based clustering with the Intel Many Integrated Core architecture. In: 9th International Conference on Application of Information and Communication Technologies, AICT 2015, 14–16 October 2015, Rostov-on-Don, Russia - Proceedings, pp. 413–417 (2015)
25. Rechkalov, T., Zymbler, M.: An approach to data mining inside PostgreSQL based on parallel implementation of UDFs. In: Kalinichenko, L.A., Manolopoulos, Y., Kuznetsov, S. O. (eds.) Selected Papers of the XIX International Conference on Data Analytics and Management in Data Intensive Domains (DAMDID/RCDL 2017), Moscow, Russia, 9–13 October 2017, CEUR Workshop Proceedings, vol. 2022, pp. 114–121. CEUR-WS.org (2017)
26. Sattler, K., Dunemann, O.: SQL database primitives for decision tree classifiers. In: Proceedings of the 2001 ACM CIKM International Conference on Information and Knowledge Management, Atlanta, Georgia, USA, 5–10 November 2001, pp. 379–386. ACM (2001)
27. Shang, X., Sattler, K.-U., Geist, I.: SQL Based Frequent Pattern Mining with FP-Growth. In: Seipel, D., Hanus, M., Geske, U., Bartenstein, O. (eds.) INAP/WLP -2004. LNCS (LNAI), vol. 3392, pp. 32–46. Springer, Heidelberg (2005). https://doi.org/10.1007/11415763_3
28. Sokolinsky, L.B.: LFU-*K*: an effective buffer management replacement algorithm. In: Lee, Y., Li, J., Whang, K.-Y., Lee, D. (eds.) DASFAA 2004. LNCS, vol. 2973, pp. 670–681. Springer, Heidelberg (2004). https://doi.org/10.1007/978-3-540-24571-1_60

Data Curation and Data Provenance Support

Data Curation Policies and Data Provenance in EUDAT Collaborative Data Infrastructure

Vasily Bunakov[1(✉)], Alexander Atamas[2], Alexia de Casanove[3], Pascal Dugénie[3], Rene van Horik[2], Simon Lambert[1], Javier Quinteros[4], and Linda Reijnhoudt[2]

[1] Science and Technology Facilities Council, Harwell, Oxfordshire, UK
{vasily.bunakov, simon.lambert}@stfc.ac.uk
[2] Data Archiving and Networked Services (DANS), The Hague, Netherlands
{alexander.atamas, rene.van.horik, linda.reijnhoudt}@dans.knaw.nl
[3] CINES, Montpellier, France
{casanove, dugenie}@cines.fr
[4] GFZ German Research Centre for Geoscience, Potsdam, Germany
javier@gfz-potsdam.de

Abstract. The work outlines the development of a data curation and data provenance framework in the EUDAT Collaborative Data Infrastructure. Practical use cases are described, as well as results of defining and implementing data curation policies and data provenance patterns.

Keywords: Data curation · Data provenance · E-infrastructures Long-term digital preservation · Policies

1 Introduction

EUDAT Collaborative Data Infrastructure (CDI) [1] is a European e-infrastructure of data services and information resources in support of research. This infrastructure and its services have been developed in close collaboration with over 50 research communities spanning across many different scientific disciplines, with more than 20 major European research organizations, data centres and computing centres involved. Researchers, research communities and service providers can use EUDAT data services to manage research data according to their own needs.

The EUDAT services offering [19] has emerged as a result of two consecutive FP7 and Horizon 2020 projects, with the actual services focused on different aspects of data management and data use, and supported by a variety of information technology stacks.

Data curation (or digital curation) is the selection, preservation, maintenance, collection and archiving of digital assets and hence is the essential part of research data management. Sensible data curation requires establishing and developing long-term repositories of digital assets for their current and future use by researchers and wider society. Collaborative data infrastructures like EUDAT that span across the borders should play a significant role in research data curation.

L. Kalinichenko et al. (Eds.): DAMDID/RCDL 2017, CCIS 822, pp. 249–263, 2018.
https://doi.org/10.1007/978-3-319-96553-6_18

An important aspect of data curation is data provenance that refers to processes, entities, activities and actors that allow reasoning over data origin, data movement and data transformation. The significance of data provenance in data infrastructures such as EUDAT is caused by a substantial complexity of data workflows that involve multiple data sources, multiple (and interacting) services, and multiple agents, both human and software.

Historically, EUDAT services have been built with only a few considerations for conscious data curation and data provenance, with secure and controlled access to data being one of the major initial goals to achieve. Other aspects of data curation and data provenance started playing a more prominent role when services matured to production stage and became a part of an operational collaborative infrastructure. Specifically, operational requirements of B2SAFE service (that currently offers what long-term digital preservation projects typically call "bit-level" preservation), as well as automated data transfers across interrelated services have made it essential to systematically explore the topic of data curation in EUDAT.

The decision was made to formulate the core approach to data curation with the involvement of two prominent unrelated research communities with substantial amounts of data to manage and then, using these two use cases as a proof-of-concept for clearly formulated data curation activities, get other user communities involved.

Another decision made was to reuse the outputs of the SCAPE project [2] and Research Data Alliance Practical Policy Working Group [3] in order to set up a reasonable data curation framework for EUDAT.

This work expands on the earlier published effort on data curation and policy modelling [20, 21] with the addition of specific data provenance considerations and the actual data policies implementation using the EUDAT data infrastructure. It also offers a consistent vision of the role of data curation and data provenance in research e-infrastructures.

The rest of the paper outlines the core use cases, characterizes the SCAPE and RDA outputs that are deemed to be applicable in EUDAT context, explains mapping of SCAPE policy elements [4] to granular data policies in EUDAT, describes a service for provenance records generation, outlines executable policy implementation effort so far and suggests a semantic approach to modelling data policies. In the end, the role of data curation for the future of e-infrastructures is discussed.

2 HERBADROP Use Case

2.1 Motivations and Relation to EUDAT Services

The HERBADROP data pilot [12] aims to offer an archival service for long-term preservation of herbarium specimen images and to develop innovative processes for extracting information from those images. HERBADROP follows the global trend towards scalable industrial-style digitizing of herbaria specimens. It is designed as both an archival service for long-term preservation of herbarium specimen images and a tool for analysing and extracting information written on the image by using Optical Character Recognition (OCR) analysis, both supported by CINES [6].

Making the specimen images and data available online from different institutes allows cross domain research and data analysis for botanists and researchers with diverse interests (e.g. ecology, social and cultural history, climate change).

Herbaria hold large numbers of collections: approximately 22 million herbarium specimens exist as botanical reference objects in Germany, 20 million in France and about 500 million worldwide. High resolution images of these specimens require substantial bandwidth and disk space. New methods of extracting information from the specimen labels have been developed using OCR but using this technology for biological specimens is particularly complex due to the presence of biological material in the image with the text, the non-standard vocabularies, and the variable and ancient fonts. Much of the information is only available using handwritten text recognition or botanical pattern recognition which are less mature technologies than OCR [13].

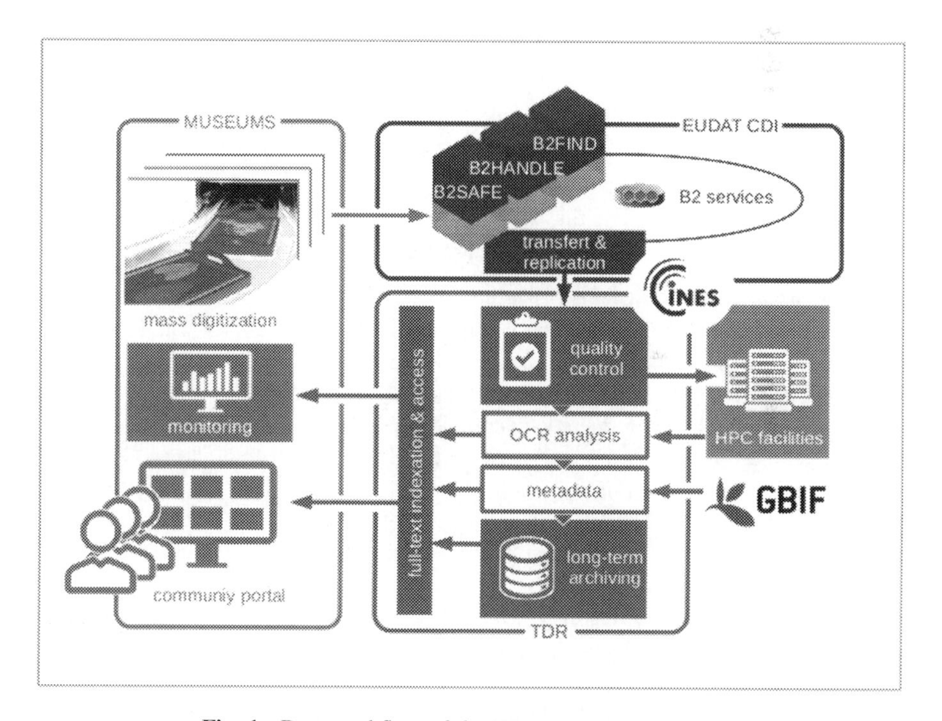

Fig. 1. Data workflow of the HERBADROP data pilot.

The proposed platform is expected to support or even substitute costly manual data input as much as possible. The platform will also curate and enrich metadata resulting from image analysis using optical character recognition (OCR) and pattern matching.

Results are exposed as platform independent Web services which can be effectively integrated into herbarium data management systems as well as metadata capture workflows. Since 2016, six European community partners have been involved. Their contribution to the pilot represents a business model that can be potentially replicated by other institutes. The partners in the HERBADROP data pilot are: Musée National d'Histoire Naturelle (MNHN) – Paris, France; Royal Botanic Garden of Edinburgh (RBGE) – United Kingdom; Botanic Garden and Botanical Museum (BGBM) – Berlin, Germany; Digitarium – Finland; Naturalis Biodiversity Center – Netherlands; Botanic Garden Meise – Belgium.

The EUDAT B2SAFE service is used in the first step of the ingestion process. Existing images of herbarium specimens along with the associated metadata (harvested from GBIF portal) are transmitted to the CINES repository using B2SAFE transfer service. The ingestion into B2SAFE is carried out in accordance with the centralized persistent identifiers (PID) management system used in EUDAT. It is envisaged that discovery and visualization of the data objects will be performed with the EUDAT B2FIND service.

The data workflow in HERBADROP is represented by Fig. 1.

2.2 Data Curation Scenarios for HERBADROP

One of the requirements from the HERBADROP communities was to implement specific use cases, such as identifying duplicates amongst specimens from the different museums. Another example of policy is long term preservation that involves a number of controls including file format verification and metadata quality. Amongst HERBADROP users, two partners of the community have proposed practical scenarios for data curation: Digitarium [14] and the Royal Botanic Garden of Edinburgh (RGBE).

Scenario Proposed by Digitarium (Finland)

Digitarium [14] planned to use Optical Character Recognition (OCR) data to generate metadata based on the label information available for the herbarium specimen. Firstly, a Natural Language Processing based system could be used to do OCR quality check and extract relevant terms. Then metadata could be either automatically generated, or manually inserted through the transcription portal [15] but with the help of OCR data.

More general for EUDAT infrastructure services, Digitarium wanted to utilize and integrate them into the whole digitisation process of natural history biological collections. The data flow goes from the beginning of the digitisation process i.e. imaging, to storage, then to transcription and analysis, until accessing. This involves data storage, high-performance computing resources, and web services in EUDAT.

Firstly, the images from the imaging station can be transferred into EUDAT storage for long-term preservation instantly or in batch. After transferring, HPC can access the images and do OCR to extract label information to generate preliminary metadata. This metadata has to be associated with corresponding images. The data can be openly accessed. However, the access rights of data have to be set up for different purposes, such as endangered species protection.

Secondly, using HTTP APIs, the images and their metadata can be accessible from EUDAT by data-owner portals. Therefore, browsing and transcribing are available. Updated metadata will be transferred back into the EUDAT B2SAFE service. Different versions of metadata have to be kept.

Thirdly, the metadata is indexed. Therefore, the data can be searched or filtered based on different terms for further scientific usages. HPC resources can be utilized also on the data for different researches.

Scenario Proposed by RBGE (the Royal Botanic Garden of Edinburgh) in Association with MNHN (Musée National d'Histoire Naturelle) – Paris
The core of the concept of HERBADROP is to harvest metadata from OCR analysis of the text that is a part of herbarium images. The choice has been to proceed to a full text analysis using a Lucene-based engine Elasticsearch [16]. The objective of this approach is to provide a powerful interface for further data curation as part of the preservation process (identifying duplicates, or inducing new taxonomic relations, etc.), see [12].

Safeguarding long-term data storage is an important precondition for reliable access to herbarium specimen information. Thanks to this pilot, it is possible to envisage long-term storage for herbarium specimen images. Moreover, the specimens will be discoverable by the entire scientific community. Thus, undescribed species stored in herbaria can be examined by experts to aid identification and discovery of new species.

Distribution information for species over time can be evaluated and these data could provide evidence of the point in time when an invasive species first occurred in a certain area. Historians could analyse herbarium data to create itineraries for historical characters. The data can be used to calibrate predictive models of the oncoming changes in biodiversity patterns under global threats. This diverse information will be useful for a wide user community including conservationists, policy makers, and politicians.

3 GEOFON Use Case

The second use case concerns GFZ, the German Research Centre for Geosciences. GFZ provides valuable seismological services in the form of a seismological infrastructure named GEOFON [7].

GFZ is one of the members of the EPOS initiative (European Plate Observatory System) [5] and, in this context, collaborates with other two seismological data centres related to EPOS (KNMI, INGV) in the EUDAT project.

Besides being one of the fastest earthquake information provider worldwide, GEOFON is also one of the largest nodes of the European Integrated Data Archive (EIDA) for seismological data under the ORFEUS umbrella [23], which is a distributed data centre established to (a) securely archive seismic waveform data and related metadata, gathered by European research infrastructures, and (b) provide transparent access to the archives by the geosciences research communities.

The internal structure of GEOFON is based on three pillars:

- A global seismic network operated in close collaboration with many partner institutions with focus on EuroMed and Indian Ocean regions. The network consists of ca. 110 high quality stations, which acquire data in real time [8].
- A global earthquake monitoring system which uses data from GEOFON and partner networks [10]. It publishes most timely earthquake information. First automatic solutions are available few minutes after the events and mostly manually revised later.
- A comprehensive seismological data archive for GFZ and partner networks, for permanent networks as well as for temporary deployments.

For some GEOFON partner networks, GEOFON acts as a data centre saving a replica of the original copy and at the same time as a data distribution centre. Additionally, data from many temporary station deployments are permanently archived at GEOFON, in particular passive seismological experiments of the GFZ Geophysical Instrument Pool Potsdam (GIPP) and the German Task Force Earthquake.

Most of data is open for public access, as well as real-time data feeds when available. However, there is a small amount of data under an embargo period, usually for a limited amount of time (3–4 years).

3.1 Data Workflow in GEOFON

GEOFON supports two scenarios for the ingestion of data into its archive: one for permanent networks and one for temporary (and most probably already finished) experiments.

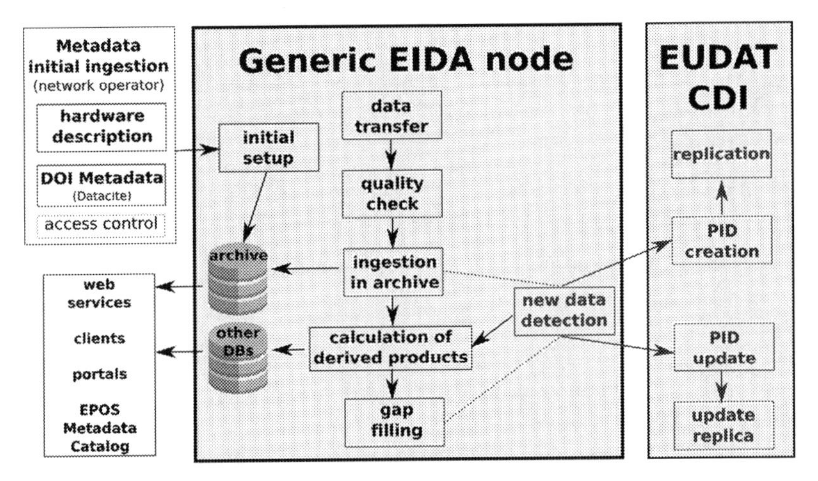

Fig. 2. Data workflow at GEOFON. It also represents the workflow from a generic seismological data centre as the ones under the EIDA/ORFEUS initiative. Boxes in black are generic activities from the data centre. Blue boxes show activities related to the EUDAT service B2SAFE, while brown boxes show the tasks related to B2HANDLE.

Usually, raw data are transmitted to the data centre with the metadata (technical hardware description) to be able to operate with them. In the case of permanent networks raw data are received continuously from the stations around the world via satellite using a protocol called SeedLink [17], a real-time data acquisition protocol which works on TCP. The packets of each individual station are always transferred in timely (FIFO) order.

In the case of temporary experiments network operators provide usually, first, the metadata needed to use the data, and in a second phase the data to be archived. Data transmission can be done as in the permanent networks case (SeedLink protocol), or can also be transmitted to the data centre by the network operator using some client-server tools provided by GEOFON, which will automatically do the first quality check of the data format. In some cases, both methods could be used.

A schematic view of the workflow at GEOFON can be seen in Fig. 2. It should be noted that this workflow is also very similar to the ones at other seismological data centres belonging to EIDA/ORFEUS. For instance, the other two data centres piloting EUDAT services (KNMI and INGV).

3.2 Service Hosting Environment with the Inclusion of EUDAT Services

Considering the workflow depicted in the previous section, GEOFON introduced some EUDAT services in order to automate and/or improve some of the tasks related to it.

Many services are being provided at GEOFON (e.g. interactive web portals, proprietary protocols to get data or derived products), with two of them (Station-WS and Dataselect) being particularly important, as they are international standards and the core services for the community upon which other services are built. Station-WS serves the information describing the hardware and everything related to the deployment, while Dataselect serves the data.

Two main EUDAT services have been integrated in the GEOFON workflow; namely, B2SAFE and B2HANDLE. The former is used to accomplish many of the Data Management tasks, while the latter is used to manage/store Persistent Identifiers (PIDs).

As the archive is stored on a partition with a particular directory structure, the B2SAFE service "mounts" the archive as an external resource in read-only mode.

One of the main requirements for the Data Policies at GEOFON was the capability to trigger processes based on the inclusion of new data. In the context of B2SAFE, this can be done by means of automatic rules which are executed under certain conditions (e.g. new data ingested).

With the proper rules we can enforce that, after new data is detected by B2SAFE, a certain set of actions is executed. For instance, the derived products can be generated and data can be replicated to a partner data centre from the EUDAT CDI, the Karlsruhe Institute of Technology (KIT). Also, as part of this replication process, persistent identifiers (PIDs) are generated for each file, so that the PID can be used to globally and univocally identify the file.

PIDs are managed and stored by means of the already mentioned service called B2HANDLE, which is based on a Handle Server and other libraries developed within the project. GFZ has a broad expertise on this type of tools and, therefore, we decided to deploy our own B2HANDLE server and work with our local instance. Each generated PID is stored with a set of key-value pairs called "PID Record". The information in the PID Record allows, among other things, to track other copies of the file in different data centres or validate its integrity by means of pre-calculated checksums.

3.3 Data Policies to Apply at GEOFON Through EUDAT Services

After the formalization of the internal workflows at GEOFON, and the inclusion of requirements from the community and the data centre, we defined a set of Data Policies to be enforced by means of the tools available within EUDAT and new developments, which could be useful for different communities.

Some of them are related to the Replication process. For instance:

- Replicate every new file in the archive to our internal backup server.
- If we are the official provider of the data in a file, replicate it to an off-site partner within the EUDAT CDI.
- Seismological data that do not belong to us but comes from our earthquake early monitoring system should be kept for 6 months only. Data still need to be replicated to the internal server.
- An automated file deletion must not be possible. In case that the system detects that a file should be deleted, an email should be sent to the appropriate operator.

Regarding the access control of the files:

- "Restricted data" must be tagged, with proper access control applied to them.
- Access restrictions can be automatically removed after a period of time (embargo period).
- Data must be able to be accessed via an HTTP API respecting the ACL (Access Control List).

Regarding automatic metadata extraction:

- Metrics derived from the data must be automatically calculated to populate some of our services when new data is ingested.
- Detailed statistics related to the data access should be available for the data owners/creators.
- In case that data are modified (e.g. correcting errors, filling gaps), this information should be available for future use (provenance information) (see Sect. 5).

Regarding the integrity of the stored data:

- A weekly process will select ~2% of the folders in our archive and verify that the synchronization is correct. The idea is that every file will be checked at least once in a year.
- Check that the data are stored in SDS format (SeisComP Directory Structure).
- Start and end time of network/station operation must be available and data outside this timespan must not be allowed.

The identified relevant policies were gradually implemented using generic EUDAT services and GEOFON-specific software.

4 Mapping of EUDAT Data Policies to SCAPE and RDA Policy Curation Frameworks

For the design and implementation of data curation actions in EUDAT, the relevant outputs of SCAPE project [2] and Practical Policy Working Group of the Research Data Alliance [3] have been identified. SCAPE outputs are perceived of high quality owing to the advanced thinking that considered long-term digital preservation policies at a granular level suitable for the machine-executable implementation. RDA Practical Policy Working Group outputs are a result of a substantial international collaborative effort including experts in iRODS platform [11], which is the technological foundation of the EUDAT B2SAFE service.

For SCAPE, we used the catalogue of preservation policy elements [4]. This is a systematized compendium of granular policies with examples of what SCAPE called "control policies" (granular statements that are easily translatable to machine-executable functions), and for the RDA Practical Policy Working Group it was their practical policy implementations report [9] that compiled a set of machine-executable functions for iRODS platform [11].

In addition to this top-down retrospective review of the SCAPE and RDA outputs, a bottom-up analysis of control policies applicable to the GEOFON and HERBADROP use case was performed, with a number of control policies identified as prime candidates for implementation in EUDAT B2SAFE. These policies are presented in Table 1.

Table 1. Candidate control policies for implementation by GEOFON and HERBADROP.

Policy category	Control policy	Policy examples
Data replication	Number and location of replicas	Data should be replicated in N locations, including in locations A and B
	Timeframe for replication	Data should be replicated within the next 24 h after the data ingestion in any particular location
	Data nodes roles	All data nodes are equivalent to read data from, but data can only be initially ingested in node X then replicated over all other nodes
Data integrity checks	The set of checksum algorithms acceptable	Checksum algorithm accepted is MD5
	Periodicity and scope of integrity checks	Calculate checksums for 2% of all data assets every week, with the aim of having the entire data collection checked annually
Data and metadata formats	Data formats accepted	BMP and PNG accepted for images
	Metadata extraction from data	Upon ingestion, file name should be extracted as metadata
	Data format check procedures acceptable	Software package X should be used for data format validation
	Minimal metadata assigned upon data release	PID is a mandatory metadata element
Data access and data reuse	Embargo rules	Embargo period of N years is applied to all PDFs and images
	The set of data licenses recommended upon data release	CC-BY license should be assigned to all data released after the embargo period ends
	Data reuse statistics collection	Number of file downloads should be collected

Then the gap analysis was performed against SCAPE policy elements, to see whether these bottom-up identified control policies allow enough coverage of the extensively defined data curation policy landscape of SCAPE project. SCAPE policy elements catalogue [4] is two-level with Guidance Policies on the top level and Policy Elements on the granular level. An example of Guidance Policy is Authenticity Policy that breaks down to Integrity, Reliability and Provenance as policy elements. Hence control policies in Data Integrity checks category from Table 1 correspond to Integrity policy element of Authenticity Policy in the SCAPE policy elements catalogue.

One noticeable gap discovered through this mapping exercise is the Digital Object lifecycle which was paid due attention to in SCAPE policy landscape but is missing in the current EUDAT considerations. This gap may be hard to address as EUDAT is a collaborative project that accumulates data from a large variety of research communities with a wide range of digital object types and lifecycles. However, this discovery should inform the future operation of EUDAT services so that they could meet all reasonable (and multi-aspect) requirements for data curation.

5 Provgen: A Web Service to Generate Provenance Data

Depending on the conceptual point of view, supplying data with enough provenance information can be seen as a specific data policy or as a track of evidence that specific data policies have been actually implemented. There is a particular challenge for sensible data provenance in data infrastructures, owing to the scale and diversity of data sources, data processing agents (services), and a variety of services implementation.

As the EUDAT services landscape becomes more complete, and also complex, much more information is generated from the execution of data workflows by different actors. Based on this, we designed and implemented a "Provenance generation service" called Provgen, which could act as a central point to generate and collect all the information generated not only from the EUDAT CDI services, but also from other external services which take part in the data workflow.

One of the main requirements of this service was to be capable of being used on almost any reasonable data workflow which could be designed and contacted by any service. Therefore, it had to be highly configurable to the different needs of the users, strongly decoupled from the other products, and with a minimum set of software dependencies.

To fulfil the requirement of flexibility to adapt to different data workflows, we designed a templating system, where templates can be loaded by the operator of the system. Templates are in Notation3 format and each template is the result of the design of a certain Provenance record type generated at a particular point in the workflow. For instance, a template for the creation of a Persistent Identifier (PID) to identify a data file can be expressed as the RDF shown in Fig. 3.

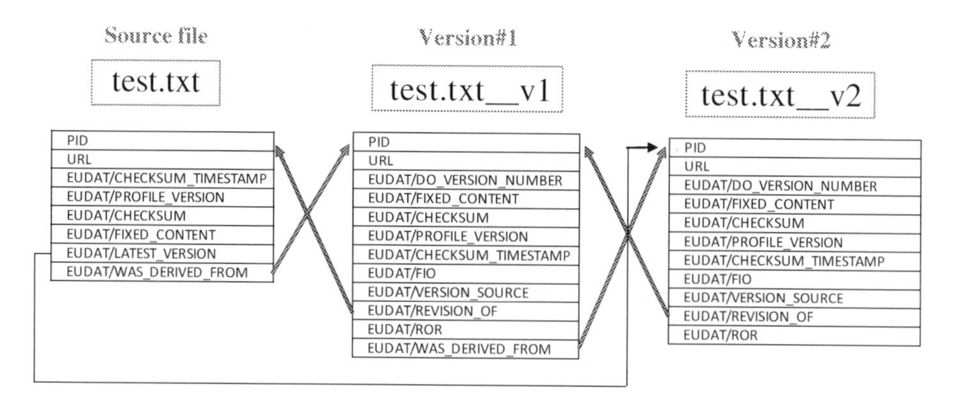

Fig. 3. B2SAFE versions cross-linking.

To decouple Provgen from other components of the EUDAT CDI and avoid software dependencies, we decided to specify and expose an API [22] which offers the following functionality over HTTP calls:

- see available templates in Provgen including the specification of each template (expected variables and their description),
- instantiate a template to generate provenance information,
- show the documentation of the system,
- see details on how Provgen was configured.

Given any particular data flow, a user of Provgen needs to do two tasks to configure it before it is available for use:

- identify the point(s) at which some Provenance information should be generated. For each of these points, a template must be generated with all the needed attributes and copied to the "templates" folder into the system,
- include in the code related to these points of the workflow a call to the Provgen API specifying the template and instantiating all the required variables.

Dublin Core and PROV [24] elements are used for the expression of RDF template, with an addition of a Provgen-specific element for the expression of persistent identifiers.

When a client needs to generate a record according to the template, it calls the API specifying the template and a set of key-value pairs. If all variables are specified and the types are correct, the user will get a Provenance record as a response to the HTTP call. Optionally, the provenance record is generated and stored on the server in a triple-store backend.

With this setup, all current and future EUDAT services (as well as external consumers) will be able to use Provgen in order to generate provenance records and provide users with a unified view of provenance information across different services in EUDAT and beyond.

More details and a longer description on Provgen component will be available in [21].

6 Implementing Data Versioning Policy in EUDAT B2SAFE

Versioning is essential for proper data curation and provenance. Software version control git-like systems are not able to deal efficiently with large size binary datasets that are one of the focuses of EUDAT B2SAFE – robust, safe and highly available service for storing large-scale data in community and departmental repositories. This is why a new service-specific versioning functionality has been designed and implemented, which can be considered a working example of an executable data policy, and a role model for other executable data policies.

B2SAFE is based on the iRODS middleware [11] and is inherently capable to operate on very large datasets distributed across multiple domains. Research communities employ B2SAFE service primarily to replicate datasets across different data centres. Information about locations of replica and their other properties is maintained and queried using Persistent Identifiers (PIDs). The so-called "PID record" keeps information as a set of key-value pairs and can store, for example, checksum, timestamp, location, etc. of a given dataset. Internally, the Handle system [26] manages PIDs in the B2SAFE system.

The versioning code is written as a custom iRODS rulebase using iRODS Rule language and is available at GitHub as a part of B2SAFE service core code [27]. In our implementation of versioning functionality for B2SAFE service, a version of digital object is understood as a timestamped copy of the object. Versions are kept in a separate directory created only for storage of versions.

As illustrated by Fig. 3, PID record of each version contains a URL of the version and a few other fields:

- EUDAT/DO_VERSION_NUMBER field keeps the version's number;
- EUDAT/FIXED_CONTENT field set to "true" meaning that content modification of the version is not allowed;
- EUDAT/CHECKSUM field containing checksum of the version file to be used for integrity validation;
- EUDAT/PROFILE_VERSION is set to one - as a legacy EUDAT value;
- EUDAT/TIMESTAMP contains a timestamp when the version has been created;
- EUDAT/FIO is a reference to the First Ingested Object (FIO) stored in the EUDAT/FIO field;
- EUDAT/VERSION_SOURCE contains a reference to direct source file which version has been created;
- EUDAT/REVISION_OF and the EUDAT/WAS_DERIVED_FROM fields refer to the previous and next version, respectively; a reference to the Repository of Records (ROR) stored in the
- EUDAT/ROR that refers to the original of the version in the Repository of Records (ROR).

An issue of practical importance is how to retrieve an older (previous) and a newer (next) version with respect to a current version. In the B2SAFE versioning service, for navigation between versions, references to older and newer versions are kept in the fields called, according to the semantics of PROV-O ontology [24], EUDAT/REVISION_OF and EUDAT/WAS_DERIVED_FROM, respectively. Hence, the new version contains a reference to the previous version and vice versa, i.e., versions are cross-linked for ease of traversing through them (as in Fig. 3). Moreover, for convenience, PID record of the source file has a reference to its latest version.

The described versioning system can create versions not only of a single data object, but also recursively versions of every data object located within a collection (optionally including sub-collections). The versioning functionality can also be applied to a file or collection of files which do not have registered PID. Furthermore, maximum number of versions of a file can be set to keep disk usage under control. By default, the B2SAFE versioning service retrieves the latest written version of a file from a specified versions repository to a given directory. A version number should be specified if an older version of the file is required.

7 Conclusion and Further Work

Analysis of data curation requirements of two use cases: HERBADROP and GEOFON has been performed, coupled with the retrospective review of the elaborated data curation policies from a dedicated EU project (SCAPE) and practical (machine-executable) policies that were the output of the dedicated RDA working group.

A set of granular control policies have been identified and implemented in two use cases, and a gap analysis of these policies has been performed against the SCAPE catalogue of policy elements.

The conceptual framework developed and requirements analyzed supported the actual implementation of data versioning policy in EUDAT B2SAFE service and the development of Provgen component [21] for generation of data provenance records.

After the set of identified policies was applied in the two use cases that have been involved in their formulation, the same policy framework could be applied in a larger number of research communities associated with EUDAT.

The scope of projects and initiatives in data curation and long-term digital preservation can be extended beyond SCAPE and RDA working groups; this specifically applies to popular functional models of digital preservation like OAIS [18] that have not been thoroughly evaluated so far for their potential application in EUDAT.

Sensible modelling of data policies and, specifically, overcoming conceptual and technological gaps between textual policy formulation and machine-executable policies require further considerations and experiments. One possible approach could be modular policy modelling supported by semantic Web technology [28].

Should the practical project opportunity arise, the body of knowledge acquired through the outlined EUDAT data curation and data provenance task can be applied to the existing or emerging data infrastructures in order to make them "data curation-centric", in opposition to the current most popular approach when data services are implemented first, with data curation requirements and implementation considered second.

For EUDAT Collaborative Data Infrastructure, this approach can be considered as addressing two major challenges:

- making a fully-fledged long-term digital preservation solution out of EUDAT B2SAFE service that looks technologically sound and supported by a distributed network of research partners with a certain level of the IT governance culture, yet lacks consistent data curation vision and enough implementation of data policies;
- replacement of "vertical" or "silos" data curation concepts, requirements and technology that are specific to particular data pilots with "horizontal" concepts, requirements and technology that will be universal across the entire data infrastructure traversing and supporting a variety of data pilots.

This vision is represented by Fig. 4 with the layers that correspond to different aspects of curation-centric data infrastructure. The right column suggests "providers" for each layer, e.g. for the operational governance, organizational structures of EUDAT CDI [1] could be used. This vision will require substantial organizational effort and sustainable sources of funding in order to be delivered and supported in the long term.

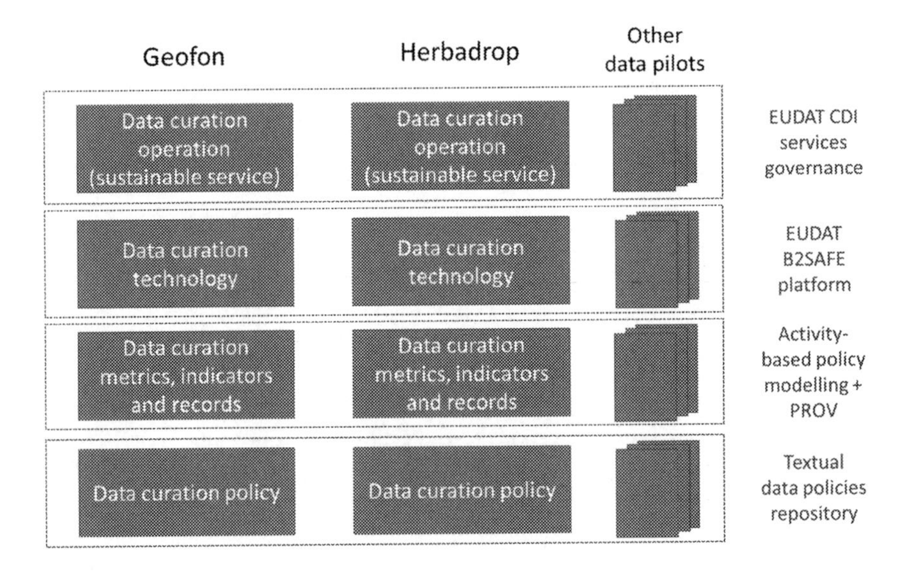

Fig. 4. Data curation-centric vision of the EUDAT Collaborative Data Infrastructure.

Acknowledgements. This work is supported by EUDAT 2020 project that receives funding from the European Union's Horizon 2020 research and innovation programme under the grant agreement No. 654065. The views expressed are those of authors and not necessarily of the project.

References

1. EUDAT Collaborative Data Infrastructure. https://www.eudat.eu/eudat-cdi
2. SCAPE: Scalable Preservation Environments. http://scape-project.eu/

3. Research Data Alliance Practical Policy Working Group. https://www.rd-alliance.org/groups/practical-policy-wg.html
4. SCAPE Catalogue of Preservation Policy Elements. http://scape-project.eu/wp-content/uploads/2014/02/SCAPE_D13.2_KB_V1.0.pdf
5. EPOS: European Plates Observing System. https://www.epos-ip.org/
6. CINES: French national IT center for higher education and research. https://www.cines.fr/en/
7. Hanka, W., Kind, R.: The GEOFON Program. Annal. Geophys. **37**(5), November 1994. https://doi.org/10.4401/ag-4196, ISSN 2037-416X
8. GEOFON Data Centre (1993): GEOFON Seismic Network. Deutsches GeoForschungsZentrum GFZ. Other/Seismic Network (1993). https://doi.org/10.14470/tr560404
9. Practical Policy Implementations Report. http://dx.doi.org/10.15497/83E1B3F9-7E17-484A-A466-B3E5775121CC
10. Hanka, W., Saul, J., Weber, B., Becker, J., Harjadi, P., Fauzi, GITEWS Seismology Group: Real-time earthquake monitoring for tsunami warning in the Indian Ocean and beyond. Nat. Hazards Earth Syst. Sci. **10**, 2611–2622 (2010). https://doi.org/10.5194/nhess-10-2611-2010
11. iRODS: Integrated Rule-Oriented Data System. https://irods.org/
12. Haston, E., Chagnoux, S., Dugénie, Herbadrop, P.: Long-term preservation of herbarium specimen images. In: Proceedings of the Second Eudat User Forum, Rome (2016)
13. Dugénie, P., Chagnoux, S.: EUDAT Data pilot Herbadrop. Second interim Herbadrop Data Pilot report (2016)
14. Digitarium: service centre for high performance digitization. http://digitarium.fi/en
15. DigiWeb + digitization platform. http://digiweb.digitarium.fi/
16. Elasticsearch search and analytics engine. https://www.elastic.co
17. SeedLink protocol and tools overview. http://ds.iris.edu/ds/nodes/dmc/services/seedlink/
18. Reference Model for an Open Archival Information System (OAIS), Recommended Practice, CCSDS 650.0-M-2 (Magenta Book). Issue 2, June 2012. CCSDS (The Consultative Committee for Space Data Systems), Washington DC (2012)
19. EUDAT services. https://www.eudat.eu/services-support
20. Bunakov, V. et al.: Data curation policies for EUDAT collaborative data infrastructure. In: Selected Papers of the XIX International Conference on Data Analytics and Management in Data Intensive Domains (DAMDID/RCDL 2017). CEUR Workshop Proceedings vol. 2022, pp. 72–78 (2017). urn:nbn:de:0074-2022-6
21. Bunakov, V., Quinteros, J., Reijnhoudt, L.: Data provenance service prototype for collaborative data infrastructure. Submitted in ALLDATA 2018: The Fourth International Conference on Big Data, Small Data, Linked Data and Open Data. http://www.iaria.org/conferences2018/ALLDATA18.html
22. Provgen API specification. https://raw.githubusercontent.com/javiquinte/provgen/master/swagger.yaml
23. ORFEUS: Observatories and Research Facilities for European Seismology. http://www.orfeus-eu.org/
24. PROV ontology. https://www.w3.org/TR/prov-o/
25. EUDAT B2SAFE service. https://www.eudat.eu/b2safe/
26. Handle.Net registry. http://handle.net
27. B2SAFE service core code. https://github.com/EUDAT-B2SAFE/B2SAFE-core/tree/versioning
28. Bunakov, V.: Data policy as activity network. In: Selected Papers of the XIX International Conference on Data Analytics and Management in Data Intensive Domains (DAMDID/RCDL 2017). CEUR Workshop Proceedings, vol. 2022, pp. 79–86 (2017). urn:nbn:de:0074-2022-6

Temporal Summaries Generation

News Timeline Generation: Accounting for Structural Aspects and Temporal Nature of News Stream

Mikhail Tikhomirov(✉) and Boris Dobrov

Lomonosov Moscow State University, Moscow, Russia
tikhomirov.mm@gmail.com, dobrov_bv@srcc.msu.ru

Abstract. The number of news articles that are published daily is larger than any person can afford to study. Correct summarization of the information allows for an easy search for the event of interest. This research was designed to address the issue of constructing annotations of news story. Standard multi-document summarization approaches are not able to extract all information relevant to the event. This is due to the fact that such approaches do not take into account the variability of the event context in time. We have implemented a system that automatically builds timeline summary. We investigated impact of three factors: query extension, accounting for temporal nature and structure of news article in form of inverted pyramid. The annotations that we generate are composed of sentences sorted in chronological order, which together contain the main details of the news story. The paper shows that taking into account the described factors positively affects the quality of the annotations created.

Keywords: Timeline summarization · Extractive summarization
Multi-document summarization · Information retrieval

1 Introduction

Due to the explosive growth of the amount of content on the Internet, the problems of extraction and automatically summarizing useful information in the incoming data stream arises. One of such problems is the summarization of news articles on an event. The news story - is a set of news reports from various sources dedicated to describing an event. Such problems are often investigated and solved by news aggregators, for example, Google.News[1] [17] or Yandex.News.[2] This is due to the fact that to work with such problems the researcher needs a huge and diverse collection of news articles.

The typical "lifetime" of the news story (the time of active discussion of the event) is usually a day or two, but not all events are so short. Some news stories have a "history" in the form of a set of previous events that occurred at different moments and are more or less related to each other. Existing multi-document summarization approaches do not take into account the fact that the context, actors, geography and other event properties can vary over time.

[1] news.google.com.
[2] news.yandex.ru.

L. Kalinichenko et al. (Eds.): DAMDID/RCDL 2017, CCIS 822, pp. 267–280, 2018.
https://doi.org/10.1007/978-3-319-96553-6_19

The fact that journalists are returning to the same events, for example, with the appearance of new data, indicates that such events are important for the society. The need for a brief summary of the event raises the problem of forming a "timeline summary". Timeline summary is a type of multi-document summary, containing the essential details of the subject matter under discussion. The construction of such annotations is a complex task, performed by journalists or analysts manually. This implies that the automation of such a process is a urgent problem.

In this paper we consider challenges and solutions for the automatic generation of temporal summaries. We consider this problem as a multi-document summarization on a query over a representative collection of news documents. The query in this case is the text of the news message. The situation corresponds to the scenario when a user would like to receive a timeline summary after reading the news document. The result should be a time-ordered list of descriptions of the key sub-events related to main event. The result consists of parts of existing sentences, since our solution refers to extractive summarization approaches.

A system was developed to automate the timeline summarization process. Experiments were conducted over a collection of 2 million Russian news for the first half of 2015. Three new factors were investigated to improve the results of constructing a timeline summary: query extension using pseudo-relevance feedback, accounting for the timing characteristics of news stories and the structure of the inverted pyramid.

This is a follow-up study of timeline summarization problem reported in previous paper [25]. In this study, we expanded the collection of standard annotations three-fold. The evaluation process was improved by dividing the collection into a training and test parts. An optimization module was added for fitting the configurations. As a result, substantial progress was achieved. Taking into account the structure of the inverted pyramid showed a significant increase in the values of metrics, which was not achieved in the previous article.

2 Related Work

2.1 Automatic Text Summarization Problem

Currently, there are quite a number of methods for automatic text summarization [3]. Some methods that use large linguistic ontologies [12, 15], that may be automatically supplemented during the analysis. Other methods are based on the statistical properties of texts [16] or machine learning [13].

During the generation of the annotations, the following problems occur [3, 7, 11]:

- Ensuring the completeness of the presentation of information, including the most up-to-date information.
- Decreasing of redundancy in the information provided.
- Ensuring the coherence and understandability of the information provided.

To ensure the completeness of the resulting annotation, it is often necessary to find links between sentences or documents [20].

To determine the redundancy in the generated annotations, various measures of similarity between sentences are used. One of the most common approaches is clustering - the selection of content groups of sentences [6]. Another approach to reduce redundancy is to compare candidate sentence with sentences that have already been included in the summary and to evaluate novel information. Example of such approach is the Maximal Marginal Relevance (MMR) [2].

The problem of ensuring the coherence of information in the summary arises both in the methods of generating the annotation [18, 19], and in the methods of evaluation, because in order to assess the connectivity and linguistic qualities of the annotation, it is necessary to perform a manual evaluation.

2.2 Timeline Summary

The problem of timeline summary construction has a number of differences from the standard summarization problem. For example, the temporal nature of events must be taken into account [9]. Also, to ensure completeness of the information provided, it is required to find documents from all sub-events of the topic under consideration.

When constructing timeline summary, data processing is mainly carried out over huge collections. In such collections, most of the information is not relevant to the user's request. This problem can be solved by using clustering methods [10, 14]. But the clustering methods have some issues. First, such a task should be solved many times over huge collections of documents, which affects the response time of the system. Secondly, the degree of closeness can be significantly smaller with standard measures of similarity for documents that describe far-in-time but related events. And, of course, it is required to identify the most characteristic objects [1, 9], for example, taking into account the structural features of the flow of documents [5, 8].

3 Statement of the Problem

3.1 General Description

The problem of constructing a timeline summary is a query-oriented. In the most general case, the user has a news document as a query. So further this problem will be considered as a problem of automatic creation of a summary on a query in the form of a text document. The output of the system is an annotation of n sentences. The connectivity between the sentences in this paper is not required. Figure 1 provides an example of a possible summary about the conflict on cemetery taken from the Interfax website.[3]

The aim of the work is to study the influence of various factors on the quality of the annotation.

[3] http://www.interfax.ru/story/215.

August 18, 2017 17:21
The case of the riots in the Khovansky cemetery is sent to the court

28 March 2017 13:39
Participants in the riots in the Khovansky cemetery were finally charged

20 January 2017 16:51
The investigation decided not to impute the murder to the defendants of the case of the fight at the Khovansky cemetery

August 30, 2016 11:59
The court dismissed the complaint about the extension of the arrest of the defendant in the case of the fight at the Khovansky cemetery

14 July 2016 13:13
For the defendants, the case of the fight in the cemetery will seek the status of victims

July 14, 2016 12:21
The police will continue to search for six suspects in the case of the fight at the Khovansky cemetery

Fig. 1. Timeline summary part about conflict at cemetery.

3.2 Mathematical Statement of the Problem

The problem described above can be formalized in the following way. Let $Q = \{q_1, q_2, \ldots, q_m\}$ be a set of queries and an associated set of reference annotations $D_g = \left\{D_g^{q_1}, D_g^{q_2}, \ldots, D_g^{q_m}\right\}$ be an associated set of reference annotations. The system generates a set of summary $D_A = \{D_A^{q_1}, D_A^{q_2}, \ldots, D_A^{q_m}\}$ in response to queries Q by algorithm A. Then the problem is reduced to maximizing the following functional:

$$\frac{\sum_{i=1}^{i=|Q|} M\left(D_A^{q_i}, D_g^{q_i}\right)}{|Q|} \rightarrow max \tag{1}$$

where M is the proximity function between the annotations. Optimization is carried out for all parameters of the algorithm.

4 Approach

4.1 Collection Processing

As mentioned earlier, the input collection contains 2 million news articles. It is not possible to work directly with such amount of information, therefore, it was decided to interact with the collection through a search engine. Search engine allows:

- Get a list of documents by text request.
- For a given document from collection, get the basic information: text, index, meta-information.

4.2 Studied Features

In this paper the following factors were investigated:

- Query extension strategy.
- Accounting for the temporal nature of news stories.
- Accounting for the structure of a news article in the form of an inverted pyramid.

4.3 Query Extension Strategy

Information that can be obtained from a query document is basically not enough to effectively build this type of annotation. This fact is a consequence of the fact that most news articles are not a general description of the event, but a discussion of some particular incident or fact.

To avoid this problem, it is necessary to use the query extension techniques. The developed algorithm uses the idea of pseudo-relevance feedback, which is widely used in information retrieval problems [21]. For the query-document, the algorithm includes the following steps:

1. The most significant K-terms are chosen on the basis of tf-idf weights forming thus the *first-level query*.
2. Further on the basis of the *first-level query* documents are retrieved.
3. The extracted cluster of documents is analyzed to find the most important terms forming thus the *second-level query*:
   a. For each document, the most significant T terms are considered.
   b. For each term, it is calculated how often it was in the top T terms on all cluster.
   c. The list of terms is sorted by frequency, the best M terms are selected.
4. Steps 2–3 are repeated (A *double query extension* process that forms a *third-level query*).
5. Output of the algorithm is a vector of N terms representing to some extent the semantics of the input document.

Note that K, T, M, N are parameters of the algorithm and they must be configured. As an example of the work of the query extension module, consider the algorithm steps on a news article about the terrorist attack in Paris (Table 1).

The table shows that a higher-level query has more significant terms for this event.

Table 1. Query extension algorithm stages example.

Entity name	Content
Initial doc	*President François Hollande of France called it a display of extraordinary "barbarism" that was "without a doubt" an act of terrorism. According to the latest information, as a result of shooting, 11 people were killed, four more are in critical condition. ...*
First-level query	Posten, Jyllands-posten, Jyllands, Herbo, Charlie, Hollande
Second-level query	Reprint, Scandal, Weekly, Caricature, Hollande, Satirical, Terrorist act, Charlie, Herbo
Third-level query (double query extension)	Journal, Muhammad, Satirical, Attack, Prophet, Terrorist act, Paris, Caricature, Hollande, Herbo, Charlie

4.4 Temporal Nature of News Stories

Since any event depends on time, the content of publications and their number also depend on the time. As an example, Fig. 2 shows a graph of the time dependence of publications on the "Earthquake in Nepal" event.

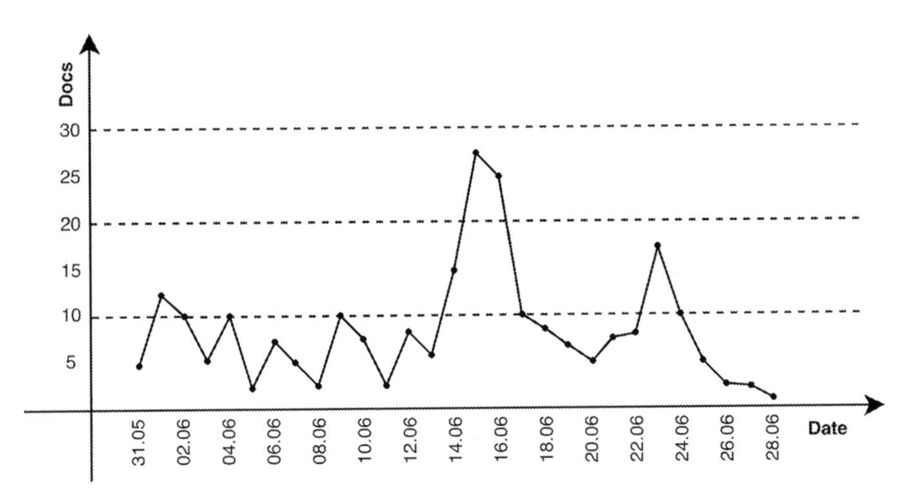

Fig. 2. Dependence of the number of publications per day for an event

To take into account this factor, for the set D of extracted documents the following procedure is undertaken:

1. The entire timeline of the event is divided into days with labels $T = \{t_1, t_2, \ldots, t_n\}$.
2. Each document receives a label from T based on the publication date D_i^t.
3. Documents published on days with a number of publications less then the $NDoc_{tr}$ threshold (2) are discarded.

$$NDoc_{tr} = 0.2 * MEAN_{top\,3}(D) \tag{2}$$

4. The output is a sorted list of collections $C = \{C_{t_1}, C_{t_2}, \ldots, C_{t_n}\}$, where each collection C_{t_i} contains only documents with the label t_i.

4.5 Inverted Pyramid

The strategy of writing a high-quality news article often relies on the structure of the "inverted pyramid" (Fig. 3). The greatest interest is the upper and lower parts of the pyramid:

- The upper part contains the most concentrated information about the event under discussion.
- The lower part may contain references to important related events in the past.

Fig. 3. Inverted pyramid on the example of an article. (https://themoscowtimes.com/articles/moscow-museum-takes-you-inside-north-korea-60240)

This structure is taken into account in two ways:

1. Inter-document feature based on the graph approach.
2. Intra-document feature, which increases the weight of sentences located in the upper and lower parts of the inverted pyramid.

Inter-Document Feature. This feature is taken into account in the following way:

1. For a set of documents D, a similarity matrix between the upper and lower parts of the documents is constructed. If the specified similarity threshold is exceeded, it is considered that there is a link between the documents D_i and D_j.
2. The importance of documents is calculated by using the LexRank algorithm over the constructed graph [4].
3. For documents whose weight is greater than a certain threshold, the previously described procedure for expanding the query is performed.

As a result, the output is a ranked list of documents D and a set of Q_D new queries, which further, together with accounting for the temporal nature of the news story, will help in sentence ranking algorithm. Among other things, document weights will also be taken into account in the ranking functions.

Intra-Document Feature. To this feature into account for the following procedure is undertaken: during the ranking of sentences, the weight of the sentence is multiplied by a coefficient that lowers the weight of sentences in the middle of the document.

Also, after described inter-document procedure, all constructed extended queries Q_D are mapped to t_i labels from T (Fig. 4).

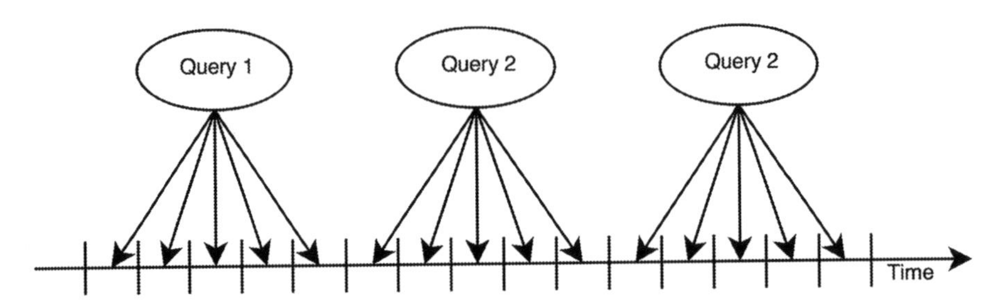

Fig. 4. Query mapping.

4.6 Similarity of Sentences

At various stages of the algorithm, there are a number of points where the measure of closeness between sentences is calculated. For these purpose a cosine measure of similarity (3) is used in all cases.

$$Sim_{cos}(S_i, S_j) = \frac{(S_i, S_j)}{|S_i| * |S_j|} \tag{3}$$

The choice of representation of a sentence plays an important role for calculating similarity. In this article we used the standard tf-idf representation. But to calculate the similarity between sentences when searching for links between documents, word2vec [24] representation was used. To achieve this, the resulting sentence vector is represented as a weighted mean of word2vec word representations. Weighing was carried out by tf-idf.

Word2vec model was trained on the entire collection of 2 million news articles. During preprocessing removal of stop words and lemmatization were applied. The width of the window was chosen to be 5, and the length of the vector was 100.

4.7 Sentence Ranking Module

This module deals with the ranking of sentences. The ranking is a modified version of the *MMR* algorithm – *MMRT* (4) taking into account all the factors described in Sect. 4.2:

$$MMRT_{s_i^t} = INC_{s_i^t} - DEC_{s_i^t} \tag{4}$$

where $INC_{s_i^t}$ – is a term describing the positive part of the formula, which depends on the similarity of the sentence to the query, the weight of the document from which the sentence is taken, and the sentence number in the document.

$$INC_{s_i^t} = (1 + \alpha * I_i) * \gamma * \lambda * Sim(Q^t, S_i^t) \tag{5}$$

$$\gamma = 1 - 0.5 * \sin\left(\frac{i * \pi}{|D_s|}\right) \tag{6}$$

The parameters α and λ are configurable parameters of the algorithm, I_i – is document weight D_s, which includes a sentence under the index i, S_i^t – is estimated sentence under the index i with label t, Q^t – query for this time label, γ – multiplier, which reduce the weight of sentences from the middle of the document.

$DEC_{S_i^t}$ is the penalty term. It depends on the similarity to the already extracted sentences:

$$DEC_{S_i^t} = (1 - \lambda) * \max_{S_j \in S} Sim(S_j, S_i^t) \tag{7}$$

where S_j is one of the extracted sentences, S is the set of all already extracted sentences.

Processing of sentences occurs in chronological order with a restriction on the maximum number of sentences per day.

4.8 System Diagram

The features described in Subsect. 4.2 are realized at various stages of the system. The general scheme of the algorithm is shown on Fig. 5.

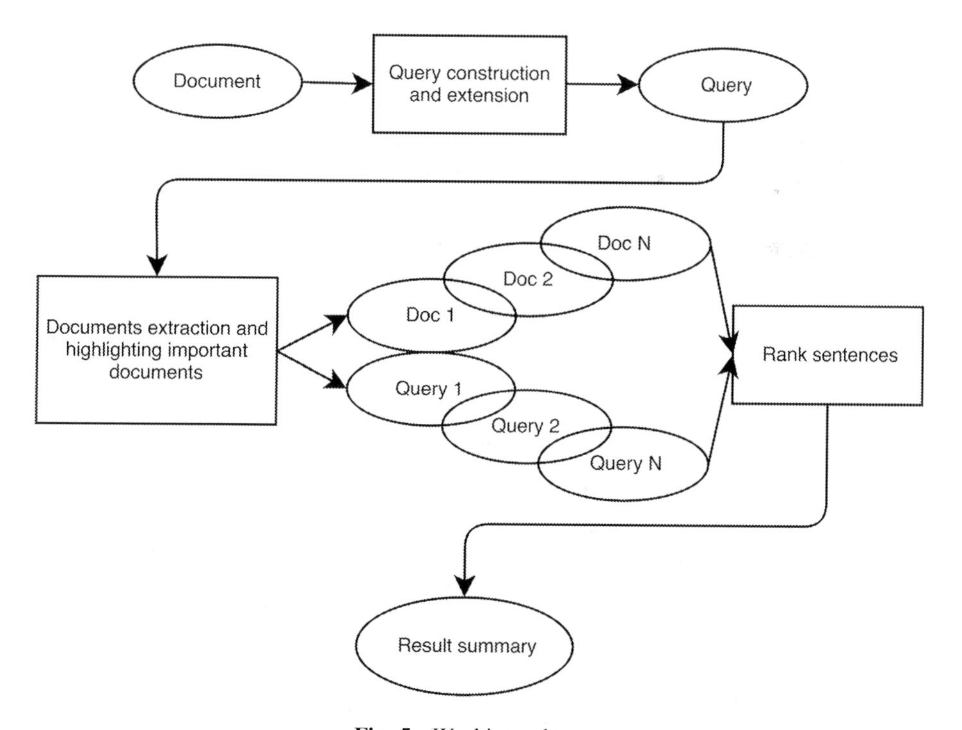

Fig. 5. Working scheme.

5 Evaluation

5.1 Metrics for Evaluation

The system was evaluated using several metrics: ROUGE-1, ROUGE-2, and Sentence Recall R^{sent}:

$$ROUGE - N = \frac{|N_A \cap N_g|}{|N_g|} \tag{8}$$

where N_A is the set of n-grams for the constructed annotations, N_g is the set of n-grams for the reference (gold) annotations.

$$R^{sent} = \frac{|S_A \equiv S_g|}{|S_g|}, \tag{9}$$

where S_A is the set of sentences from the constructed annotations, S_g is the set of sentences from the reference annotations. Operator $\equiv$ denotes the following: the result of the $|S_A \equiv S_g|$ is a subset of S_A such that their semantic equivalent is present in S_g.

5.2 Data Preparation

Since a test set of annotations is required for evaluating procedure, in the course of the research, timeline summaries were manually prepared. The procedure for the formation of such a collection was as follows:

1. At the first stage with the help of Wikipedia there high-profile events were selected, which were actively covered in the press for the beginning of 2015.
2. Further, for most of the events on the site “Interfax”, the search for the corresponding story was carried out. On the basis of documents corresponding to the story, a timeline summary was created.
3. If there is no corresponding story on the “Interfax”, the materials were studied on the topic and a timeline summary was created on the basis of the documents read.

As a result, 45 annotations on 15 news stories were created (Table 2).

5.3 Optimization of Algorithm Parameters

Since the system contains a large number of parameters (total 23 parameters), some of which are presented in Table 3, there was a need to optimize the choice of the values of these parameters.

To achieve this, the entire collection of the reference annotations was divided into train and test parts with the ratio 2 to 1. Further, the functional (1) was implemented in Python using an open hyperopt [22] package based on machine learning. This package uses the technique of Sequential model-based optimization (SMBO) [23] for the parameters selection. The parameters were trained on the training part. After that, the final evaluation of the configurations took place on the test part.

Table 2. News stories on which the reference annotations are made.

Story ID	Description
Story 1	The flight of the space probe DAWN to Ceres
Story 2	Fire in Khakassia
Story 3	Terrorist act in Paris
Story 4	The plane crash in Taiwan
Story 5	Earthquake in Nepal
Story 6	Fire on the Orel submarine.
Story 7	Attack on the synagogue in Copenhagen
Story 8	Fire in the mall «Admiral»
Story 9	Terrorist act in Sydney
Story 10	Ice Hockey Championship
Story 11	The case of Svetlana Davydova
Story 12	The Murder of Boris Nemtsov
Story 13	Dangerous coronavirus in Korea
Story 14	Protest in Yerevan
Story 15	The failure of the spaceship "Progress"

Table 3. Some system parameters.

Parameter name	Description	Opt. value
KeepL	The number of lemmas to choose when building a first-level query	6
DocCount	The maximum number of documents retrieved using a search engine	400
QuerrySize	The size of the resulting expanded query	14
TopLemms	The number of the most significant lemmas extracted in the work of the algorithm for constructing an extended query	18
MinLinkScore	The minimum value of similarity between the top and bottom of the documents to identify the reference	0.5
Lambda	The value of the parameter λ for *MMRT*	0.58
Alpha	The value of the parameter α for *MMRT*	0.45
MaxDaily AnswerSize	The maximum number of sentences per day	15
Doc Boundary	Threshold of the importance of documents for building new queries	0.61

6 Results

In order to evaluate the contribution of the considered features, a fitting and evaluation of the following 6 configurations was made:

1. **baseline** – a simple approach to summarization, without taking into account the factors considered, using MMR as ranking algorithm.

2. **querry-ex** – adding a query extension strategy feature to *baseline* (Sect. 4.3), but without double query extension.
3. **double-ex** – *querry-ex* + double query extension (Sect. 4.3).
4. **temporal** – *double-ex* + accounting for the temporal nature of news stories (Sect. 4.4).
5. **importance** – *temporal* + accounting for the structure of a news article in the form of an inverted pyramid, when tf-idf representation is used (Sect. 4.5).
6. **w2v-imp** – *importance,* but using w2v for computing sentence similarity when accounting for the structure of a news article (Sect. 4.6).

The result of evaluation of the configurations can be seen in Table 4. This table shows that each of the features considered gives a positive contribution to the quality of generation of timeline summary. As an example of the final annotation, one can consider a fragment of the annotation on the previously mentioned incident of the crash in Taiwan in Table 5.

Table 4. Evaluation results.

	R1	*R2*	R^{sent}
baseline	0.320	0.127	0.162
querry-ex	0.326	0.133	0.175
double-ex	0.356	0.176	0.233
temporal	0.384	0.178	0.240
importance	0.399	0.176	0.246
w2v-imp	**0.403**	**0.181**	**0.254**

Table 5. The generated timeline summary fragment about the plane crash in Taiwan.

Date	Sentence
11.02.2015	Transasia Airways will pay relatives of victims of a plane crash in Taiwan for 470 thousand
11.02.2015	The tragedy in Taiwan, one-fifth of the pilots of Taiwan's Transasia airline have not passed the proficiency test
12.02.2015	Rescuers completed the search operation for victims of the crash of an airline to Transasia Airways, which crashed on February 4 in Taiwan
01.07.2015	Crew crashed in Taiwan aircraft Transasia Airways shut off the engines after a loss of power
02.07.2015	The Transasia plane crashed on February 4 in Taiwan, because the pilot accidentally turned off the running engine when the second engine stalled

7 Conclusions and Future Work

In this article we presented an approach for building a timeline summary. The conducted research shows that the problem of constructing the timeline summary differs from the standard MDS problem. The effectiveness of using the following features was shown:

- Query extension strategy.
- Accounting for the temporal nature of news stories.
- Accounting for the structure of a news article in the form of an inverted pyramid.

Extending the query, as expected, has a positive effect on the event representation discussed in the document. But the interesting fact is that re-extension the query (*double query extension*) has a much greater effect. This is because the documents that are retrieved on the first-level query are not sufficient for a good presentation of the event.

The fact that accounting for the temporal nature of news stories improves the quality of the annotation is an obvious consequence of the fact that news stories and events have temporal characteristics.

Taking into account the structure of the inverted pyramid gives an improvement. Increase the values of metrics on the *w2v-imp* configuration means that the correctness of the recognized links between the documents plays a significant role. This fact raises challenges for future research.

Using structural features of news articles make it possible to obtain information, the use of which can significantly improve the quality of generated annotations.

References

1. Binh, T.G., Alrifai, M., Quoc Nguyen, D.: Predicting relevant news events for timeline summaries. In: Proceedings of the 22nd International Conference on World Wide Web, pp. 91–92. ACM (2013)
2. Carbonell, J., Goldstein, J.: The use of MMR, diversity-based reranking for reordering documents and producing summaries. In: Proceedings of the 21st Annual International ACM SIGIR Conference on Research and Development in Information Retrieval, pp. 335–336. ACM (1998)
3. Dang, H.T.: Overview of DUC 2006. In: Proceedings of the Document Understanding Workshop, Presented at HLT-NAACL 2006 (2006). http://duc.nist.gov/pubs/2006papers/duc2006.pdf
4. Erkan, G., Radev, D.R.: Lexrank: graph-based lexical centrality as salience in text summarization. J. Artif. Intell. Res. **22**, 457–479 (2004)
5. Hu, P., Huang, M.L., Zhu, X.Y.: Exploring the interactions of storylines from informative news events. J. Comput. Sci. Technol. **29**(3), 502–518 (2014)
6. Radev, D., Jing, H., Budzikowska, M.: Centroid-based summarization of multiple docuemtns: sentence extraction, utility-based evaluation, and user studies. In: Proceedings of the 2000 NAACL-ANLP Workshop on Automatic Summarization, Seattle, pp. 21–30 (2000)

7. Radev, D., McKeown, K., Hovy, E.: Introduction to the special issue on summarization. Comput. Linguist. **28**(4), 399–408 (2002)
8. Shahaf, D., Guestrin, C.: Connecting two (or less) dots: discovering structure in news articles. ACM Trans. Knowl. Discov. Data (TKDD) **5**(4), 24–54 (2012)
9. Tran, G., Alrifai, M., Herder, E.: Timeline summarization from relevant headlines. In: Hanbury, A., Kazai, G., Rauber, A., Fuhr, N. (eds.) ECIR 2015. LNCS, vol. 9022, pp. 245–256. Springer, Cham (2015). https://doi.org/10.1007/978-3-319-16354-3_26
10. Yan, R., et al.: Evolutionary timeline summarization: a balanced optimization framework via iterative substitution. In: Proceedings of the 34th International ACM SIGIR Conference on Research and Development in Information Retrieval, Beijing, China, 24–28 July 2011, pp. 745–754. ACM (2011). https://doi.org/10.1145/2009916.2010016
11. Wu, Z., Lei, L., Li, G., Huang, H., Zheng, C., Chen, E., Xu, G.: A topic modeling based approach to novel document automatic summarization. Expert Syst. Appl. **84**, 12–23 (2017)
12. Hennig, L., Umbrath, W., Wetzker, R.: An ontology-based approach to text summarization. In: Proceedings of the 2008 IEEE/WIC/ACM International Conference on Web Intelligence and Intelligent Agent Technology, vol. 3, pp. 291–294 (2008)
13. Nallapati, R., Zhou, B., Gulcehre, C., Xiang, B.: Abstractive text summarization using sequence-to-sequence RNNs and beyond. arXiv preprint arXiv:1602.06023 (2016)
14. Wei, T., Lu, Y., Chang, H., Zhou, Q., Bao, X.: A semantic approach for text clustering using WordNet and lexical chains. Expert Syst. Appl. **42**(4), 2264–2275 (2015)
15. Allahyari, M., Pouriyeh, S., Assefi, M., Safaei, S., Trippe, E.D., Gutierrez, J.B., Kochut, K.: Text summarization techniques: a brief survey. arXiv preprint arXiv:1707.02268 (2017)
16. Rush, A.M., Chopra, S., Weston, J.: A neural attention model for abstractive sentence summarization. arXiv preprint arXiv:1509.00685 (2015)
17. Hertzfeld, A.: Introducing Google News Timeline. https://news.googleblog.com/2009/04/introducing-google-news-timeline.html. Accessed 10 Jan 2018
18. Christensen, J., Mausam, S.S., Soderland, S., Etzioni, O.: Towards Coherent Multi-Document Summarization. In: HLT-NAACL, pp. 1163–1173 (2013)
19. Nishikawa, H., Arita, K., Tanaka, K., Hirao, T., Makino, T., Matsuo, Y.: Learning to generate coherent summary with discriminative hidden semi-markov model. In: COLING, pp. 1648–1659 (2014)
20. Barzilay, R., Elhadad, M.: Using lexical chains for text summarization. In: Advances in Automatic Text Summarization, pp. 111–121 (1999)
21. Jiang, L., Mitamura, T., Yu, S.I., Hauptmann, A.G.: Zero-example event search using multimodal pseudo relevance feedback. In: Proceedings of International Conference on Multimedia Retrieval, p. 297 (2014)
22. Bergstra, J., Komer, B., Eliasmith, C., Yamins, D., Cox, D.D.: Hyperopt: a python library for model selection and hyperparameter optimization. Comput. Sci. Discov. **8**(1), 014008 (2015)
23. Hutter, F., Hoos, Holger H., Leyton-Brown, K.: Sequential model-based optimization for general algorithm configuration. In: Coello, C.A. (ed.) LION 2011. LNCS, vol. 6683, pp. 507–523. Springer, Heidelberg (2011). https://doi.org/10.1007/978-3-642-25566-3_40
24. Goldberg, Y., Levy, O.: word2vec Explained: deriving Mikolov et al.'s negative-sampling word-embedding method. arXiv preprint arXiv:1402.3722 (2014)
25. Tikhomirov, M.M., Dobrov, B.V.: Using news corpora for temporal summary formation (in Russian). In: Selected Papers of the XIX International Conference on Data Analytics and Management in Data Intensive Domains (DAMDID/RCDL 2017), CEUR Workshop Proceedings, Moscow, Russia, vol. 2022, pp. 165–171 (2017)

Author Index